Handbook of Evolution

Edited by
Franz M. Wuketits
Francisco J. Ayala

Related Titles

Journal

Evolutionary Anthropology

2005, 6 Issues
ISSN 1060-1538

S. Stinson, B. Bogin, R. Huss-Ashmore, D. O'Rourke (Eds.)

Human Biology: An Evolutionary and Biocultural Perspective

2000
ISBN 0-471-13746-4

K. M. Weiss, A. V. Buchanan

Genetics and the Logic of Evolution

2004
ISBN 0-471-23805-8

I. K. Bradbury

The Biosphere, Second Edition

1999
ISBN 0-471-98549-X

K. J. McNamara, J. Long

The Evolution Revolution

1998
ISBN 0-471-97407-2

Handbook of Evolution

Volume 2:
The Evolution of Living Systems
(Including Hominids)

Edited by
Franz M. Wuketits and Francisco J. Ayala

WILEY-VCH

WILEY-VCH Verlag GmbH & Co. KGaA

Editors

Prof. Dr. Franz M. Wuketits
Institut für Wissenschaftstheorie
Universität Wien
Sensengasse 8
1090 Wien
Austria

Prof. Dr. Francisco J. Ayala
Ecology and Evolutionary Biology
University of California, Irvine
321 Steinhaus Hall
Irvine, CA 92697
USA

Library of Congress Card No.: applied for
A catalogue record for this book is available from the British Library.

Bibliographic information published by Die Deutsche Bibliothek
Die Deutsche Bibliothek lists this publication in the Deutsche Nationalbibliografie; detailed bibliographic data is available in the internet at http://dnb.ddb.de.

© 2005 Wiley-VCH Verlag GmbH & Co. KGaA, Weinheim

Printed in the Federal Republic of Germany
Printed on acid-free paper

Cover Design SCHULZ Grafik-Design, Fußgönheim
Composition Manuela Treindl, Laaber
Printing Strauss GmbH, Mörlenbach
Bookbinding Litges & Dopf Buchbinderei GmbH, Heppenheim

ISBN-13 978-3-527-30838-5
ISBN-10 3-527-30838-5

Preface

The present volume of the *Handbook* is devoted to biological evolution, i.e., the evolution of living systems, including hominids. This is an enormous topic. The evolution of organisms has been studied by means of various concepts and methods applied in different fields of scientific research, virtually comprising all biological disciplines, from comparative anatomy to genetics, from biogeography to molecular biology. In addition, evolutionists make use of results in disciplines like biophysics, biochemistry, geology, climatology, etc. Thus, evolutionary biology relies on a vast amount of data and can be regarded as an extensive interdisciplinary approach. However, it should not be misunderstood as a mere accumulation of "facts". First and foremost, it is a comprehensive theory of life's history on the Earth and the overwhelming diversity of organisms, both extant and extinct.

The editors and authors of this volume aim to give a synopsis of the most important data, concepts and theoretical reflections in evolutionary biology. The intention is not to give a full account of our present knowledge of biological evolution in all its details. This would not be possible, anyway, in one volume. What is probably more interesting to the reader is a presentation of some basic ideas and "facts", and, even more, of the implications of evolutionary thinking for a comprehensive modern world view. In this sense, the eight chapters of the volume are to be understood as parts of a mosaic. Their authors present and discuss different aspects of the biological theory of evolution, tackle problems and controversial issues, and point to "open questions". As the reader will notice, the theory of biological evolution consists of many "subtheories" and is not yet a finished case (as science never is). While nobody can seriously doubt that evolution occurs, the questions "Why?" and "How?" still require some deeper analyses and leave room for speculation. The conceptual framework for evolutionary biology is Charles Darwin's theory of natural selection. However, as the altmeister himself was well aware, the theory needed some improvement. In the meantime, much work has indeed been done and the results of evolutionary research are astounding. However, we cannot yet rest our case.

Chapter 1 presents a synopsis of biological evolution. It summarizes the empirical evidence for evolution, the genetic basis of evolutionary change, the meaning of selection (in its different expressions), ideas of mode and tempo in evolution, and molecular evolution. In Chapter 2, Michael Ruse critically reviews some evolution controversies (including the sociobiology debate and the controversy surrounding

Handbook of Evolution, Vol. 2: The Evolution of Living Systems (Including Hominids)
Edited by Franz M. Wuketits and Francisco J. Ayala
Copyright © 2005 Wiley-VCH Verlag GmbH & Co. KGaA, Weinheim
ISBN: 3-527-30838-5

evolutionism versus creationism). Edward O. Wilson's contribution (Chapter 3) is a discussion of the evolutionary meaning of complex social systems. Although not an original paper (it was first published in 1991), it is still fresh and lucid, and helps our understanding of some important aspects of biological evolution. Chapter 4 is a brief presentation of the history and philosophy of evolutionary thinking in biology. It reconstructs the very idea of evolution, the main (philosophical) obstacles to evolutionism, and Darwin's theory and its ramifications in the 20th century. In Chapter 5, Gerd B. Müller gives an account of the comparatively new discipline of evolutionary developmental biology, its empirical background, agenda and concepts, research programs, and its meaning for a deeper understanding of evolutionary change. Chapter 6, by Winfried Henke, provides a comprehensive compilation and discussion of current ideas of human evolution. This chapter can also serve as an introduction to biological systematics and taxonomy, and its relevance in the context of evolutionary theory. In Chapter 7, Peter J. Richerson and his co-authors pose the question whether environmental variability and environmental change were major motors of human evolution, and reach some interesting conclusions. The human impact on evolutionary change is the topic of Chapter 8. Bernhard Verbeek discusses the question to what extent the emergence and evolutionary development of humans has changed the face of our planet.

Clearly, evolutionary biology includes many other aspects that are of some interest. However, we hope that the selection of topics presented and discussed in these chapters will serve as a guide to one of the most fascinating adventures of our intellectual history, and will invite readers to further reflection and discussion. Moreover, we hope that this volume will give the impression of evolutionary biology as a discipline "in flux". It is not by chance that human evolution is given so much space in the present volume. The future of our own and many other species will depend, to a considerable extent, on our understanding of – and attitude towards – evolutionary history and the meaning of biodiversity. We want to encourage readers to think about the importance of evolution and evolutionary biology in our institutions of higher education.

Our thanks go to the contributors of this volume who, despite their many other obligations, found time and energy to write and – in some cases – even to rewrite "their chapters". Also, we want to express our thanks to the staff of Wiley-VCH (Weinheim), particularly to Dr. Waltraut Wüst for her characteristic patience and constant encouragement.

May 2005

Franz M. Wuketits, Vienna (Austria)
Francisco J. Ayala, Irvine, CA (USA)

Contents

Handbook of Evolution, Vol. 2: The Evolution of Living Systems (Including Hominids)
Edited by Franz M. Wuketits and Francisco J. Ayala
Copyright © 2005 Wiley-VCH Verlag GmbH & Co. KGaA, Weinheim
ISBN: 3-527-30838-5

List of Contributors

Prof. Dr. Francisco J. Ayala
Ecology and Evolutionary Biology
University of California, Irvine
321 Steinhaus Hall
Irvine, CA 92697
USA

Prof. Dr. Robert L. Bettinger
Department of Anthropology
University of California, Davis
One Shields Avenue
Davis, CA 95616
USA

Prof. Dr. Robert Boyd
Department of Anthropology
University of California, Los Angeles
375 Portola Plaza
Los Angeles, CA 90095
USA

Prof. Dr. Winfried Henke
Institute of Anthropology
Johannes Gutenberg University Mainz
55099 Mainz
Germany

Prof. Dr. Dr. Gerd B. Müller
Universität Wien
Biozentrum
Althanstrasse 14
1090 Wien
Austria

Prof. Dr. Peter Richerson
Department of Environmental
Science and Policy
University of California, Davis
One Shields Avenue
Davis, CA 95616-8573
USA

Prof. Dr. Michael Ruse
Department of Philosophy
Florida State University
156B Dodd Hall
Tallahassee, FL 32306-1500
USA

Prof. Dr. Bernhard Verbeek
University of Dortmund, FB 3, Biology
Otto-Hahn-Strasse 6
44227 Dortmund
Germany

Prof. Edward O. Wilson
Museum of Comparative Zoology
Harvard University
Cambridge, MA 02138
USA

Prof. Dr. Franz M. Wuketits
Institut für Wissenschaftstheorie
Universität Wien
Sensengasse 8
1090 Wien
Austria

Handbook of Evolution, Vol. 2: The Evolution of Living Systems (Including Hominids)
Edited by Franz M. Wuketits and Francisco J. Ayala
Copyright © 2005 Wiley-VCH Verlag GmbH & Co. KGaA, Weinheim
ISBN: 3-527-30838-5

1
The Evolution of Organisms: A Synopsis

Francisco J. Ayala

1.1
Biological Diversity and Evolution

The diversity of life is staggering. More than two million existing species of plants and animals have been named and described: many more remain to be discovered, at least ten million according to most estimates. What is impressive is not just the numbers but also the incredible heterogeneity in size, shape, and ways of life: from lowly bacteria, less than one thousandth of a millimeter in diameter, to the stately sequoias of California, rising 100 meters above the ground and weighing several thousand tons; from microorganisms living in the hot springs of Yellowstone National Park at temperatures near the boiling point of water, some like *Pyrolobus fumarii* able to grow at more than 100 °C, to fungi and algae thriving on the ice masses of Antarctica and in saline pools at −23 °C; from the strange wormlike creatures discovered in dark ocean depths thousands of meters below the surface, to spiders and larkspur existing on Mt. Everest more than 6600 meters above sea level.

These variations on life are the outcome of the evolutionary process. All organisms are related by descent from common ancestors. Humans and other mammals are descended from shrew-like creatures that lived more than 150 million years ago; mammals, birds, reptiles, amphibians, and fishes share as ancestors small worm-like creatures that lived in the world's oceans 600 million years ago; plants and animals are derived from bacteria-like microorganisms that originated more than three billion years ago. Because of biological evolution, lineages of organisms change through time; diversity arises because lineages that descend from common ancestors diverge through the generations as they become adapted to different ways of life.

Charles Darwin argued that organisms come about by evolution, and he provided a scientific explanation, essentially correct but incomplete, of how evolution occurs and why it is that organisms have features – such as wings, eyes, and kidneys – clearly structured to serve specific functions. Natural selection was the fundamental concept in his explanation. Genetics, a science born in the 20th century, revealed in detail how natural selection works and led to the development of the modern theory of evolution. Since the 1960s a related scientific discipline, molecular biology,

Handbook of Evolution, Vol. 2: The Evolution of Living Systems (Including Hominids)
Edited by Franz M. Wuketits and Francisco J. Ayala
Copyright © 2005 Wiley-VCH Verlag GmbH & Co. KGaA, Weinheim
ISBN: 3-527-30838-5

has enormously advanced knowledge of biological evolution and has made it possible to investigate detailed problems that seemed completely out of reach a few years earlier – for example, how similar the genes of humans, chimpanzees, and gorillas are (they differ in about 2% of their DNA).

The evolution of organisms, i.e., their common descent with modification from simple ancestors that lived many million years ago, is at the core of genetics, biochemistry, neurobiology, physiology, ecology, and other biological disciplines and makes sense of the emergence of new infectious diseases and other matters of public health. The evolution of organisms is universally accepted by biological scientists. "Nothing in biology makes sense except in the light of evolution" (Dobzhansky 1973).

I present a summary of some central tenets of the theory of biological evolution. The processes by which planets, stars, galaxies, and the universe form and change over time are a type of 'evolution', but in a different sense. In both instances there is change over time, but the processes are quite different, and I do not discuss the evolution of the universe. The evolution of the hominids, the lineage that leads to our own species is treated separately in this volume.

A distinction must be drawn at the outset between the questions (1) *whether* and (2) *how* biological evolution happened. The first refers to the finding, now supported by an overwhelming body of evidence, that descent with modification has occurred during some 3.5 billion years of earth's history. The second refers to the theory explaining how those changes came about. The mechanisms accounting for these changes are still undergoing investigation; the currently favored theory is an extensively modified version of Darwinian natural selection.

Natural selection was proposed by Darwin primarily to account for the adaptive organization of living beings: it is a process that promotes or maintains adaptation and thus gives the appearance of purpose or design. Evolutionary change through time and evolutionary diversification (multiplication of species) are not directly promoted by natural selection, but they often ensue as byproducts of natural selection as it fosters adaptation to different environments. Darwin's theory of natural selection is summarized in the *Origin of Species* as follows:

> As many more individuals are produced than can possibly survive, there must in every case be a struggle for existence, either one individual with another of the same species, or with the individuals of distinct species, or with the physical conditions of life. ... Can it, then, be thought improbable, seeing that variations useful to man have undoubtedly occurred, that other variations useful in some way to each being in the great and complex battle of life, should sometimes occur in the course of thousands of generations? If such do occur, can we doubt (remembering that many more individuals are born than can possibly survive) that individuals having any advantage, however slight, over others, would have the best chance of surviving and of procreating their kind? On the other hand, we may feel sure that any variation in the least degree injurious would be rigidly destroyed. This preservation of favorable variations and the rejection of injurious variations, I call Natural Selection.

The most serious difficulty facing Darwin's evolutionary theory was the lack of an adequate theory of inheritance that would account for the preservation through the generations of the variations on which natural selection was supposed to act. Contemporary theories of 'blending inheritance' proposed that the characteristics of parents became averaged in the offspring. As Darwin became aware, blending inheritance could not account for the conservation of variations, because halving the differences among variant offspring would rapidly reduce the original variation to the average of the already existing characteristics.

Mendelian genetics provided the missing link in Darwin's argument. About the time the *Origin of Species* was published, the Augustinian monk Gregor Mendel was performing a long series of experiments with peas in the garden of his monastery in Brünn (now Brno, Czech Republic). These experiments and the analysis of their results are an example of masterly scientific method. Mendel's theory accounts for biological inheritance through particulate factors (genes) inherited one from each parent, which do not mix or blend but segregate in the formation of the sex cells, or gametes. Mendel's discoveries, however, remained unknown to Darwin and, indeed, did not become generally known until 1900, when they were simultaneously rediscovered by three scientists.

The rediscovery in 1900 of Mendel's theory of heredity ushered in an emphasis on the role of heredity in evolution. In the 1920s and 1930s the theoretical work of geneticists demonstrated, first, that continuous variation (in such characteristics as size, number of eggs laid, and the like) could be explained by Mendel's laws; and second, that natural selection acting cumulatively on small variations could yield major evolutionary changes in form and function. Distinguished members of this group of theoretical geneticists were R. A. Fisher and J. B. S. Haldane in Britain and Sewall Wright in the United States. Their work had a limited impact on contemporary biologists because it was almost exclusively theoretical, formulated in mathematical language, and with little empirical corroboration. A major breakthrough came in 1937 with the publication of *Genetics and the Origin of Species* by Theodosius Dobzhansky, who advanced a reasonably comprehensive account of the evolutionary process in genetic terms, laced with experimental evidence supporting the theoretical argument. Other writers who importantly contributed to the formulation of the synthetic theory were the zoologists Ernst Mayr and Sir Julian Huxley, the paleontologist George G. Simpson, and the botanist George Ledyard Stebbins. By 1950, acceptance of Darwin's theory of evolution by natural selection was universal among biologists, and the synthetic theory had become widely adopted.

Since 1950, the most important line of investigation has been the application of molecular biology to evolutionary studies. In 1953 James Watson and Francis Crick discovered the structure of DNA (deoxyribonucleic acid), the hereditary material contained in the chromosomes of every cell's nucleus. The genetic information is contained within the sequence of components (nucleotides) that make up the long chain-like DNA molecules, very much in the same manner as semantic information is contained in the sequence of letters in an English text. This information determines the sequence of amino acids in the proteins, including the enzymes that carry out the organism's life processes. Comparisons of the amino acid sequences of proteins in different species provides quantitatively precise measures of species divergence, a considerable improvement over the typically qualitative evaluations obtained by comparative anatomy and other evolutionary subdisciplines.

1.2
Evolutionary Theory

Three different, though related, issues have been the main subjects of evolutionary investigations: (1) the fact of evolution; that is, that organisms are related by common descent with modification; (2) evolutionary history; that is, the details of when lineages split from one another and of the changes that occurred in each lineage; and (3) the mechanisms or processes by which evolutionary change occurs.

The fact of evolution is the most fundamental issue and the one established with utmost certainty. Darwin gathered much evidence in its support, but evidence has accumulated continuously ever since, derived from all biological disciplines. The second and third issues go much beyond the general affirmation that organisms evolve. The theory of evolution seeks to ascertain the evolutionary relationships between particular organisms and the events of evolutionary history, as well as to explain how and why evolution takes place. These are matters of active scientific investigation. Many conclusions are well established; for example, that the chimpanzee and gorilla are more closely related to humans than any of these three species is to the baboon or other monkeys; and that natural selection explains the adaptive configuration of such features as the human eye and the wings of birds. Some other matters are less certain, others are conjectural, and still others – such as precisely when life originated on earth and the characteristics of the first living things – remain largely unresolved.

1.3
The Evidence for Evolution: Paleontology

That organisms are related by common descent with modification has been demonstrated by evidence from paleontology, comparative anatomy, biogeography, embryology, biochemistry, molecular genetics, and other biological disciplines. The idea first emerged from observations of systematic changes in the succession of fossil remains found in a sequence of layered rocks. Such layers are now known to have a cumulative thickness of many scores of kilometers and to represent at least 3.5 billion years of geological time. The general sequence of fossils from bottom upward in layered rocks had been recognized before Darwin perceived that the observed progression of biological forms strongly implied common descent. The farther back into the past one looked, the less the fossils resembled recent forms, the more the various lineages merged, and the broader the implications of a common ancestry appeared.

Paleontology, however, was still a rudimentary science in Darwin's time, and large parts of the geological succession of stratified rocks were unknown or inadequately studied. Darwin, therefore, worried about the rarity of truly intermediate forms. Although gaps in the paleontological record remain even now, many have been filled by the researches of paleontologists since Darwin's time. Hundreds of thousands of fossil organisms found in well dated rock sequences represent a succession of forms through time and manifest many evolutionary transitions. Microbial life of the simplest type (i.e., prokaryotes, which are cells whose nuclear matter is not bounded by a nuclear membrane) was already in existence more than three billion years ago. The oldest evidence suggesting the existence of more complex organisms (i.e., eukaryotic cells with a true nucleus) has been discovered in fossils that had been sealed in flinty rocks approximately 1.4 billion years old. More advanced forms like true algae, fungi, higher plants, and animals have been found only in younger geological strata. Table 1.1 presents the order in which progressively complex forms of life appeared.

The sequence of observed forms and the fact that all except the first are constructed from the same basic cellular type strongly imply that all the major categories of life (including plants, algae, and fungi) have a common ancestry in the first eukaryotic cell. Moreover, there have been so many discoveries of intermediate forms between fish and amphibians, between amphibians and reptiles, between reptiles and mammals, and even along the primate line of descent from apes to humans, that it is often difficult to identify categorically along the line when the transition occurs from one to another particular genus or from one to another particular species. Nearly all fossils can be regarded as intermediates in some sense; they are life forms that come between ancestral forms that preceded them and those that followed.

The fossil record thus provides compelling evidence of systematic change through time – of descent with modification. From this consistent body of evidence it can be predicted that no reversals will be found in future paleontological studies. That is, amphibians will not appear before fishes nor mammals before reptiles, and no complex life will occur in the geological record before the oldest eukaryotic cells.

Table 1.1 Order in which progressively complex forms of animal life appeared.

Life form	*Millions of years since first known appearance (approximate)*
Microbial (prokaryotic cells)	3500
Complex (eukaryotic cells)	1400
First multicellular animals	670
Shell-bearing animals	540
Vertebrates (simple fishes)	490
Amphibians	350
Reptiles	310
Mammals	200
Nonhuman primates	60
Earliest apes	25
Earliest hominids	6
Homo sapiens (modern humans)	0.15 (150 000 years)

That prediction has been upheld by the evidence that has accumulated thus far: no reversals have been found.

1.3.1
Comparative Anatomy, Biogeography, Embryology

Inferences about common descent derived from paleontology have been reinforced by comparative anatomy. The skeletons of humans, dogs, whales, and bats are strikingly similar, despite the different ways of life led by these animals and the diversity of environments in which they have flourished. The correspondence, bone by bone, can be observed in every part of the body, including the limbs: a person writes, a dog runs, a whale swims, and a bird flies with structures built of the same bones (Figure 1.1). Scientists call such structures homologous and have concurred that they are best explained by common descent. Comparative anatomists investigate such homologies, not only in bone structure but also in other parts of the body as well, working out relationships from degrees of similarity. Their conclusions provide important inferences about the details of evolutionary history that can be tested by comparisons with the sequence of ancestral forms in the paleontological record.

The mammalian ear and jaw offer an example in which paleontology and comparative anatomy combine to show common ancestry through transitional stages. The lower jaws of mammals contain only one bone, whereas those of reptiles have several. The other bones in the reptile jaw are homologous with bones now found in the mammalian ear. What function could these bones have had during intermediate stages? Paleontologists have now discovered intermediate forms of mammal-like reptiles (Therapsida) with a double jaw joint – one composed of the bones that persist in mammalian jaws, the other consisting of bones that eventually became the hammer and anvil of the mammalian ear. Similar examples are numerous.

Biogeography also has contributed evidence for common descent. The diversity of life is stupendous. Approximately 250 thousand species of living plants, 100 thousand species of fungi, and 1.5 million species of animals and microorganisms have been described and named, each occupying its own peculiar ecological setting or niche, and the census is far from complete. Some species, such as human beings and our companion the dog, can live under a wide range of environmental conditions. Others are amazingly specialized. One species of the fungus *Laboulbenia* grows exclusively on the rear portion of the covering wings of a single species of beetle (*Aphaenops cronei*) found only in some caves of southern France. The larvae of the fly *Drosophila carcinophila* can develop only in specialized grooves beneath the flaps of the third pair of oral appendages of the land crab *Gecarcinus ruricola*, which is found only on certain Caribbean islands.

How can we make intelligible the colossal diversity of living beings and the existence of such extraordinary, seemingly whimsical creatures as *Laboulbenia*, *Drosophila carcinophila*, and others? Why are island groups like the Galapagos inhabited by forms similar to those on the nearest mainland but belonging to different species? Why is the indigenous life so different on different continents?

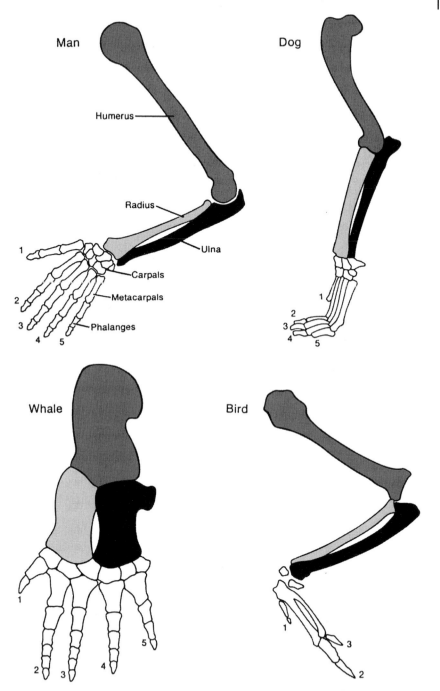

Figure 1.1 The forelimbs of four species of vertebrate animals. Although the limbs are used for different functions (writing, running, swimming, flying), the bones and their organization are similar, reflecting their derivation from a common ancestor.

Creationists contend that the curious facts of biogeography result from the occurrence of special creation events. A scientific hypothesis proposes that biological diversity results from an evolutionary process whereby the descendants of local or migrant predecessors became adapted to their diverse environments. A testable corollary of this hypothesis is that present forms and local fossils should show homologous attributes indicating how one is derived from the other. Also, there should be evidence that forms without an established local ancestry had migrated into the locality. Wherever such tests have been carried out, these conditions have been confirmed. A good example is provided by the mammalian populations of North and South America, where strikingly different endemic forms evolved in isolation until the emergence of the isthmus of Panama approximately three million years ago. Thereafter, the armadillo, porcupine, and opossum – mammals of South American origin – were able to migrate to North America along with many other species of plants and animals, while the placental mountain lion and other North American species made their way across the isthmus to the south.

The evidence that Darwin found for the influence of geographical distribution on the evolution of organisms has become stronger with advancing knowledge. For example, approximately 2000 species of flies belonging to the genus *Drosophila* are now found throughout the world. About one-quarter of them live only in Hawaii. More than a thousand species of snails and other land mollusks are also only found in Hawaii. The natural explanation for the occurrence of such great diversity among closely similar forms is that the differences resulted from adaptive colonization of isolated environments by animals with a common ancestry. The Hawaiian islands are far from and were never attached to any mainland or other islands, and they have had few colonizers. Organisms that reached these islands found many unoccupied ecological niches where they could then undergo separate evolutionary diversifications. No mammals other than one bat species lived on the Hawaiian islands when the first human settlers arrived; very many other kinds of plants and animals were also absent. The scientific explanation is that these kinds of organisms never reached the islands because of their great geographic isolation, while those that reached there multiplied in kind, because of the absence of related organisms that would compete for resources.

The vagaries of biogeography cannot be attributed to environmental peculiarities alone. The Hawaiian islands are no better than other Pacific islands for the survival of *Drosophila*, nor are they less hospitable than other parts of the world for many organisms not indigenous to them. For example, pigs and goats multiplied in Hawaii after their introduction by humans. The general observation is that all sorts of organisms are absent from places well suited to their occupancy. The animals and plants vary from continent to continent and from island to island in a distribution pattern consistent with colonization and evolutionary change, rather than being simply responsive to the conditions of place.

Embryology, the study of biological development from the time of conception, is another source of independent evidence for common descent. Barnacles, for instance, are sedentary crustaceans with little apparent similarity to such other crustaceans as lobsters, shrimps, or copepods. Yet barnacles pass through a free-swimming larval stage, in which they look unmistakably like other crustacean larvae. The similarity of larval stages supports the conclusion that all crustaceans have homologous parts and a common ancestry. Human and other mammalian embryos pass through a stage during which they have unmistakable but useless grooves similar to gill slits found in fishes – evidence that they and the other vertebrates shared remote ancestors that respired with the aid of gills.

1.3.2
Molecular Biology

The substantiation of common descent that emerges from all the foregoing lines of evidence has been validated and reinforced by the discoveries of molecular biology, a biological discipline that emerged in the mid-20th century. This new discipline has unveiled the nature of the hereditary material and the workings of organisms at the level of enzymes and other molecules. Molecular biology provides very detailed and convincing evidence for biological evolution.

The hereditary material, DNA, and the enzymes that govern all life processes hold information about an organism's ancestry. This information has made it possible to reconstruct evolutionary events that were previously unknown and to confirm and adjust the view of events that already were known. The precision with which events of evolution can be reconstructed is one reason the evidence from molecular biology is so compelling. Another reason is that molecular evolution has shown all living organisms, from bacteria to humans, to be related by descent from common ancestors.

The molecular components of organisms exhibit a remarkable uniformity in the nature of the components as well as in the ways in which they are assembled and used. In all bacteria, plants, animals, and humans, the DNA is made up of the same four component nucleotides, although many other nucleotides exist, and all of the various proteins are synthesized from different combinations and sequences of the same 20 amino acids, although several hundred other amino acids do exist. The genetic code by which the information contained in the nuclear DNA is translated into proteins is everywhere the same. Similar metabolic pathways are used by the most diverse organisms to produce energy and to construct the cell components.

This unity reveals the genetic continuity and common ancestry of all organisms. There is no other rational way to account for their molecular uniformity when numerous alternative structures are equally likely. The genetic code may serve as an example. Each particular sequence of three nucleotides in the nuclear DNA acts as a pattern, or code, for the production of exactly the same amino acid in all organisms. This is no more necessary than it is for a language to use a particular combination of letters to represent a particular reality. If it is found that many different combinations of letters, such as 'planet', 'tree', 'woman', are used with identical meanings in a number of different books, one can be sure that the languages used in those books are of common origin.

Genes and proteins are long molecules that contain information in the sequence of their components in much the same way as sentences of the English language contain information in the sequence of their letters and words. The sequences that make up the genes are passed on from parents to offspring, identical except for occasional changes introduced by mutations. The similarity between the sequences is evidence that they derive from a common ancestral sequence; the differences reflect the time elapsed since they started to diverge and thus make it possible to reconstruct the evolutionary history of the organisms that carry the sequences. To illustrate, assume that two books are being compared; both books are 200 pages long and contain the same number of chapters. Closer examination reveals that the two books are identical page for page and word for word, except that an occasional word – say one in 100 – is different. The two books cannot have been written independently; either one has been copied from the other or both have been copied, directly or indirectly, from the same original book. Similarly, if each nucleotide is represented by one letter, the complete sequence of nucleotides in the DNA of a higher organism would require several hundred books of hundreds of pages, with several thousand letters on each page. When the 'pages' (or sequence of nucleotides) in these 'books' (organisms) are examined one by one, the correspondence in the 'letters' (nucleotides) gives unmistakable evidence of common origin.

Two arguments attest to evolution. Using the alphabet analogy, the first argument says that languages that use the same dictionary – the same genetic code and the same 20 amino acids – cannot be of independent origin. The second argument, concerning similarity in the sequence of nucleotides in the DNA or the sequence of amino acids in the proteins, says that books with very similar texts cannot be of independent origin.

The evidence of evolution revealed by molecular biology goes one step further. The degree of similarity in the sequence of nucleotides or of amino acids can be precisely quantified. For example, cytochrome c (a protein) of humans and chimpanzees consists of the same 104 amino acids in exactly the same order; but differs from that of rhesus monkeys by one amino acid, from that of horses by 11 amino acids, and from that of tuna by 21 amino acids. The degree of similarity reflects the recentness of common ancestry. Thus, the inferences from comparative anatomy and other disciplines concerning evolutionary history can be tested in molecular studies of DNA and proteins by examining their sequences of nucleotides and amino acids.

The authority of this kind of test is overwhelming; each of the thousands of genes and thousands of proteins contained in an organism provides an independent test of that organism's evolutionary history. Not all possible tests have been performed, but many hundreds have been done, and not one has given evidence contrary to evolution. There is probably no other notion in any field of science that has been as extensively tested and as thoroughly corroborated as the evolutionary origin of living organisms. There is no reason to doubt the evolutionary theory of the origin of organisms any more than to doubt the heliocentric theory of the rotations of the planets around the sun.

1.4
Genetics of Evolution

The central argument of Darwin's theory of evolution starts from the existence of hereditary variation. Experience with animal and plant breeding demonstrates that variations can be developed that are useful to humans. So, reasoned Darwin, variations must occur in nature that are favorable or useful in some way to the organism itself in the struggle for existence. Favorable variations are ones that increase chances for survival and procreation. These advantageous variations are preserved and multiplied from generation to generation at the expense of less advantageous ones in the process known as natural selection. The outcome of the process is an organism that is well adapted to its environment, and evolution often occurs as a consequence.

Biological evolution is the process of change and diversification of living things over time, and it affects all aspects of their lives – morphology, physiology, behavior, and ecology. Underlying these changes are changes in the hereditary materials. Hence, in genetic terms, evolution consists of changes in the organism's hereditary makeup.

Natural selection, then, can be defined as the differential reproduction of alternative hereditary variants, determined by the fact that some variants increase the likelihood that the organisms having them will survive and reproduce more successfully than will organisms carrying alternative variants. Selection may be due to differences in survival, in fertility, in rate of development, in mating success, or in any other aspect of the life cycle. All of these differences can be incorporated under the term 'differential reproduction', because all result in natural selection to the extent that they affect the number of progeny an organism leaves.

Darwin explained that competition for limited resources results in the survival of the most effective competitors. But natural selection may occur not only as a result of competition but also as an effect of some aspect of the physical environment, such as inclement weather. Moreover, natural selection would occur even if all the members of a population died at the same age, if some of them produced more offspring than others. Natural selection is quantified by a measure called Darwinian fitness, or relative fitness. Fitness in this sense is the relative probability that a hereditary characteristic will be reproduced; that is, the degree of fitness is a measure of the reproductive efficiency of the characteristic.

Evolution can be seen as a two-step process. First, hereditary variation takes place; second, selection occurs of those genetic variants that will be passed on most effectively to the following generations. Hereditary variation also entails two mechanisms: the spontaneous mutation of one variant to another, and the sexual process that recombines those variants to form a multitude of variations.

The gene pool of a species is the sum total of all of the genes and combinations of genes that occur in all organisms of the same species. The necessity of hereditary variation for evolutionary change to occur can be understood in terms of the gene pool. Assume, for instance, that at the gene locus that codes for the human MN blood groups, there is no variation; only the *M* form exists in all individuals.

Evolution of the MN blood groups cannot take place in such a population, since the allelic frequencies have no opportunity to change from generation to generation. On the other hand, in populations in which both forms *M* and *N* are present, evolutionary change is possible.

The more genetic variation that exists in a population, the greater the opportunity for evolution to occur. As the number of genes that are variable increases and as the number of forms of each gene becomes greater, the likelihood that some forms will change in frequency at the expense of their alternates grows. The British geneticist R. A. Fisher mathematically demonstrated a direct correlation between the amount of genetic variation in a population and the rate of evolutionary change by natural selection. This demonstration is embodied in his fundamental theorem of natural selection: "The rate of increase in fitness of any organism at any time is equal to its genetic variance in fitness at that time." This theorem has been confirmed experimentally. Because a population's potential for evolving is determined by its genetic variation, evolutionists are interested in discovering the extent of such variation in natural populations. Techniques for determining genetic variation have been used to investigate numerous species of plants and animals. Typically, insects and other invertebrates are more varied genetically than mammals and other vertebrates; and plants bred by outcrossing exhibit more variation than those bred by self-pollination. But the amount of genetic variation is in any case astounding. Every individual represents a unique genetic configuration that will never be repeated again. The enormous reservoir of genetic variation in natural populations provides virtually unlimited opportunities for evolutionary change in response to environmental constraints and the needs of the organisms, beyond the new variations that arise every generation by the process of mutation, which I discuss next.

1.4.1
Variation and Mutation

There are more than two million known species, which are widely diverse in size, shape, and ways of life, as well as in the DNA sequences that contain their genetic information. What has produced the pervasive genetic variation within natural populations and the genetic differences among species?

The information encoded in the nucleotide sequence of DNA is, as a rule, faithfully reproduced during replication, so that each replication results in two DNA molecules that are identical to each other and to the parent molecule. But, occasionally 'mistakes', or mutations, occur in the DNA molecule during replication, so that daughter cells differ from the parent cells in at least one of the letters in the DNA sequence. A mutation first appears on a single cell of an organism, but it is passed on to all cells descended from the first. Mutations can be classified into two categories: gene, or point, mutations, which affect only a few letters (nucleotides) within a gene; and chromosomal mutations, which either change the number of chromosomes or change the number or arrangement of genes on a chromosome. (Chromosomes are the elongated structures that store the DNA of each cell.)

Gene mutations can occur spontaneously; that is, without being intentionally caused by humans. They can also be artificially induced by ultraviolet light, X irradiation, and other high-frequency types of radiation, as well as by exposure to certain mutagenic chemicals, such as mustard gas. The consequences of gene mutations may range from negligible to lethal. Some have a small or undetectable effect on the organism's ability to survive and reproduce, because no essential biological functions are altered. But when the active site of an enzyme or some other essential function is affected, the impact may be severe.

Newly arisen mutations are more likely to be harmful than beneficial to their carriers, because mutations are random events with respect to adaptation; that is, their occurrence is independent of any possible consequences. Harmful mutations are eliminated or kept in check by natural selection. Occasionally, however, a new mutation may increase the organism's adaptation. The probability of such an event's happening is greater when organisms colonize a new territory or when environmental changes confront a population with new challenges. In these situations, the established adaptation of a population is less than optimal, and there is greater opportunity for new mutations to be better adaptive. This is so because the consequences of mutations depend on the environment. Increased melanin pigmentation may be advantageous to inhabitants of tropical Africa, where dark skin protects them from the sun's ultraviolet radiation; but it is not beneficial in Scandinavia, where the intensity of sunlight is low and light skin facilitates the synthesis of vitamin D.

Mutation rates are low, but new mutants appear continuously in nature, because there are many individuals in every species and many genes in every individual. The process of mutation provides each generation with many new genetic variations. More important yet is the storage of variation, the results of past mutations, in each organism. Thus, it is not surprising to see that when new environmental challenges arise, species are able to adapt to them. More than 200 insect and rodent species, for example, have developed resistance to the pesticide DDT in different parts of the world where spraying has been intense. Although the insects had never before encountered this synthetic compound, they adapted to it rapidly by means of mutations that allowed them to survive in its presence. Similarly, many species of moths and butterflies in industrialized regions have shown an increase in the frequency of individuals with dark wings in response to environmental pollution, an adaptation known as industrial melanism. The examples can be multiplied at will.

1.4.2
Genetic Equilibrium and Genetic Change

The genetic variation present in natural populations of organisms is sorted out in new ways in each generation by the process of sexual reproduction. But heredity by itself does not change gene frequencies. This principle is formally stated by the Hardy–Weinberg law, an algebraic equation that describes the genetic equilibrium in a population. In the simplest instance, there are two alleles, *A* and *a*, at a gene locus, so that three genotypes are possible, *AA*, *Aa*, and *aa*. If the frequencies of the two alleles in the population are p and q, respectively, the equilibrium frequencies of the three genotypes are given by $(p + q)^2 = p^2 + 2\,p\,q + q^2$ for *AA*, *Aa*, and *aa*, respectively. The genotype equilibrium frequencies for any number of alleles are derived in the same way.

The Hardy–Weinberg law assumes that gene frequencies remain constant from generation to generation – that there is no gene mutation or natural selection and that populations are very large. But these assumptions are not correct; indeed, if they were, evolution could not occur. Why, then, is the Hardy–Weinberg law significant if its assumptions do not hold true in nature? The answer is that the Hardy–Weinberg law plays in evolutionary studies a role similar to that of Newton's first law of motion in mechanics. Newton's first law says that a body not acted upon by a net external force remains at rest or maintains a constant velocity. In fact, there are always external forces acting upon physical objects (gravity, for example), but the first law provides the starting point for the application of other laws. Similarly, organisms are subject to mutation, selection, and other processes that change gene frequencies, and the effects of these processes are calculated by using the Hardy–Weinberg law as the starting point. There are four processes of gene frequency change: mutation, migration, drift, and natural selection.

The allelic variations that make evolution possible are generated by the process of mutation; but new mutations change gene frequencies very slowly, since mutation rates are low. Moreover, gene mutations are reversible. Changes in gene frequencies due to mutation occur, therefore, very slowly. In any case, allelic frequencies usually are not in mutational equilibrium, because some alleles are favored over others by natural selection. The equilibrium frequencies are then decided by the interaction between mutation and selection, with selection usually having the greater consequence.

I have already discussed the process of mutation. Migration, or gene flow, takes place when individuals migrate from one population to another and interbreed with its members. The genetic makeup of populations changes locally whenever different populations intermingle. In general, the greater the difference in allele frequencies between the resident and the migrant individuals and the larger the number of migrants, the greater effect the migrants have in changing the genetic constitution of the resident population.

Moreover, gene frequencies can change from one generation to another by a process of pure chance known as genetic drift. This occurs because populations are finite in numbers, and thus the frequency of a gene may change in the following generation by accidents of sampling, just as it is possible to get more or fewer than 50 heads in 100 throws of a coin simply by chance. The magnitude of the gene frequency changes due to genetic drift is inversely related to the size of the population: the larger the number of reproducing individuals, the smaller the effects of genetic drift. The effects of genetic drift in changing gene frequencies from one generation to the next are small in most natural populations, which generally consist of thousands of reproducing individuals. The effects over many generations are more important. Genetic drift can have important evolutionary consequences when a new population becomes established by only a few individuals, as in the colonization of islands and lakes. This is one reason why species in neighboring islands, such as those in the Hawaiian archipelago, are often more heterogeneous than species in comparable continental areas adjacent to one another.

1.5
Natural Selection

The phrase 'natural selection' was used by Darwin to refer to any reproductive bias favoring some hereditary variants over others. He proposed that natural selection promotes the adaptation of organisms to the environments in which they live because the organisms carrying such useful variants would leave more descendants than those lacking them. The modern concept of natural selection derives directly from Darwin's but is defined precisely in mathematical terms as a statistical bias favoring some genetic variants over their alternates (the measure to quantify natural selection is called fitness). Hereditary variants, favorable or not to the organisms, arise by mutation. Unfavorable ones are eventually eliminated by natural selection; their carriers leave no descendants or leave fewer than those carrying alternative variants. Favorable mutations accumulate over the generations. The process continues indefinitely because the environments that organisms live in are forever changing. Environments change physically – in their climate, physical configuration, and so on – but also biologically, because the predators, parasites, and competitors with which an organism interacts are themselves evolving.

If mutation, migration, and drift were the only processes of evolutionary change, the organization of living things would gradually disintegrate, because they are random processes with respect to adaptation. These three processes change gene frequencies without regard for the consequences that such changes may have on the ability of the organisms to survive and reproduce. The effects of such processes alone would be analogous to those of a mechanic who changed parts in an automobile engine at random, with no regard for the role of the parts in the engine. Natural selection keeps the disorganizing effects of mutation and other processes in check, because it multiplies beneficial mutations and eliminates harmful ones. But natural selection accounts not only for the preservation and improvement of the organization of living beings but also for their diversity. In different localities or in different circumstances, natural selection favors different traits, precisely those that make the organisms well adapted to their particular circumstances and ways of life.

The effects of natural selection can be studied by measuring the ensuing changes in gene frequencies; but they can also be explored by examining changes on the observable characteristics – or phenotypes – of individuals in a population. Distribution scales of phenotypic traits such as height, weight, number of progeny, or longevity typically show greater numbers of individuals with intermediate values and fewer and fewer toward the extremes (the so-called normal distribution). When individuals with intermediate phenotypes are favored and extreme phenotypes are selected against, the selection is said to be stabilizing. The range and distribution of phenotypes then remains approximately the same from one generation to another. Stabilizing selection is very common. The individuals that survive and reproduce more successfully are those that have intermediate phenotypic values. Mortality among newborn infants, for example, is highest when they are either very small or very large; infants of intermediate size have a greater chance of surviving.

But the distribution of phenotypes in a population sometimes changes systematically in a particular direction. The physical and biological aspects of the environment are continuously changing, and over long periods of time the changes may be substantial. The climate and even the configuration of the land or water vary incessantly. Changes also take place in the biotic conditions; that is, in the other organisms present, whether predators, prey, parasites, or competitors. Genetic changes occur as a consequence, because the genotypic fitnesses may be shifted so that different sets of variants are favored. The opportunity for directional selection also arises when organisms colonize new environments where the conditions are different from those of their original habitat. The process of directional selection often takes place in spurts. The replacement of one genetic constitution for another changes the genotypic fitnesses of genes for other traits, which in turn stimulates additional changes, and so on in a cascade of consequences.

The nearly universal success of artificial selection and the rapid response of natural populations to new environmental challenges are evidence that existing variation provides the necessary materials for directional selection, as Darwin already explained. More generally, human actions have been an important stimulus to this type of selection. Mankind transforms the environments of many organisms, which rapidly respond to the new environmental challenges through directional selection. Well known instances are the many cases of insect resistance to pesticides, synthetic substances not present in the natural environment. Whenever a new insecticide is first applied to control a pest, the results are encouraging, because a small amount of the insecticide is sufficient to bring the pest organism under control. As time passes, however, the amount required to achieve a certain level of control must be increased again and again, until finally it becomes ineffective or economically impractical. This occurs because organisms become resistant to the pesticide through directional selection. Resistance of the housefly, *Musca domestica*, to DDT was first reported in 1947. Resistance to one or more pesticides has now been recorded in more than 100 species of insects.

Sustained directional selection leads to major changes in morphology and ways of life over geologic time. Evolutionary changes that persist in a more-or-less continuous fashion over long periods of time are known as evolutionary trends. Directional evolutionary changes increased the cranial capacity of the human lineage from *Australopithecus afarensis*, human ancestors who lived four million years ago, with a small brain weighing somewhat less than one pound, to *Homo sapiens*, modern humans with a brain three and a half times as large. The more-or-less gradual increase in size during the evolution of the horse family from 50 million years ago to modern times is another of the many well studied examples of directional selection.

Two or more divergent traits in an environment may be favored simultaneously, which is called diversifying selection. No natural environment is homogeneous; rather, the environment of any plant or animal population is a mosaic consisting of more-or-less dissimilar subenvironments. There is heterogeneity with respect to climate, food resources, and living space. Also, the heterogeneity may be temporal, with change occurring over time, as well as spatial, with dissimilarity found in different areas. Species cope with environmental heterogeneity in diverse ways. One strategy is the selection of a generalist genotype that is well adapted to all of the subenvironments encountered by the species. Another strategy is genetic polymorphism, the selection of a diversified gene pool that yields different genetic makeups, each adapted to a specific subenvironment.

1.6
Sexual Selection and Kin Selection

One important factor in reproduction is mutual attraction between the sexes. The males and females of many animal species are fairly similar in size and shape except for the sexual organs and secondary sexual characteristics such as the mammary glands of female mammals. There are, however, species in which the sexes exhibit striking dimorphism. Particularly in birds and mammals, the males are often larger and stronger, more brightly colored, or endowed with conspicuous adornments. But bright colors make animals more visible to predators; for example, the long plumage of peacocks and birds of paradise and the enormous antlers of aged male deer are cumbersome loads at best. Darwin knew that natural selection could not be expected to favor the evolution of disadvantageous traits, and he was able to offer a solution to this problem. He proposed that such traits arise by 'sexual selection', which "depends not on a struggle for existence in relation to other organic beings or to external conditions, but on a struggle between the individuals of one sex, generally the males, for the possession of the other sex." Thus, the colored plumage of the males in some bird species makes them more attractive to their females, which more than compensates for their increased visibility to potential predators. Sexual selection is a topic of intensive research at present.

The apparent altruistic behavior of many animals is, like some manifestations of sexual selection, a trait that at first seems incompatible with the theory of natural selection. Altruism is a form of behavior that benefits other individuals at the expense of the one that performs the action; the fitness of the altruist is diminished by its behavior, whereas individuals that act selfishly benefit from it at no cost to themselves. Accordingly, it might be expected that natural selection would foster the development of selfish behavior and eliminate altruism. This conclusion is not so compelling when it is noticed that the beneficiaries of altruistic behavior are usually relatives. They all carry the same genes, including the genes that promote altruistic behavior. Altruism may evolve by kin selection, which is simply a type of natural selection in which relatives are taken into consideration when evaluating an individual's fitness.

Kin selection is explained as follows. Natural selection favors genes that increase the reproductive success of their carriers, but it is not necessary that all individuals with a given genetic makeup have higher reproductive success. It suffices that carriers of the genotype reproduce more successfully on average than those possessing alternative genotypes. A parent shares half its genes with each progeny, so a gene that promotes parental altruism is favored by selection if the behavior's cost to the parent is less than half its average benefits to the progeny. Such a gene will be more likely to increase in frequency through the generations than an alternative gene that does not promote parental care. The parent spends some energy caring for the progeny because it increases the reproductive success of the parent's genes. But kin selection extends beyond the relationship between parents and their offspring. It facilitates the development of altruistic behavior when the energy invested, or the risk incurred, by an individual is compensated in excess by the benefits ensuing to relatives.

In many species of primates (as well as in other animals), altruism also occurs among unrelated individuals when the behavior is reciprocal and the altruist's costs are smaller than the benefits to the recipient. This reciprocal altruism is found, for example, in the mutual grooming of chimpanzees as they clean each other of lice and other pests. Another example appears in flocks of birds that post sentinels to warn of danger. One crow sitting in a tree watching for predators, while the rest of the flock forages, incurs a small loss by not feeding, but this is well compensated by the protection it receives when it itself forages and others of the flock stand guard.

1.7
The Origin of Species

Darwin sought to explain the splendid diversity of the living world: thousands of organisms of the most diverse kinds, from lowly worms to spectacular birds of paradise, from yeasts and molds to oaks and orchids. His *Origin of Species* is a sustained argument showing that the diversity of organisms and their characteristics can be explained as the result of natural processes. As Darwin noted, different species may come about as the result of gradual adaptation to environments that are continuously changing in time and differ from place to place. Natural selection favors different characteristics in different situations.

In everyday experience we identify different kinds of organisms by their appearance. Everyone knows that people belong to the human species and are different from cats and dogs, which in turn are different from each other. There are differences among people, as well as among cats and dogs; but individuals of the same species are considerably more similar among themselves than they are to individuals of other species. But there is more to it than that: a bulldog, a terrier, and a golden retriever are very different in appearance, but they are all dogs because they can interbreed. People can also interbreed with one another, and so can cats, but people cannot interbreed with dogs or cats, nor these with each other. It is, then, clear that although species are usually identified by appearance, there is something basic, of great biological significance, behind similarity of appearance; namely, that individuals of a species are able to interbreed with one another but not with members of other species. This is expressed in the following definition: species are groups of interbreeding natural populations that are reproductively isolated from other such groups.

The ability to interbreed is of great evolutionary importance, because it determines that species are independent evolutionary units. Genetic changes originate in single individuals; they can spread by natural selection to all members of the species but not to individuals of other species. Thus, individuals of a species share a common gene pool that is not shared by individuals of other species, because they are reproductively isolated.

Although the criterion for deciding whether individuals belong to the same species is clear, there may be ambiguity in practice for two reasons. One is lack of knowledge: it may not be known for certain whether individuals living in different sites below to the same species, because it is not known whether they can naturally interbreed. The other reason for ambiguity is rooted in the nature of evolution as a gradual process. Two geographically separate populations that at one time were members of the same species later may have diverged into two different species. Since the process is gradual, there is no particular point at which it is possible to say that the two populations have become two different species. It is an interesting curiosity that some antievolutionists have referred to the existence of species intermediates as evidence against evolution; quite the contrary, such intermediates are precisely expected.

A similar kind of ambiguity arises with respect to organisms living at different times. There is no way to test whether or not today's humans could interbreed with those who lived thousands of years ago. It seems reasonable that living people, or living cats, would be able to interbreed with people, or cats, exactly like those that lived a few generations earlier. But what about the ancestors removed by one thousand or one million generations? The ancestors of modern humans that lived one million years ago (about 50 thousand generations) are classified in the species *Homo erectus*, whereas present-day humans are classified in a different species, *Homo sapiens*, because those ancestors were quite different from us in appearance and thus it seems reasonable to conclude that interbreeding could not have occurred with modern-like humans. But there is no exact time at which *Homo erectus* became *Homo sapiens*. It would not be appropriate to classify remote human ancestors and modern humans in the same species just because the changes from one generation to the next are small. It is useful to distinguish between the two groups by means of different species names, just as it is useful to give different names to childhood, adolescence, and adulthood, even though there is no one moment at which an individual passes from one to the next. Biologists distinguish species in organisms that lived at different times by means of a commonsense rule: if two organisms differ from each other about as much as two living individuals belonging to two different species differ today, they are classified into separate species and given different names.

Bacteria and blue-green algae do not reproduce sexually, but by fission. Organisms that lack sexual reproduction are classified into different species according to criteria such as external morphology, chemical and physiological properties, and genetic constitution. The definition of species given above applies only to organisms able to interbreed.

Since species are groups of populations reproductively isolated form one another, asking about the origin of a species is equivalent to asking how reproductive isolation arises between populations. Two theories have been advanced to answer this question. One theory considers isolation as an accidental byproduct of genetic divergence. Populations that become genetically less and less alike (as a consequence, for example, of adaptation to different environments) may eventually be unable to interbreed because their gene pools are disharmonious. The other theory regards isolation as a product of natural selection. Whenever hybrid individuals are less fit than nonhybrids, natural selection directly promotes the development of reproductive isolation. This occurs because genetic variants interfering with hybridization have greater fitness than those favoring hybridization, given that the latter are often present in poorly fit hybrids. Scientists have shown that these two theories of the origin of reproductive isolation are not mutually exclusive.

1.7.1
Adaptive Radiation

The geographic separation of populations derived from common ancestors may continue long enough that the populations become completely differentiated species before coming together again. As the separated populations continue evolving independently, morphological differences may arise. Examples of adaptive radiation are common in archipelagos far removed from the mainland. The Galapagos Islands are about 600 miles off the west coast of South America. When Darwin arrived there in 1835, he discovered many species not found anywhere else in the world – for example, 14 species of finch (known as Darwin's finches). These passerine birds have adapted to a diversity of habitats and diets, some feeding mostly on plants, others exclusively on insects. The various shapes of their bills are clearly adapted to probing, grasping, biting, or crushing – the diverse ways in which these different Galapagos species obtain their food. The explanation for such diversity (which is not found in finches from the continental mainland) is that the ancestor of Galapagos finches arrived in the islands before other kinds of birds and encountered an abundance of unoccupied ecological opportunities. The finches underwent adaptive radiation, evolving a variety of species with ways of life capable of exploiting niches that in continental faunas are exploited by different kinds of birds.

Striking examples of adaptive radiation occur in the Hawaiian Islands. The archipelago consists of several volcanic islands, ranging from less than one million to more than ten million years in age, far away from any continent or other large islands. An astounding number of plant and animal species of certain kinds exist in the islands while many other kinds are lacking. Among the species that have evolved in the islands, there are about two dozen (about one-third of them now extinct) of honeycreepers, birds of the family Drepanididae, all derived from a single immigrant form. In fact, all but one of Hawaii's 71 native bird species are endemic; that is, they have evolved there and are found nowhere else. More than 90% of the native species of the hundreds of flowering plants, land mollusks, and insects in Hawaii are also endemic, as are two-thirds of the 168 species of ferns. About one-fourth of the world's total number of known species of *Drosophila* flies (more than 500) are native Hawaiian species. The species of *Drosophila* in Hawaii have diverged by adaptive radiation from one or a few colonizers, which encountered an assortment of ecological opportunities that in other lands are occupied by different groups of flies or insects.

1.7.2
Rapid or Quantum Speciation

Examples of rapid speciation are also known in many organisms. Instances of rapid speciation are sometimes called quantum or saltational speciation. An important form of quantum speciation occurs as a result of polyploidy, which is the multiplication of entire sets of chromosomes. This can happen in a single generation, for example, if meiosis fails so that an individual's gametes have two sets of chromosomes, rather than only one. If a male and a female gamete, each with two sets of chromosomes, combine, the resulting individual will have four sets, rather than two sets, of chromosomes. A typical (diploid) organism carries in the nucleus of each cell two sets of chromosomes, one inherited from each parent; a polyploid organism has several sets of chromosomes. Many cultivated plants are polyploid: bananas have three sets of chromosomes, potatoes have four, bread wheat has six, some strawberries have eight. All major groups of plants have natural polyploid species, but they are most common among flowering plants (angiosperms), of which about 47% are polyploids.

In animals, polyploidy is relatively rare because it disrupts the balance between chromosomes involved in the determination of sex. But polyploid species are found in hermaphroditic animals (individuals having both male and female organs), which include snails and earthworms, as well as in forms with parthenogenetic females (which produce viable progeny without fertilization), such as some beetles, sow bugs, goldfish, and salamanders.

1.8
Evolutionary History

It is possible to look at two sides of evolution: one, called anagenesis, refers to changes that occur within a lineage; the other, called cladogenesis, refers to the split of a lineage into two or more separate lineages. Anagenetic evolution has, over the last four million years more than tripled the size of the human brain; in the lineage of the horse, it has reduced the number of toes from four to one. Cladogenetic evolution has produced the extraordinary diversity of the living world, with more than two million species of animals, plants, fungi, and microorganisms.

The evolution of all living organisms, or of a subset of them, can be represented as a tree, with branches that divide into two or more as time progresses. Such trees are called phylogenies. Their branches represent evolving lineages, some of which eventually die out while others have persisted to the present time. Evolutionists are interested in the history of life and hence in the topology, or configuration, of evolution's trees. They also want to know the anagenetic changes along lineages and the timing of important events.

Tree relationships are ascertained by means of several complementary sources of evidence. First, there is the fossil record, which provides definitive evidence of relationships among some groups of organisms, but is far from complete and often seriously deficient. Second, there is comparative anatomy, the comparative study of living forms; and the related disciplines of comparative embryology, cytology, ethology, biogeography, and others. In recent years the comparative study of informational macromolecules – proteins and nucleic acids – has become a powerful tool for the study of evolution's history. We saw earlier how the results from these disciplines demonstrate that evolution has occurred. Advanced methods have now been developed to reconstruct evolution's history.

These methods make it possible to identify whether the correspondence of features in different organisms is due to inheritance from a common ancestor, which is called homology. The forelimbs of humans, whales, dogs, and bats are homologous. The skeletons of these limbs are all constructed of bones arranged according to the same pattern because they derive from an ancestor with similarly arranged forelimbs (Figure 1.1). Correspondence of features due to similarity of function but not related to common descent is termed analogy. The wings of birds and of flies are analogous. Their wings are not modified versions of a structure present in a common ancestor but rather have developed independently as adaptations enabling them to fly.

Homology can be recognized not only between different organisms but also between repetitive structures of the same organism. This has been called serial homology. There is serial homology, for example, between the arms and legs of humans, among the seven cervical vertebrae of mammals, and among the branches or leaves of a tree. The jointed appendages of arthropods are elaborate examples of serial homology. Crayfish have 19 pairs of appendages, all built according to the same basic pattern but serving diverse functions – sensing, chewing, food handling, walking, mating, egg carrying, and swimming. Serial homologies are not useful in reconstructing the phylogenetic relationships of organisms, but they are an important dimension of the evolutionary process.

Relationships in some sense akin to those between serial homologs exist at the molecular level between genes and proteins derived from ancestral gene duplications. The genes coding the various hemoglobin chains are an example. About 500 million years ago a chromosome segment carrying the gene encoding hemoglobin became duplicated, so that the genes in the different segments thereafter evolved in somewhat different ways, one eventually giving rise to the modern gene coding for α hemoglobin, the other for β hemoglobin. The β hemoglobin gene became duplicated again about 200 million years ago, giving rise to the γ (fetal) hemoglobin gene. The α, β, and γ hemoglobin genes are homologous: similarities in their DNA sequences occur because they are modified descendants of a single ancestral sequence.

1.9
Punctuational Evolution

Morphological evolution is a more-or-less gradual process, as shown by the fossil record. Major evolutionary changes are usually due to a building up over the ages of relatively small changes. But the fossil record is often discontinuous, so that it fails to manifest the gradual transition from one form to another. Fossil strata are separated by sharp boundaries; accumulation of fossils within a geologic deposit (stratum) is fairly constant over time, but the transition from one stratum to another may involve gaps of tens of thousands of years. Different species, characterized by small but discontinuous morphological changes, typically appear at the boundaries between strata, whereas the fossils within a stratum exhibit little morphological variation. This is not to say that the transition from one stratum to another always involves sudden changes in morphology; on the contrary, fossil forms often persist virtually unchanged through several geologic strata, each representing millions of years.

Paleontologists attributed the apparent morphological discontinuities in the fossil record to the discontinuity of the sediments; that is, to substantial time gaps encompassed in the boundaries between strata. The assumption was that, if the fossil deposits were more continuous, they would show a more gradual transition of forms. Even so, morphological evolution would not always keep progressing gradually, because some forms, at least, remain unchanged for extremely long times. Examples are the lineages known as living fossils: the lamp shell *Lingula*, a genus of brachiopod that appears to have remained essentially unchanged since the Ordovician Period, some 450 million years ago; or the tuatara (*Sphenodon punctatus*), a reptile that has shown little morphological evolution for nearly 200 million years since the early Mesozoic.

According to some paleontologists, however, the frequent discontinuities in the fossil record are not artifacts created by gaps in the record, but rather reflect the true nature of morphological evolution, which happens in sudden bursts associated with the formation of new species. The lack of morphological evolution, or stasis, of lineages such as *Lingula* and *Sphenodon* is in turn due to lack of speciation within those lineages. The proposition that morphological evolution is jerky, with most morphological change occurring during the brief speciation events and virtually no change during the subsequent existence of the species, is known as the punctuated equilibrium model of morphological evolution. Most evolutionists, however, agree that evolution is on the whole gradual, rather than discontinuous.

1.10
Molecular Evolution

Molecular biology has made possible the comparative study of proteins and nucleic acid, DNA, which is the repository of hereditary (and therefore, evolutionary) information. The relationship of proteins to DNA is so immediate that they closely reflect the hereditary information. This reflection is not perfect, because the genetic code is redundant and, consequently, some differences in DNA do not yield differences in the synthesized proteins. Moreover, it is not complete, because a large fraction of the DNA (about 90% in many organisms) does not code for proteins. Nevertheless, proteins are so closely related to the information contained in the DNA that they, as well as the nucleic acids, are called informational macromolecules.

Nucleic acids and proteins are linear molecules made up of sequences of units – nucleotides in nucleic acids, amino acids in proteins – which retain considerable amounts of evolutionary information. Comparing two macromolecules establishes the number of their units that are different. Because evolution usually occurs by changing one unit at a time, the number of differences is an indication of the recentness of common ancestry. Changes in evolutionary rates may create difficulties, but macromolecular studies have two notable advantages over comparative anatomy and other classical disciplines. One is that the information is more readily quantifiable. The number of units that are different is precisely established when the sequence of units is known for a given macromolecule in different organisms. The other advantage is that comparisons can be made even between very different sorts of organisms. There is very little that comparative anatomy can say when organisms as diverse as yeasts, pine trees, and human beings are compared; but there are homologous DNA and protein molecules that can be compared among all three (Figure 1.2).

Informational macromolecules provide information not only about the topology of evolutionary history (that is, the configuration of evolutionary trees), but also about the amount of genetic change that has occurred in any given branch. It might seem at first that determining the number of changes in a branch would be impossible for proteins and nucleic acids, because it would require comparison of molecules from organisms that lived in the past with those from living organisms. But this determination can actually be made using elaborate methods developed by scientists who investigate the evolution of DNA and proteins.

One conspicuous attribute of molecular evolution is that differences between homologous molecules can readily be quantified and expressed as, for example, proportions of nucleotides or amino acids that have changed. Rates of evolutionary change can therefore be more precisely established with respect to DNA or proteins than with respect to morphological traits. Studies of molecular evolution rates have led to the proposition that macromolecules evolve as fairly accurate 'clocks'. If the rate of evolution of a protein or gene were approximately the same in the evolutionary lineages leading to different species, proteins and DNA sequences would provide a molecular clock of evolution. The sequences could then be used to reconstruct not

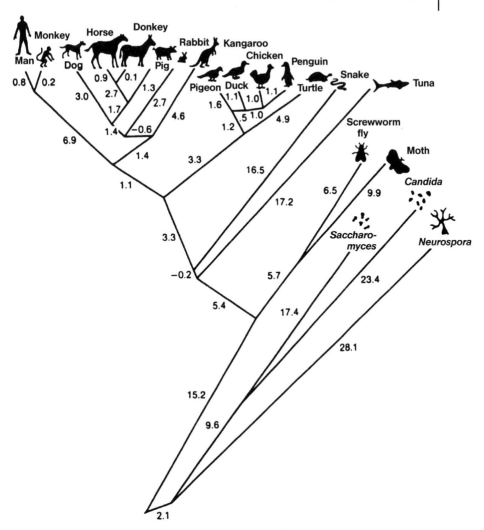

Figure 1.2 Evolutionary tree of 20 diverse organisms, ranging from the human, through the horse, chicken, and snake, to moth and yeast. This tree is based on the amino acid sequences of a small protein, cytochrome c. Although the evolutionary relationships are not all accurate, it is remarkable that they can be determined by examining a single protein. A more nearly correct evolutionary tree can be constructed by combining data for several proteins.

only the topology of the evolutionary tree but also the time when the various branching events occurred.

The molecular evolutionary clock is not a metronomic clock, like a watch or other timepiece that measures time exactly, but a stochastic clock like that of radioactive decay. In a stochastic clock, the probability of a certain amount of change is constant, although some variation occurs in the actual amount of change.

Over fairly long periods of time, a stochastic clock is quite accurate. The enormous potential of the molecular evolutionary clock lies in the fact that each gene or protein is a separate clock. Each clock 'ticks' at a different rate – the rate of evolution characteristic of a particular gene or protein, but each of the thousands and thousands of genes or proteins provides an independent measure of the same evolutionary events.

Evolutionists have found that the amount of variation observed in the evolution of DNA and proteins is greater than is expected from a stochastic clock; in other words, the clock is inaccurate. The discrepancies in evolutionary rates along different lineages are not excessively large, however. It turns out that it is possible to time phylogenetic events with as much accuracy as may be desired; but more genes or proteins (about two to four times as many) must be examined than would be required if the clock were stochastically accurate. The average rates obtained for several DNA sequences or proteins taken together become a fairly precise clock, particularly when many species are investigated (Figure 1.3).

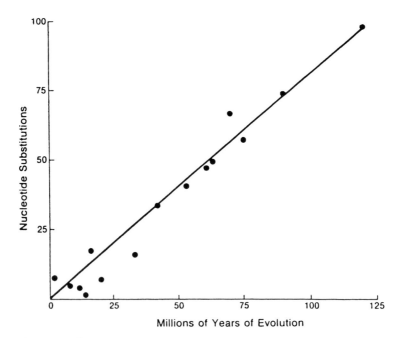

Figure 1.3 Differences in DNA composition vs. paleontological time. The total number of differences for seven proteins (cytochrome *c*, fibrinopeptides A and B, hemoglobins α and β, myoglobin, and insulin C-peptide) were calculated for comparisons between pairs of species whose ancestors diverged at the time indicated. The solid line was drawn from the origin to the outermost point. The fit between the observed number of differences and the expected number (as determined by the solid line) is fairly good in general. However, in the primates (points below the diagonal at lower left) protein evolution seems to have occurred at a slower rate than in most other organisms.

2
The Evolution Controversies: An Overview

Michael Ruse

2.1
Introduction

Evolutionary theorizing has been controversial from its birth in the eighteenth century. No sooner had people like the Englishman Erasmus Darwin (1794–1796) and the Frenchman Jean-Baptiste de Lamarck (1809) started putting forwards ideas of upward organic development, than the critics started to launch their counter-attacks. The greatest biologist of the age, George Cuvier (1813), was a leader of the critique, arguing that evolution is a false, pernicious doctrine, religiously un-acceptable, politically dangerous, and ideologically suspect. Things did not change during the nineteenth century. Charles Darwin, the grandson of Erasmus, rightly known as the father of evolution, published his great work *On the Origin of Species*, in which he argued for the mechanism of natural selection, in 1859. Only a year later, Darwin's great supporter Thomas Henry Huxley found himself debating the forces of reaction and status quo: first the comparative anatomist and paleontologist Richard Owen, and then publicly the Bishop of Oxford, Samuel Wilberforce (Ruse 2000b). In the twentieth century, evolution continued to be controversial. Notoriously in 1929 in the state of Tennessee a school teacher, John Thomas Scopes, was prosecuted for teaching evolution, and although his conviction was overturned on appeal, it had a chilling effect on evolutionary ideas, particularly in schools (Larson 1997). It was not indeed until the 1960s and 1970s that evolution started to find its way back into textbooks in the United States. In the last part of the twentieth century, evolution certainly throve in universities and museums, but still it continued a source of much controversy. There was a court case in the state of Arkansas in 1981, and then as the millennium drew to an end, more and more voices were raised against evolution. Noteworthy were such people as the Berkeley lawyer Phillip Johnson and the Notre Dame philosopher Alvin Plantinga (Ruse 2000a). But even within the halls of evolutionism itself there was much controversy, with people like the Harvard entomologist and sociobiologist Edward O. Wilson (1975) on the one side arguing for an extension of evolutionary ideas to our own species, and his colleagues at Harvard the geneticist Richard Lewontin (1977) and the paleontologist

Handbook of Evolution, Vol. 2: The Evolution of Living Systems (Including Hominids)
Edited by Franz M. Wuketits and Francisco J. Ayala
Copyright © 2005 Wiley-VCH Verlag GmbH & Co. KGaA, Weinheim
ISBN: 3-527-30838-5

Stephen Jay Gould (1980) on the other side arguing that such extensions are scientifically unwarranted and morally dubious.

In this brief overview I cannot hope to cover all the controversies that follow evolutionists at the beginning of a new century. I therefore confine my discussion to some five controversies, hoping thereby to give the reader a good sense not only of the debates, but also of the excitement that surrounds one of the most important, if controversial, ideas of our time. I start with a vigorous debate about the history of Darwinism itself. This is revealing not just for the past but also for our understanding of the present. Next, I move to some of the debate and discussions about origin-of-life studies. I then move to central issues to do with natural selection. This is followed by extensions to humankind. I conclude with the attacks from without, particularly by today's enthusiasts for so-called Intelligent Design.

2.2
The History of Evolutionism

I begin with discussion of a controversy that is not so much about or within evolutionary theory itself, but rather about the history of evolutionism. As was said at the beginning, evolutionary theory is a child of the eighteenth century (Ruse 1999a). It was then that discoveries of the fossil record and of strange organisms and the like started to persuade people to think that the Genesis-based account of origins is not correct (Rudwick 1972). Rather, we must suppose that organisms came by a natural process of development from primitive forms. However, from the beginning, all would agree evolutionary theory was very much more than just a straight scientific theory. The early evolutionists, Erasmus Darwin in Britain and Jean-Baptiste de Lamarck in France, had strong ideological axes to grind (Ruse 1996, 1999b). They were enthusiasts of the chief secular philosophy or ideology of the age: progress. They believed that humans, unaided, can improve their knowledge, and through this their social conditions. Darwin and Lamarck and other early evolutionists took this ideology and read it into the biological world. Metaphorically, they saw evolutionary development as a progressive rise from the primitive to the complex, from the monad to the man, as they used to say. Then generally, in good circular fashion, they read their biological progressivism back into culture, thinking that they had justified their overall ideology!

Expectedly, the great critics of evolution – notably Georges Cuvier (1813) in France and Adam Sedgwick (1833) in England – opposed not only the factual basis of evolution, which to be frank was not very great, but also the ideological underpinnings. Cuvier and Sedgwick and others were keen Protestants, and as such they saw progressivism striking directly at their central theological commitment to a Providential God: to a world where we humans can effect no lasting changes on our own but can seek salvation only by appealing to God for His mercy and his grace.

Evolutionary theory in those early days was therefore less a genuine scientific theory and more a kind of pseudoscience or secular religion. And this is the way that it persisted at least until the time of Charles Darwin. In the *Origin of Species,* published in 1859, Darwin put forward a much firmer (than hitherto understood) empirical base for evolutionary change, appealing right across the spectrum of the biological world: instinct, paleontology, biogeography, anatomy, embryology, systematics, and more. Thus he supported his belief in a gradual process of descent. Conversely, he used this belief in a gradual process of descent to illuminate problems in the various subdisciplines of the biological sciences. In short, it was Darwin's hope that the *Origin of Species* would provide a new paradigm of research – biology under the umbrella of evolution (Ruse 1999a).

However, this was not to be. Darwin's great supporters (notably Thomas Henry Huxley) had little interest in making a functioning professional science out of Darwinism. They were not interested in evolutionary problems at an experimental level, being much more concerned with broad-scale issues to do with morphology and paleontology (Desmond 1994, 1997). And as all agree, for a long while after the *Origin* was published, evolutionary biology became not very Darwinian at all. Indeed, it owed far more to the transcendental idealistic morphology coming out of Germany, reaching back ultimately to the Naturphilosophen, than to anything to be found in the *Origin of Species* (Richards 1992; Nyhart 1995). It is at this point that we start to enter into controversy. I argue that what happened after the *Origin* was little more than a disaster from the point of view of science. Evolutionary biology was hijacked by others, notably Ernst Haeckel in Germany and Herbert Spencer and his followers in England (and then in America). It remained as it has long been: a kind secular religion, an alternative to Christianity. As a science, it became (or rather stayed) thoroughly second-rate.

All agree that now there was a science of evolutionary biology, but it was a kind of evolutionary morphology, which was increasingly pushed out of the universities, if indeed it ever got in them, and became a museum subject. Evolution was much more a kind of popular science than anything else, and because it turned out that trying to trace phylogenies using evolutionary morphology and embryology ended in paradox more often than not, the best young biologists of the age turned increasingly away from evolutionary problems. They took up other issues like cytology, embryology, and ultimately in the twentieth century, genetics. People like William Bateson, the English Mendelian, and Thomas Hunt Morgan, the great American fruit fly geneticist, started life as evolutionary biologists but found that the subject was getting them nowhere. So they turned to richer and more fertile fields.

I argue that this situation continued until the 1920s and even into the 1930s. Evolutionary biology was a second-rate subject, not one for a top-quality university mind. Students were steered away from it. It was only after the population geneticists, notably R. A. Fisher and J. B. S. Haldane in England and Sewall Wright in America, had put genetics on a firm evolutionary basis from a theoretical point of view that it was possible to develop evolutionary theory as a university-based, mature, professional science (Provine 1971). This finally occurred in the 1930s and 1940s, thanks particularly to the labors of such people as Theodosius Dobzhansky and his followers in America and to E. B. Ford and others in Britain. By the 1950s, one did at last have the kind of selection-based, mature theory that Charles Darwin had hoped for.

The question now is one of interpretation. Am I correct in my claims that what happened after Darwin until the 1930s was indeed a disaster for evolutionary theory? (The case is made fully in Ruse 1996.) The eminent British historian of evolutionary biology, Peter Bowler, disagrees strongly. He has detailed very carefully the work of the evolutionary morphologists in the latter decades of the nineteenth century and the first decades of the twentieth century (Bowler 1984, 1996). He shows rightly that there was a great deal of work that went on, and he argues that this was interesting and significant. More importantly, he argues that it was because of this work that evolutionary biology was able to take off and professionalize in the way that it did in the late 1920s and 1930s. He believes that there is indeed a continuity, even though he himself has stressed that much of the work that was done around the turn of the century was not very Darwinian. He feels that whatever the regard for Darwin's work, a basis was provided for the advances that were to come.

I am obviously not the person to make a disinterested evaluation of this debate. I would simply say from my corner that I see no evidence to back Bowler's claim. I believe that the evolutionists of the 1930s had themselves to break with what had gone before, and that they themselves set out deliberately to make a theory, not only at the conceptual level, but also at the social level. One finds that few if any of them had any interest in morphology. Even the paleontologist of the group, G. G. Simpson (1944), succeeded as a modern evolutionist because he was breaking away from the old fashions of searching desperately for phylogenies. He was developing and embracing a much newer, dynamic, causal picture of evolutionary change. Simpson in his way was just as much a revolutionary as the others.

Probably the time is not yet fully ripe to decide if the truth lies with me or with Bowler. Although much work has been done on the history of evolutionary theory, very little has been done at the meta-level on this grand a range. Most studies on evolutionary history are micro-studies on particular events or particular people. Few other than Bowler or myself – who are now both long past the years of PhD thesis writing, an exercise that encourages micro-studies – have attempted to look at the broader picture and to offer overall interpretations. No doubt more work will be done on this, and nothing stimulates inquiry more than dispute. Quite probably, both Bowler and I will come through with our positions modified. However, at the moment we have an ongoing and as-yet-unresolved controversy about the nature of evolutionary theory. All agree that Darwin was crucially important in providing a theory of evolution through natural selection. All agree that, although Darwin was crucially important, his ideas were not picked up after the publication of the *Origin* in the 1860s. It was indeed not until the 1920s and 1930s that people became truly Darwinian and the modern theory of evolution was developed. All agree that between Darwin and the 1930s a great deal of non-Darwinian biology was done, which drew heavily on German morphology and other roots. The question is Where does one go from here? Does one want to argue as does Bowler that this intermediary science was good quality science albeit non-Darwinian, and that it led in an important way to the modern theory of evolution? Or does one argue as I do that this non-Darwinian science was an absolute disaster for the history of evolutionary theory? That it was second-rate, if science at all, and often was used simply as a vehicle for quasi-religious ideas, and that the development to the modern theory in the 1920s and 1930s took place only because the new practitioners turned their backs on or were ignorant of what had gone before? That they have reached back rather across the years to Darwin himself?

Here we have two very different pictures of evolutionary theory. In its history, time and more research will tell which is correct.

2.3
Origin of Life

If you are an evolutionist, that is to say, if you believe that all organisms came gradually by natural processes from one or a few forms (as Darwin said), then it is almost inevitable that you will start to think about ultimate origins. You will start to ask questions about how life itself came into being. Was it something that was seeded from elsewhere, or was it something that developed naturally in some fashion here on earth? Certainly the early evolutionists, Erasmus Darwin and Jean-Baptiste de Lamarck, thought long and hard about ultimate origins, both of them subscribing to some form of the venerable doctrine of 'spontaneous generation' (Farley 1977). Lamarck, for instance, believed that in pools of warm mud, lightening and heat and so forth would stir things up. Worms and the like would come about instantaneously, and thus life would be on its path upward. This was a belief shared by other early evolutionists through to the first part of the nineteenth century. You find, for instance, that Robert Chambers (the author of *Vestiges of the Natural History of Creation*) thought that somehow life comes about instantaneously, and he was forever looking for lifelike forms in the inorganic world that he believed might be generated into living beings.

At the time that Darwin wrote the *Origin,* however, the whole notion of spontaneous generation was coming under very heavy criticism, and as is well known the French biologist Louis Pasteur in the 1860s was to drive the final nail into its coffin. Darwin, a very skilled scientist, knew that he could only get into trouble if he subscribed to anything like spontaneous generation, and yet he knew he had no alternative. Wisely therefore, he realized that the best course of action was to say nothing at all, and so in the *Origin* the vital question of the origin of life is significant mainly by its total absence. Darwin said nothing because he had nothing to say, and in many respects his strategy was very successful, because any question of life's origin never became one of the focal points of debate after the *Origin* (Ruse 1982).

It was, however, a problem that would not go away, especially if spontaneous generation no longer seemed to be a viable option. It is generally agreed that the major breakthroughs on this topic came about in the 1920s, when independently the English biochemist and geneticist J. B. S. Haldane (1929) and the Russian biochemist A. I. Oparin (1924) put forward ideas about life's origins. They suggested that it would be something that came about naturally but in a sequential form – not an instantaneous appearance of fully functioning life as was supposed by supporters of spontaneous generation. But there would have been a kind of organic evolution leading up to functioning primitive life forms. These ideas of Haldane and Oparin received considerable support in the 1950s, when researchers at the University of Chicago, Harold Urey and Stanley Miller (1953), showed how it would be possible to make some of life's basic building blocks (amino acids) naturally from inorganic compounds – naturally, under experimental condition similar to what might have been expected back when life was supposedly formed.

Much work has gone on since then, although no one could say that the questions have been answered with unambiguous success. Indeed, the whole matter of the origin of life is today one of some considerable controversy (Ruse 2000b). On the one hand, we have some of the leading researchers admitting that we really do not yet know how life could come about. One prominent researcher, using the metaphor of a detective story, writes "We must conclude that we have identified some important suspects and, in each case, we have some ideas about the method they might have used. However, we are very far from knowing whodunit" (Orgel 1998, p. 495). On the other hand, we find critics who are contemptuous of supposed claims about the supposed origins of life. For instance, the eminent philosopher of religion Alvin Plantinga states that, in his opinion, we are, if anything, further away from solutions now than we were in the past. He speaks of hypotheses about the origin of life as for "the most part mere arrogant bluster", adding that "given our present state of knowledge, I believe it is vastly less probable, on our present evidence, than is its denial" (Plantinga 1991, p. 20). Is this really so?

It is useful to break the Oparin–Haldane hypothesis into a number of steps (Freeman and Herron 1998). We begin first with the question of the conditions required for the natural creation of organic molecules, those that constitute the ultimate building blocks of life. If conditions back when life is supposed to have begun – some 3.5 billion years ago, if we read the fossil record right – are like conditions today, it simply would not be possible for primitive building blocks of life to be created and persist. The oxygen atmosphere that we have today would preclude that. However, it is thought that, around the time that life is supposed to have started, in fact the atmosphere on the earth was oxygen-free, somewhat laden with other gases such as methane, ammonia, carbon dioxide, and hydrogen sulfide. These are just the ingredients that we need for making elementary molecules. Then, second, there is the production of complex molecules from inorganic substances. Mention has already been made of the work of Miller and Urey, and it is still thought that their work holds up well. There would have been precisely the kinds of radiation, electrical storms, etc. that would have given rise to primitive building blocks, particularly amino acids.

Next in the Oparin–Haldane sequence you have the question of linking these primitive molecules into long macromolecules, to make the substances we find in cells today: proteins and nucleic acids. The problem is not so much the joining of the molecules – this happens quite easily – but keeping them joined long enough for them to start functioning. Quite possibly what happened was that the molecules adhered to clays and built up long chains. This has been simulated. But how then do we move on the fourth step, to get the chain molecules to replicate themselves? Some suggest that the clays continue to play an important role. Crystals repeat themselves, building copies on templates although sometimes with mistakes built in. These mistakes then get repeated. Perhaps the organic molecules piggybacking on the crystals got repeated; then mistakes got built in, and eventually the molecules could themselves take over reproduction, dropping their supports. Possibly, however, the molecules themselves could move directly to making proteins. It is thought that the key molecule here would be RNA, because in some organisms it is the only nucleic acid. It is capable of replicating itself and also of acting as a model for other cellular building blocks.

There are going to be more steps after this. For instance, one has to make a globular cell. It was here that Oparin (1968) spent most of his life's work, as did a number of Americans, notably Sidney Fox (1988). They were showing how one could make self-contained spheres that can maintain and reproduce themselves. Possibly the organic molecules would work within these spheres. Later it is thought that other events had to occur. Most researchers today subscribe to the ideas of the American researcher Lynn Margulis (1970, 1993), who has argued that, to make more complex cells, one needed a kind of symbiotic relationship between more primitive cells. The cell parts today, like mitochondria, perhaps once had an independent existence. Then they were incorporated within other cells and now remain there as part of a united whole.

There is more after this. For instance, one needs the development of sexuality and things of this nature. I really do not think that even the greatest optimist could say that we are anywhere close to having a full and adequate picture of life's origins. There are far too many gaps waiting to be filled in and unknowns that have not been properly addressed and answered. On the other hand, this said, huge advances have been made in the past half century on this problem of life's origin. Moreover, this is an area that is not stagnating, but rather where a great deal of effort is being put into exploring the various links. It is not a topic that scientists avoid because they have no ideas at all about how to tackle the problems. If anything, it is a more vigorous area than it has ever been before.

It seems therefore that it would be foolish indeed to conclude with the critics that this is an area of great weakness for the evolutionist. It is simply not true, as Plantinga claims, that we are further back now than we were before. Indeed, to the contrary, thanks to the coming of molecular biology we now are starting to have a full appreciation of the great problems that are involved in coming to an adequate theory of life's origins. Recognizing the magnitude of problems is the first step toward solving them. Philosophers are notoriously unreliable when making forecasts about the future of science. But my suspicion is that in the next half century we will see even greater advances than before, and without confidently predicting that it will be possible to create life anew in the test tube in the laboratory, my suspicion is that we will at some point in the foreseeable future start to get really solid answers to the origin of life. It is no longer the black hole for evolution that it was at the time for Charles Darwin. when he wrote the *Origin of Species*.

2.4
Natural Selection

Let me move now to the central issue of modern evolutionary theory, namely the Darwinian mechanism of natural selection. Charles Darwin argued that more organisms are born than can possibly survive and reproduce. There is consequently a struggle for existence. Then following this, given the naturally occurring variation in populations, there will be something analogous to the breeders' form of artificial selection. This Darwin called natural selection: more organisms are born than can survive and reproduce, and those that are successful or 'fitter' succeed because of variations they have that the losers do not have. Overall, given enough time, this leads to evolutionary change. But it is important to note that for the Darwinian, this change is of a particular kind; namely, it is change in the direction of adaptation. Organisms have characteristics that enable them to survive and reproduce – characteristics like the eye or hand. It is the essence of Darwinism that evolution produces organisms that are not simply thrown together, but which are tightly integrated or functioning thanks to their adaptive nature (Ruse 1999a).

There have always been critics of natural selection, and these critics continue vocal today. Particularly strident have been Richard Lewontin and Stephen Jay Gould, who in a famous, or notorious, paper "The Spandrels of San Marco" argued that Darwinian evolutionists much over-emphasize the extent to which the world is tightly adapted (Gould and Lewontin 1979). These critics argued to the contrary that, just as in medieval cathedrals one found areas at the tops of pillars – so-called spandrels – that have no direct function, so also in looking at organisms one should recognize that much of their nature is not particularly adaptive. Therefore, although natural selection undoubtedly occurs and is an important evolutionary mechanism, it is nothing like as important or significant as Darwinians claim.

What would the critics put in the place of natural selection? There is no one answer, and to a certain extent what they imply is that really things happen more or less randomly or by happenstance. Perhaps the process of genetic drift, in which the population frequencies of genes just float up or down by chance, is an important mechanism. Also, frequently suggested are so-called developmental constraints. It is argued that simply as part of the development and architecture of organisms many characteristics happen, not for any good reason but just because they do (Gould 1984). Often highlighted by the critics of natural selection is the four-limbedness of vertebrates. Why, ask people like Gould and Lewontin, should there be four limbs for vertebrates rather than six as for insects or even eight as for arachnids; or perhaps five as for starfish?

Darwinians have been combating these arguments with some vigor. John Maynard Smith (1981), the doyen of English evolutionists, argues that the four-limbedness of vertebrates may not have any direct adaptive function at the moment, but certainly was rooted in adaptive advantage in the past. The earliest vertebrates were aquatic and needed two limbs in front and two limbs in back to move up and down in the water just as planes today move up and down in the air. In fact, argues Maynard Smith, the earliest vertebrates were not necessarily four-limbed, and the number four was something that came about precisely because it proved to be of adaptive advantage to the possessors. Likewise, argue the Darwinians, other mechanisms seem not to have been greatly significant. Genetic drift recently has been much criticized by a number of theoreticians who suggest that it is highly improbable that it was ever very significant – certainly not in any major way in the course of evolution (Coyne et al. 1997).

In a sense, the jury is still out on the issue of the power and ubiquity of selection, although it is fair to say that most practicing evolutionists rely very heavily on natural selection. Indeed, they would argue that, unless they did, they simply could not do their science. If one wants to put things down to other random factors, then this is a kind of negative argument that leads to no new research hypotheses. Talking of constraints or drift is simply not very helpful. There is an asymmetry here between selection and its rivals. However, this is not to deny that – taken over long periods of time, when one starts to look at macro events rather than micro events – one might see certain patterns emerging that were not directly fueled by natural selection, patterns that reflect other factors or constraints on the course of evolution. This particularly was the hypothesis of the late Jack Sepkoski, a paleontologist who argued that, if one looks at the course of evolutionary change over millions of years, one starts to see certain patterns emerging – patterns that, although not negating natural selection, do not depend on it directly.

In particular, Sepkoski was initially interested in the rise of organisms during the Cambrian period, some five or six hundred million years ago. What Sepkoski (1976, 1978, 1979, 1984) argued was that, with the coming of sexuality, perhaps one got a kind of exponential growth as organisms were able to move into new empty ecological niches. And then, as these niches got filled up, the growth slowed. So one gets the kind of sigmoid pattern one associates with such growth. In fact, Sepkoski was able to map the numbers of families of marine organisms extant at that time, and he showed that, roughly speaking, one does get such a growth as one expects. More than this: Sepkoski was then able to add other more subtle factors to his analysis, showing in particular that once the Cambrian explosion was over, then the Cambrian organisms, if anything, started to decline somewhat. But now new forms took over and one got a kind of sigmoid growth on the back of the earlier ones. Indeed, this has happened even a third time among animals. Thus, Sepkoski felt able to model (using computers and vast databases of extinct taxa) the history of life in the past six or seven hundred million years.

I stress that Sepkoski certainly did not look upon his work as being in any sense anti- Darwinian. It was rather neutrally non-Darwinian, because he was doing was trying to show broad patterns that would not be captured directly by natural selection. However, even here, in a sense, Sepkoski went on to appeal to Darwinism. He certainly thought that the new waves of organisms, building on the earlier growth patterns, came about because the organisms developed new adaptations of an altogether innovative kind. One has here an example of what the British ultra-Darwinian Richard Dawkins refers to as the evolution of evolvability. Organisms grew in great numbers during the Cambrian and then tailed off, but then new forms and new adaptations were developed so life could take off again. Moreover, this has happened no less than three times in all, giving the kind of picture that we find in the fossil record. (When it came to plants, Sepkoski believed that perhaps one even had a fourth wave. This might have been a function of the evolution of the angiosperms, a new kind of life form that was able to build on previous patterns.)

We have here therefore a case of an extension of Darwinism, but by no means a refutation of Darwinism. (I see analogies between what Sepkoski attempted in evolutionary history and what Bowler and I attempt in the history of evolutionary theory. Like Sepkoski, Bowler and I both think that when you look at the broader picture, new patterns emerge – patterns that could not be deduced from the micro-studies on their own. That Bowler and I differ in our findings and interpretations is another matter. Sepkoski and his colleague David Raup also thought that extinction is periodic (Sepkoski and Raup 1986a, b). Some agreed, some did not. We all concur that it is only by looking at the big picture that these issues arise.)

What about Stephen Jay Gould's supposed new theory of punctuated equilibria? Gould has argued that he has found aspects of the fossil record, which are not only beyond Darwinism but which are in some sense anti-Darwinian (Eldredge and Gould 1972; Gould and Eldredge 1977; Ruse 1989). It is an essential tenet of Darwinian theory that evolutionary change must be gradual, yet Gould argues that in the fossil record one finds long periods of inaction, or what he refers to as stasis, followed by sharp changes in form from one kind to another. This apparently supports a whole new vision or theory of evolution – non-adaptive, jerky, em-phasizing constraints. And this theory of punctuated equilibria, as Gould calls it, is in some important sense not just non-Darwinian but anti-Darwinian. It cannot be incorporated within the Darwinian picture.

However, conventional Darwinians are not certain that any of this is very significant:

> Gould occupies a rather curious position, particularly on his side of the Atlantic. Because of the excellence of his essays, he has come to be seen by non-biologists as the preeminent evolutionary theorist. In contrast, the evolutionary biologists with whom I have discussed his work tend to see him as a man whose ideas are so confused as to be hardly worth bothering with, but as one who should not be publically criticised because he is at least on our side against the creationists. All this would not matter, were it not that he is giving non-biologists a largely false picture of the state of evolutionary biology. (Maynard Smith 1995, p. 46)

Darwinians point out that what may be a sharp change in the fossil record could take many hundred of generations in real life, and this could be quite time enough for natural selection to work on the random variation produced by genetics (Ruse 1982). Orthodox Darwinians have never claimed that change, although gradual, is always at a steady rate. Rather, it has always been thought, from Darwin on, that sometimes one will have times of rapid change and sometimes periods when nothing much happens. After all, selection is going to change things only if there is a good reason for this to happen. Hence, although conventional Darwinians allow that Gould has done service in pointing to the uneven patterns in the fossil record, inasmuch as punctuated equilibria is taken to be a whole new theory of evolutionary change or to demand new mechanisms, Gould is stepping way beyond his competence and the evidence given to us by nature. There is no need for a theory beyond Darwinism.

I would not want to say that all of these controversies around selection are completely concluded. Amongst nonbiologists (philosophers, for example) there is much sympathy for a critique of adaptationism. However, in the working community of evolutionary biologists, it seems fair to say that almost all who are actually doing ongoing research feel that Darwinism functions as a viable paradigm. It provides new problems and sets new standards. At this time, there is little need to look elsewhere. (Although I shall suggest that there was more to the story than this, I would entertain seriously the suggestion that Lewontin and Gould show the dreadful effects of the turn-of-the-century rejection of Darwinism. They have not fully internalized the revolution of the 1920s and 1930s. If this is so, then relating back to the dispute between Bowler and myself, you might say that he is certainly correct in saying that the non-Darwinism did have a lasting effect. But I am correct in saying that the non-Darwinism did not feed into the new Darwinism. Moreover, I am correct in seeing the turn-of-the-century non-Darwinism as an absolute disaster. It was a disaster at the turn of one century and it is a disaster at the turn of another century.)

2.5
Human Sociobiology

It has always been the claim of the traditional Darwinian that his or her theory applies to humankind. Erasmus Darwin and Jean-Baptiste de Lamarck both thought that evolution applies to humans, and the same was true of Charles Darwin. In fact, the very first records we have of Darwin speculating about natural selection are about human brains and human intelligence. In the *Descent of Man* written some 12 years after the *Origin of Species*, Darwin argued strongly not only that humans have evolved from other organisms, but that human nature is in many respects something molded by natural selection

For many years however, people held back from exploring these ideas. There was the growth of the social sciences that naturally did not want to be swallowed by the biological sciences. Then also there were the terrible events in the first part of the twentieth century, particularly as occurred in Nazi Germany in the 1930s. People grew even less willing to see human nature as in some sense a function of or dictated by biology. But times do change, and by the 1960s and 1970s an increasing number of evolutionary biologists (and others) were wondering whether the time had come to look in some detail at the state of human nature from a biological perspective (Degler 1991). They were encouraged in doing this by the fact that the 1960s saw a great development of our evolutionary understanding of social behavior. It was realized that organisms not only fight and compete against each other, but that very widely they cooperate. The reasons for this are fairly simple: enlightened self-interest dictates that one is better off if one works together with one's group than if one is constantly competing. Although there are costs to cooperation, or as biologists call it, altruism, the costs are well out-paid by the benefits from such cooperation or altruism (Ruse 1985).

In 1975 the Harvard entomologist Edward O. Wilson brought together these new ideas about social behavior – what he called sociobiology – and incorporated them into a magnificent overview of the whole subject: *Sociobiology: The New Synthesis*. What made Wilson's work controversial, however, was the fact that he applied the ideas of animal sociobiology to human nature. He argued that just as ants and lions and chimpanzees are animals that function the way they do because of their biology, so also in many major respects we should see that humans are products of evolution through natural selection. Much that we do in culture is dictated by or determined by our evolutionary past.

The sorts of things that Wilson talked about in *Sociobiology: The New Synthesis* and in a later more popular exposition of the human case (*On Human Nature*), included family structures, parent–child relationships, and sexual encounters. Wilson argued that the fact that males so often are dominant in societies, whereas women are more subservient and tend to be the gatherers as opposed as to the hunters, is not simply a function of male oppression, but also in many major respects a function of our biology. There are good reasons why males should be aggressive and compete for females, whereas females should be somewhat more choosey – to use a word of Wilson's which gave great offence – to be much more 'coy'. Females are the ones left holding babies, whereas males can impregnate a great many partners. In other words, sexual relationships are asymmetrical, an imbalance that follows directly from our biology. Wilson offered similar arguments for such things as aggression, social status, politics, and life (including ethics) generally. Noteworthy was Wilson's belief that religion is something that has no basis in reality but is an adaptation put in place by our biology to make us good social animals.

These ideas were attacked, viciously at times, by a whole range of people (Kitcher 1985). Naturally enough, feminists and others did not like Wilson's claims about the supposedly natural relationship between the sexes. Social scientists were made very uncomfortable by what they saw as the territorial ambitions of evolutionary biologists. If sociobiology worked, then what place would there be in the curriculum for psychology and sociology? And above all, the radical biologists disliked Wilson's ideas. These critics felt that his views were just a prescription for racism and worse. They saw human sociobiology as being little more than an echo of the vile doctrines of Germany of the 1930s. Prominent amongst these critics were Lewontin and Gould (Lewontin 1977, 1991; Gould 1980, 1981; Lewontin et al. 1984).

Much of the critical objection was directed less at Wilson's Darwinian extensions to humankind and more at the very core of Darwinism. I have already looked in the last section at some of the objections made against natural selection, and it is right and proper to see these objections in the context of an overall critique of Darwinism – a critique intending to forestall its application to humankind. However, the critics also had specific objections to the human sociobiological program. They argued that there is simply no evidence for the genetic basis of such things as sexual differences, that these can be explained simply in terms of culture and politics. Likewise, they argued that things like aggression have no ultimate basis in biology, nor does ethics. At best, something like moral behavior is a cultural phenomenon, if not indeed a reflection of certain material or otherwise objective entities.

There is no doubt that the critics made some telling points. The work of Wilson and his like is frequently speculative, even verging into sloppy. Moreover, the motives of the human sociobiologist are often based more in ideology than in anything to be found in the empirical world. However, there does seem to be an area where (over the last 25 years) there have been significant scientific advances. Now, even if the battle is not finished, it is at least dying out. During the past quarter century, fundamental advances have been made by molecular biologists and others in understanding the nature of the human genetic blueprint (the genome). We now do know beyond reasonable doubt that much of our behavior is rooted in our biology. This does not in itself mean that it is adaptive, but this a first step toward such a conclusion.

Take sexual orientation, the putative causes of which were very speculative at the time that Wilson wrote (something that led to much scorn and criticism). It used to be thought that sexual orientation is all a question of the environment (Ruse 1988b). Classically, Freud (1905) argued that homosexual men are produced by a combination of dominant mothers and weak or distant fathers. Now however, we are starting to get really solid evidence suggesting that (at least some) homosexuality is directly under the control of the genes. The way that you think about sex, and the objects toward whom you are attracted, are indeed a matter of one's DNA rather anything in one's upbringing. The nature of sexual encounters is no less a function of our evolutionary heritage than are the tools of such encounters (LeVay 1993; Hamer and Copeland 1994).

What about the question of adaptive advantage? Here also one finds that major advances have been made. One of the most stunning successes of human sociobiology stems from the work of the Canadian evolutionary psychologists Martin Daly and Margo Wilson (1988) at McMaster University. They took as their subject their problem of homicide. Why is it that people kill other people, and who does what to whom? They were particularly interested in the issue of parents killing children. This would seem to be blatantly nonadaptive and to go against the Darwinian paradigm. Excluding infanticide, which raises issues of its own, Daly and Wilson hypothesized that the parents who kill children are almost always social parents and but rarely biological parents. In other words, Daly and Wilson suggested that it is step-parents who do the killing of children rather than so-called natural parents. I should say that in making this hypothesis they were not sailing in entirely untested waters. It is a very well documented fact in the animal world that males moving in on pregnant females or on females with litters will first and foremost bring on an abortion or kill all the offspring. The female is thus brought into heat, and the new male can impregnate her. And there will be no rivals making demands on her when the next batch of children is born. Daly and Wilson suggested that perhaps humans are similar.

They tested this hypothesis and found that it was stunningly confirmed. A child is more than 100 times more likely to be killed by a step-father than by a biological father. (This does not mean that most step-fathers kill their step-children. This does not follow. The point is that one is looking at comparative figures.) What made the Daly–Wilson finding even more stunning was that the relative differential held both in Hamilton, Ontario, Canada, and in Detroit, Michigan, U.S.A. The murder rate in the United States is at least five times higher than it is in Canada, and Detroit even more so. But although the absolute numbers are much increased in Detroit, proportionately the same ratio of step-parents (as opposed to biological parents) kill children. What made the work of Daly and Wilson truly convincing was that this was hardly a conclusion built in before they set out in their research. They found when they went to the authorities for figures to check their hypothesis there were no such figures. When it came to family violence, agencies had been reluctant or indifferent to distinguishing between step-parents and natural parents. They thought it was either irrelevant or would be prejudicial to reconstituted families. However, once the statistics were kept in a proper manner, the differences emerged dramatically.

As Aristotle said, one swallow does not make a summer. But this is far from the only swallow. There is now a really large and ever-growing body of evidence suggesting that human sociobiology is something with a strong grounding in empirical reality. Here therefore is a place where an evolutionary controversy is, perhaps is not yet concluded, but well on the way to closure. The interest now is in developing human sociobiology in broader and more subtle ways, so that we can further tease out the effects of genes as opposed to the environment. (Although I have been talking about biology, let me stress that no one wants to argue that the environment or culture is a matter of causal indifference. The reason why I speak English rather than German is because I was brought up in England rather than in Germany. This is a cultural phenomenon. The reason why I speak at all, rather than just grunt, is because I am a human being. This is a biological phenomenon. The task now for sociobiologists is that of teasing apart the relative causal factors in human nature and seeing which are really significant and which are not.)

2.6
The Creationist Controversy

In 1981 in the State of Arkansas the American fundamentalists, better known today as Creationists, masterminded the passing of a law that insisted on the teaching of the Bible taken literally in biology classes in the state. This law was soon overturned and ruled unconstitutional, but the Creationism controversy has continued since then (Ruse 1988a; Hull and Ruse 1998). Indeed, as we start the twenty-first century, it thrives as never before. A wide range of thinkers – some biologists but many from medicine, philosophy, law, and elsewhere – argue that evolutionary theory, Darwinism in particular, is on very shaky grounds indeed (Johnson 1991). They believe that one ought rather subscribe to an alternative position. These days,

Creationists are cautious about parading the early chapters of Genesis taken absolutely literally, although one suspects that most of them do in fact subscribe to literal interpretations. However, what is often said now is that one ought to prefer something called Intelligent Design over Darwinism. That is to say, one should accept that at some point or points, some being – which may or may not have been the Christian God (one knows the Creationists think it was) – intervened in creation. Organisms were produced miraculously, in one step, with the full glory that one finds about them today.

On what basis do the Intelligent Design theorists make their case? Probably the most sophisticated argument has come from the Lehigh biochemist Michael Behe, in his *Darwin's Black Box*. He argues that, at the biochemical level particularly, one finds examples of what he refers to as 'irreducible complexity': organism parts are so complex that there is simply no way that they could have been put together by natural selection. Behe argues that many organic systems are rather like mouse traps, which need five parts all functioning together to work. Take away one part, for instance the spring, and the mouse trap simply does not work. Behe argues that this means that it could never have been produced by selective processes, because the essence of selection is gradualness. Likewise, at the biochemical level, Behe argues that one finds that parts have to be put together at one instant, to get them to work. Behe cites the example of blood clotting, for which we have a kind of cascade process with one part of the process kicking in and then another, and then a third until the whole thing is over. Take away one part of the process, argues Behe, and nothing works. Analogous to the mouse trap example, he suggests that natural selection could not possibly have assembled all the constituent parts and then put them together in one spot at the same time. Here then we have irreducible complexity, which cannot be explained by natural selection. The only other reasonable hypothesis is one of intervention of some kind, or what Behe and his like call Intelligent Design.

This argument has been much criticized by scientists and others. The blood clotting example is thought by experts in the field to be a paradigm of how something was in fact manufactured sequentially (Doolittle 1997). The various parts of the cascade are all similar, and apparently not all are truly necessary. It is thought therefore that one started with very primitive forms and then things got more sophisticated by duplicating various parts and modifying them slowly. In other words, Darwinians argue that selection could indeed produce something like blood clotting or other sophisticated processes at the biochemical level. (The Krebs cycle – a process necessary for producing available energy for the functioning body – has been a recent counterexample of the evolutionists (Miller 1999).)

More fundamentally, critics argue that Behe misunderstands the way in which evolution works (Ruse 2000a). Perhaps all the parts are put together now, and to move one part would cease to make for functioning. It does not follow that all parts were assembled at one fell swoop. Suppose, to counter the mousetrap example, one looked at a stone bridge with no mortar. If one just saw it as it stands at the moment, one could not understand how the keystone in the centre was inserted. If one built up the walls of the bridge and sides, and tried then to move towards the middle everything would just keep falling down. However, we know that first the bridge maker makes a structure underneath the arch – perhaps scaffolding or a mound of earth. The stones are laid on this, and the bridge completed. Then, once everything is in place and the stones pushing against each other, one can remove the scaffolding underneath and the bridge remains standing. So likewise this is a common practice in evolution. We have already seen it suggested that clays might have worked this way for the earliest functioning molecules. There is therefore no need to invoke an alternative non-Darwinian mechanism.

In any case, argue the Darwinians, someone like Behe is even in worse trouble theologically than he is scientifically. Suppose one does in fact conclude that a Being – a Designer – was necessary for making the functioning, complex parts of the body. How then can one deny that this Being was responsible for the not-very-complex, indeed very-simple-but-absolutely-horrendous mistakes that sometimes occur in the functioning body. Why, for instance, is the Designer not responsible for sickle-cell anemia, which is brought about by the change of just one base in a DNA molecule? Here, the philosopher–mathematician William Dembski has intervened, arguing that through a filter process one can get the designer off the hook. Dembski (1988a, b) argues that, when an event occurs in biology, one has three choices. First, one might put it down to the working of law. So, if a blue-eyed child is born to blue-eyed parents, one simply explains this as the workings of Mendel's laws. But if one cannot put something down to law, for instance one has random mutation, then one puts this down to chance. Sickle-cell anemia is to be put down to chance and is to be blamed on no one. If neither law nor chance explain things, then it is appropriate to invoke Intelligent Design. Hence, argues Dembski, one can give the Designer the credit for something like the blood-clotting cascade but can exonerate the Designer from the problems of the sickle-cell anemia malfunctioning gene.

Critics do not find this response convincing. Are law, chance, and design quite as logically distinct as Dembski thinks? Why should one not say that, even though the malmutation came about by what we would regard as chance, in some sense it could still be a function of a Designer. Why should the Designer's responsibility be evaded at this point? After all, even though one might want to say that the malmutation came about in a random way, one could still insist that ultimately the malmutation was caused by law. It was certainly not an event outside law. In fact, we now know a great deal about the causes behind mutations, including bad mutations. So the critics of Intelligent Design argue that, even though one might speak of a malmutation as being a function of chance, at the same time it is probably something that is covered by law. And, if indeed one is prepared to invoke an Intelligent Designer for one set of events, then why should one not be forced to invoke an Intelligent Designer for all other sets of events?

Despite these strong criticisms, I doubt very much that Intelligent Design or Creationism generally is going to go away in a hurry. It seems to be thriving more at the moment than ever before. However popular though it may be, and socially and pedagogically threatening as it truly is, it does not seem to be a matter of great intellectual import or concern to the orthodox Darwinian.

2.7
Conclusions

I have run through a number of controversies. I suggest that Darwinism emerges looking stronger than ever. This is not to say that all problems are resolved. They never are when one has interesting, forward-looking science. No doubt new advances, both in theory and in empirical finding, will throw up new challenges. For instance, now molecular biologists are suggesting that maybe the history of life is less that of a tree and more that of a lattice. Perhaps through viruses genes can be transferred from one branch of the living world to another. If true, surely this will be significant in the course of evolution and possibly also for the mechanisms. But these are matters for the future. As things stand at the moment, controversial though evolutionary theory in general and Darwinian theory in particular may be, I would argue that it is a mature, thriving, and very successful branch of modern science.

2.8
References

BEHE, M. (1996) *Darwin's Black Box: The Biochemical Challenge to Evolution.* New York: Free Press.

BOWLER, P. (1984) *Evolution: The History of an Idea.* Berkeley: University of California Press.

BOWLER, P. (1996) *Life's Splendid Drama.* Chicago, IL: University of Chicago Press.

CHAMBERS, R. (1844) *Vestiges of the Natural History of Creation.* London: Churchill.

COYNE, J. A., BARTON, N. H., TURELLI, M. (1997) Perspective: a critique of Sewall Wright's shifting balance theory of evolution. *Evolution* 51, 643–671.

CUVIER, G. (1813) (trans. Kerr, R.) *Essay on the Theory of the Earth.* Edinburgh: W. Blackwood.

DALY, M., WILSON, M. (1988) *Homicide.* New York: De Gruyter.

DARWIN, C. (1859) *On the Origin of Species.* London: John Murray.

DARWIN, C. (1871) *The Descent of Man.* London: John Murray.

DARWIN, E. (1794–1796) *Zoonomia; or, The Laws of Organic Life.* London: J. Johnson.

DAWKINS, R. (1988) The evolution of evolvability. In: LANGTON, C. G. (Ed.), *Artificial Life.* Redwood City, CA: Addison-Wesley.

DEGLER, C. N. (1991) In: *Search of Human Nature: The Decline and Revival of Darwinism in American Social Thought.* New York: Oxford University Press.

DEMBSKI, W. (1998a) *The Design Inference: Eliminating Chance through Small Probabilities.* Cambridge: Cambridge University Press.

DEMBSKI, W. (Ed.) (1998b) *Mere Creation: Science, Faith and Intelligent Design.* Downers Grove, IL: Intervarsity Press.

DESMOND, A. (1994) *Huxley, the Devil's Disciple.* London: Michael Joseph.

DESMOND, A. (**1997**) *Huxley, Evolution's High Priest*. London: Michael Joseph.

DOOLITTLE, R. F. (**1997**) A delicate balance. *Boston Review* **22**, 28–29.

ELDREDGE, N., GOULD, S. J. (**1972**) Punctuated equilibria: an alternative to phyletic gradualism. In: SCHOPF, T. J. M. (Ed.), *Models in Paleobiology*. San Francisco: Freeman, Cooper, 82–115.

FARLEY, J. (**1977**) *The Spontaneous Generation Controversy from Descartes to Oparin*. Baltimore: The Johns Hopkins University Press.

FOX, S. W. (**1988**) *The Emergence of Life: Darwinian Evolution from the Inside*. New York: Basic Books.

FREEMAN, S., HERRON, J. C. (**1998**) *Evolutionary Analysis*. Englewood Cliffs, NJ: Prentice-Hall.

FREUD, S. ([1905] **1955**) Three Essays on the Theory of Sexuality. In: STRACHEY, J. (Ed.), *The Standard Edition of the Complete Psychological Works of Sigmund Freud* 7. London: Hogarth, 125–243.

GOULD, S. J. (**1980**) Sociobiology and the theory of natural selection. In: BARLOW, G., SILVERBERG, J. (Eds.), *Sociobiology: Beyond Nature/Nurture?* Boulder, CO: Westview, 257–269.

GOULD, S. J. (**1981**) *The Mismeasure of Man*. New York: Norton.

GOULD, S. J. (**1984**) Morphological channeling by structural constraint: convergence in styles of dwarfing and giantism in Cerion, with a description of two new fossil species and a report on the discovery of the largest Cerion. *Paleobiology* **10**, 172–194.

GOULD, S. J., ELDREDGE, N. (**1977**) Punctuated equilibria: the tempo and mode of evolution reconsidered. *Paleobiology* **3**, 115–151.

GOULD, S. J., LEWONTIN, R. C. (**1979**) The spandrels of San Marco and the Panglossian paradigm: a critique of the adaptationist program. *Proc. Roy. Soc. London, Series B: Biol. Sci.* **205**, 581–598.

HALDANE, J. B. S. (**1929**) The origin of life. *Rationalist Annual*, 1–10.

HAMER, D., COPELAND, P. (**1994**) *The Science of Desire: The Search for the Gay Gene and the Biology of Behavior*. New York: Simon and Schuster.

HULL, D. L., RUSE, M. (Eds.) (**1998**) *Readings in the Philosophy of Biology: Oxford Readings in Philosophy*. Oxford: Oxford University Press.

JOHNSON, P. E. (**1991**) *Darwin on Trial*. Washington, DC: Regnery Gateway.

KITCHER, P. (1985) *Vaulting Ambition*. Cambridge, MA: MIT Press.

LAMARCK, J.-B. ([1809] **1963**) (trans. ELLIOT, H.) *Zoological Philosophy*. New York: Hafner.

LARSON, E. J. (**1997**) *Summer for the Gods: The Scopes Trial and America's Continuing Debate over Science and Religion*. New York: Basic Books.

LEVAY, S. (**1993**) *The Sexual Brain*. Cambridge, MA: MIT Press.

LEWONTIN, R. C. (**1977**) Sociobiology: a caricature of Darwinism. In: SUPPE, F., ASQUITH, P. (Eds.), *PSA 1976*. East Lansing, MI: Philosophy of Science Association.

LEWONTIN, R. C. (**1991**) *Biology as Ideology: The Doctrine of DNA*. Toronto: Anansi.

LEWONTIN, R. C., ROSE, S., KAMIN, L. J. (**1984**) *Not in Our Genes: Biology, Ideology and Human Nature*. New York: Pantheon.

MARGULIS, L. (**1970**) *Origin of Eukaryotic Cells.* New Haven, CT: Yale University Press.

MARGULIS, L. (**1993**) *Symbiosis in Cell Evolution.* New York: Freeman.

MAYNARD SMITH, J. (1981) Did Darwin get it right? *London Review of Books* **3** (11), 10–11.

MAYNARD SMITH, J. (**1995**) Genes, memes, and minds. *New York Review of Books* **42** (19), 46–48.

MILLER, K. (**1999**) *Finding Darwin's God.* New York: Harper and Row.

MILLER, S. L. (**1953**) A production of amino acids under possible primitive earth conditions. *Science* **117**, 528–529.

NYHART, L. K. (1995) *Biology Takes Form: Animal Morphology and the German Universities.* Chicago, IL: University of Chicago Press.

OPARIN, A. ([1924] **1967**) (trans. SYNGE, A.) The origin of life (Originally published as Proishkhozhdenie zhizni [1928]). In: BERNAL, J. D. (Ed.), *The Origin of Life.* Cleveland, OH: World, 199–234.

OPARIN, A. (**1968**) *The Origin and Initial Development of Life (NASA TTF-488).* Washington, DC: GPO.

PENNOCK, R. (**1998**) *Tower of Babel: Scientific Evidence and the New Creationism.* Cambridge, MA: MIT Press.

PLANTINGA, A. (**1991**) Evolution, neutrality, and antecedent probability: a reply to Van Till and McMullin. *Christian Scholar's Review* **21**, 80–109.

PROVINE, W. B. (**1971**) *The Origins of Theoretical Population Genetics.* Chicago: University of Chicago Press.

RICHARDS, R. J. (**1992**) *The Meaning of Evolution: The Morphological Construction and Ideological Reconstruction of Darwin's Theory.* Chicago: University of Chicago Press.

RUDWICK, M. J. S. (**1972**) *The Meaning of Fossils.* New York: Science History Publications.

RUSE, M. (**1982**) *Darwinism Defended: A Guide to the Evolution Controversies.* Reading, MA: Benjamin/Cummings.

RUSE, M. (**1985**) *Sociobiology: Sense or Nonsense?* 2nd ed. Dordrecht: Reidel.

RUSE, M. (Ed.) (**1988a**) *But is it Science? The Philosophical Question in the Creation/Evolution Controversy.* Buffalo, NY: Prometheus.

RUSE, M. (**1988b**) *Homosexuality: A Philosophical Inquiry.* Oxford: Blackwell.

RUSE, M. (**1989**) *The Darwinian Paradigm: Essays on its History, Philosophy and Religious Implications.* London: Routledge.

RUSE, M. (**1996**) *Monad to Man: The Concept of Progress in Evolutionary Biology.* Cambridge, MA: Harvard University Press.

RUSE, M. (**1999a**) *The Darwinian Revolution: Science Red in Tooth and Claw,* 2nd ed. Chicago: University of Chicago Press.

RUSE, M. (**1999b**) *Mystery of Mysteries: Is Evolution a Social Construction?* Cambridge, MA: Harvard University Press.

RUSE, M. (**2000a**) *Can a Darwinian be a Christian? The Relationship between Science and Religion.* Cambridge: Cambridge University Press.

RUSE, M. (**2000b**) *The Evolution Wars.* Santa Barbara, CA: ABC-CLIO.

SEDGWICK, A. (**1833**) *A Discourse on the Studies of the University*. London: Parker.

SEPKOSKI, J. J. JR. (**1976**) Species diversity in the Phanerozoic: species-area effects. *Paleobiology* **2**, 298–303.

SEPKOSKI, J. J. JR. (**1978**) A kinetic model of Phanerozoic taxonomic diversity. I. Analysis of marine orders. *Paleobiology* **4**, 223–251.

SEPKOSKI, J. J. JR. (**1979**) A kinetic model of Phanerozoic taxonomic diversity. II. Early Paleozoic families and multiple equilibria. *Paleobiology* **5**, 222–252.

SEPKOSKI, J. J. JR. (**1984**) A kinetic model of Phanerozoic taxonomic diversity. III. Post-Paleozoic families and mass extinctions. *Paleobiology* **10**, 246–267.

SEPKOSKI, J. J. JR., RAUP, D. M. (**1986a**) Periodicity in marine extinction events. In: ELLIOTT, D. (Ed.), *Dynamics of Extinction*. New York: Wiley, 3–36.

SEPKOSKI, J. J. JR., RAUP, D. M. (**1986b**) Was there 26-Myr periodicity of extinctions? *Nature* **321**, 533.

SIMPSON, G. G. (**1944**) *Tempo and Mode in Evolution*. New York: Columbia University Press.

WILSON, E. O. (**1975**) *Sociobiology: The New Synthesis*. Cambridge, MA: Harvard University Press.

WILSON, E. O. (**1978**) *On Human Nature*. Cambridge, MA: Cambridge University Press.

3
The Effects of Complex Social Life on Evolution and Biodiversity*

Edward O. Wilson

Abstract

Social vertebrates display faster chromosomal evolution, faster species turnover, and higher levels of allelic heterozygosity than nonsocial vertebrates. At least some species also display substantially more genetic differentiation among geographically spaced social groups. The explanation for this pattern may be that matrilines stay together over multiple generations to compose such groups, and a relatively small number of males inseminate them. In contrast, higher social insects, comprising the termites, ants, eusocial bees, and eusocial wasps, have slower rates of evolution, similar to that of other, related groups of insects. The explanation appears to be that their population structure is very different from that of vertebrates: from the view-point of genetics their colonies are actually individuals (superorganisms), and the population equivalent to the local vertebrate society (= population) is the population of colonies. Moreover, outbreeding is extensive. The four major assemblages of highly social organisms – the colonial invertebrates, eusocial insects, nonhuman social mammals, and man – are characterized for the most part by great biomass, interspecific competitive superiority, and low species diversity. Competitive superiority in colonial invertebrates and eusocial insects over nonsocial animals is achieved substantially by social means. Lower species diversity on the other hand is due substantially to large organism size, where invertebrate clones and insect colonies are the equivalent of the organisms composing nonsocial species.

* This contribution is a re-release. First published in Oikos 61, 13–18, Copenhagen 1991. Wiley-VCH thanks the author for his permission.

Handbook of Evolution, Vol. 2: The Evolution of Living Systems (Including Hominids)
Edited by Franz M. Wuketits and Francisco J. Ayala
Copyright © 2005 Wiley-VCH Verlag GmbH & Co. KGaA, Weinheim
ISBN: 3-527-30838-5

3.1
Introduction

The usual form of sociobiological analysis addresses the prerequisites in heredity and the environment for the origin and evolution of sociality. In this article I will reverse the procedure and pose the question of how sociality affects evolution, biological diversity, and certain aspects of the environment.

3.2
Sociality and Rate of Evolution

Bush et al. (1977) found that among 225 genera of vertebrates, speciation and chromosomal evolution have proceeded faster in genera whose species are organized into troops or harems, such as those of horses and many primates. They found the same trend in genera with limited vagility, patchy distribution, and strong individual territoriality, a category which includes (for example) many rodents. The criteria used by these authors were: (1) rate of chromosome change inferred from phylogenies of living species; (2) rate of appearance of new species, inferred from phylogenies of living species and fossils; and (3) rate of generic extinctions, hence turnover, in monophyletic clades (species and their descendants).

Marzluff and Dial (1990) obtained what might seem at first a result contrary to that of Bush et al.: hyperdiverse living groups of vertebrates, that is, families with many genera and genera with many species, are not significantly more social than less diverse groups. However, the two sets of results do not conflict. The data reviewed by Marzluff and Dial represent the standing crops of genera and species, while those reviewed by Bush et al. refer to turnover rates. It is possible to have very high turnover – new taxa replacing old ones – and hence rapid evolution overall, but achieve only a small standing crop base. The turnover is due to the dynamism of the evolutionary process, while the standing diversity is constrained by population size at equilibrium (K) and hence the number of species that can be packed together in the same community.

This brings us to the question of the causes of rapid evolution in social vertebrates. Bush et al. suggested that social groups, by staying clear of one another through behavioral repulsion, separately constitute demes with low effective breeding size. The groups are more prone to allele and chromosome type fixation by genetic drift. Thus, in accordance with Sewall Wright's shifting balance hypothesis, these socially demarcated populations might "experiment" with temporarily less adaptive combinations. If a new, favorable combination were to be struck (in Wright's metaphor, the population crosses the adaptive valley to a new adaptive peak), divergent evolution among the populations would be accelerated. And if the populations were further isolated by a geographic barrier, such as a deep valley or riverine corridor forest, the divergence might lead to full species formation, with intrinsic isolating mechanisms thereafter preventing free genetic exchange with other, sister populations. Finally, in taxonomic groups where only a few species can coexist, sympatry of the newly formed species would tend to result in the extinction of those less well adapted – in other words increase the rate of turnover.

Unfortunately for this neat explanation, Melnick (1987, 1988) noted from his research on rhesus macaques and that of earlier investigators on mammals generally that mammalian societies apparently exchange too many individuals to allow protracted evolution based on sustained genetic drift. For at least the rhesus macaques of Dunga Gali, Pakistan, Melnick also excluded the possibility that divergence among local troops is due to fission and formation of new troops along matrilines, in other words, females and their descendents.

Nonetheless, the data on social vertebrates confirm a high degree of differentiation among social groupings of mammals, just as Bush et al. stressed. Melnick offers the following alternative explanation to inbreeding and genetic drift. Matrilines stay together to compose each troop over multiple generations. Only a small number of males inseminate the females, as a result of dominance interactions, and they soon leave, as do their sons, without inbreeding. The total result is a strong genetic divergence among the troops, even when these groups are stationed only a short distance apart. If geographical isolation occurred, speciation might soon follow. One can imagine that intergroup selection is also an accelerating force, with certain genetic combinations prevailing over others in territorial disputes and takeovers.

In a parallel study, Nevo et al. (1984) found that among 127 vertebrate species surveyed with isozyme electrophoresis, social species possessed higher levels of genic heterozygosity than nonsocial species. Nevo's own work on *Spalax* mole rats of the Middle East indicates that heterozygosity in local populations increases with the level of environmental stress, in particular physical stress associated with arid environments. Nevo has suggested (pers. comm.) that social organization is a form of stress that could promote higher degrees of behavioral polymorphism and hence allelic heterozygosity. But it also seems likely that chance variation arising among geographically proximate groups, as postulated by Bush et al. and Melnick, would result in elevated genic diversity within groups through gene flow, even when gene flow between groups is moderately restricted across a few generations at a time.

When we turn to the second great pinnacle of social evolution, the eusocial insects (Wilson 1975), we encounter a radically different situation from that of the vertebrates. The rate of evolution at the level of the species and above is slower by as much as an order of magnitude. Slow evolution, at least slow macroevolution, is a trait held in common with most other insects. All of the 28 orders of insects alive in the Cretaceous Period are still alive today, with several extending all the way back to the late Paleozoic, whereas only one Cretaceous mammalian order, the Marsupialia, still persists. Within the single insect order Hymenoptera, 25 of the 36 families present in the Cretaceous are alive, whereas this is true for only one mammalian family, the Didelphidae or opossums. In contrast to the fast-evolving mammals, the early Tertiary faunas of social insects have a decidedly modern cast, down to the level of the genus and even the species group. A species of the contemporary stingless bee genus *Trigona* has been found in Late Cretaceous amber (Michener and Grimaldi 1988). No fewer than 56% of the ant genera of the early Oligocene Baltic amber still survive. A modern facies is even more evident in the Dominican amber, which is apparently early Miocene in age. here 34 genera or 92% of the total 37 are extant. Furthermore, the majority of species belong to living species groups, and a few are difficult to separate from living ants at the species level (Wilson 1985). The eusocial insects (ants, bees, wasps, and termites, characterized by a nonreproductive worker caste) appear to be about as conservative in evolution as their closest relatives among the solitary insects (Wilson 1990).

If the rapid evolution and high levels of heterozygosity of social mammals are to be explained as the impact of social organization on population structure, as the evidence seems to indicate, how are we to account for the complete lack of such correspondence in the social insects? The answer is that a colony of social insects, unlike a troop of macaques or herd of elephants, is not ordinarily a population, at least not in the ordinary sense applied in population genetics. Rather, it is the counterpart of a vertebrate organism, with the queen as the reproductive system and the workers as the rest of the body. The queen is inseminated by one to several males at the start of the life on the colony, or in the case of termites, is accompanied by a long-lived consort male. The offspring workers rarely reproduce on their own while the queen still lives, and in the vast majority of species, virgin primary reproductives leave the nest to start their own colonies alone and elsewhere. Soon after departure they mate with individuals from other nests, often in the midst of large panmictic swarms drawn from many other nests. In short, gene flow appears to be as free and random as it is in solitary insects. The equivalent in the social insects of a conventional local breeding population of vertebrates or solitary insects is the population of colonies.

There are exceptions to the population structure of the social insects just characterized. Some ant species in the genera *Formica, Linepithema* (formerly part of *Iridomyrmex*), *Myrmica, Pheidole,* and *Pristomyrmex* are organized into super-colonies or unicolonial populations, in which there are no colonial boundaries and the local population is just one great sprawling colony serviced by large numbers of queens (Hölldobler and Wilson 1990). Much of the mating occurs locally, within the territories of the supercolonies. These aggregations are thus structured somewhat like vertebrate society-populations, and have a similar potential for rapid evolution. Yet they do not present the appearance of much cumulative change. They are typically closely related to species that form ordinary colonies, and in fact may represent evolutionary dead ends that do not evolve much further on their own. The supercolony species are also poor in standing diversity compared with their ordinary relatives. Here the reason seems clear. Supercolony species typically dominate the local environment, aggressively excluding colonies of many of the related species. They are also broad generalists, preying on arthropods, scavenging corpses, and collecting honeydew excreta from aphids and other homopterous insects. It is hard to imagine two recently evolved sister species with supercolonies that overlap in the same habitat.

3.3
Sociality, Success, and Dominance

In considering the status of any group of organisms, biologists intuitively use the concepts of success, dominance, and diversity – and just as often conflate them. For the purposes of evaluating sociality in insects with greater precision, I recently proposed the following definitions (Wilson 1990):

Success

Longevity of the entire clade (a species and all its descendents) through geological time. Paleontologists often use the average duration of genera or families as a measure of a still higher taxon, such as the class Anthozoa or phylum Porifera. What they mean is the clade that starts with the ancestral species first displaying the diagnostic traits of the genus or family, plus all of the species descended from that species. What they exclude is pseudoextinction, the "termination" of a line by evolution of new diagnostic traits without extinction of the populations bearing the ancestral traits. A species evolving radically different traits may be said to give rise to a second species. Thus there are two chronospecies, the ancestor and the descendent, but the line has not gone extinct; the two forms are counted together as part of the same clade. Clade longevity is intuitively the best criterion of success, because it represents the simplest and most direct measure of species-level selection over long periods of time. It is a pure Darwinian criterion of the evolutionary process.

Dominance

Relative abundance, especially as it effects the appropriation of biomass and energy and impacts the populations and evolution of the rest of the biota. Dominance, as I have used the word (Wilson, 1990), is thus not merely population density but the degree to which the species influence the remainder of the community. The influence in general will certainly be correlated with density, and all other things being equal it must be identical to density. But other things are very seldom equal. Influence will rise as organism size increases, hence more biomass is appropriated. It will rise as basal metabolism increases, resulting in the appropriation of more free energy. It will rise with the number of links in the food webs, hence the number of species affected. For parasites it will rise with the degree of virulence. For free-living organisms generally it will rise as more of the physical environment id disturbed, by soil excavation during nest-building, by adventitious destruction of vegetation during trail construction and feeding, and so on. How these varied concomitants of population density and their impact are to be generally measured is a problem not easily resolved, but the usefulness of refining the concept of dominance in particular cases is clear enough.

Speciosity

The number of species, which in turn can be usefully separated into the geographically expanding components of diversity (Magurran 1988): alpha (the number of species occurring together within a given habitat), beta (the rate of turnover in species along a transect in passing from one habitat to another), and gamma (the number for all of the sampled localities taken together). The greater the alpha and beta diversity, the greater the adaptive radiation.

Adaptive radiation
The array of niches occupied by the various species of a clade, especially those that coexist in the same area simultaneously.

Geographic range
The entire area occupied by a clade at a particular time.

Success, which is an evolutionary quality, and dominance, which is an ecological trait, are correlated. All mathematical models to date have confirmed the strong intuitive impression that the risk of extinction decreases with maximum population size (K) and increases with the temporal coefficient of variation in population size (CV). Pimm et al. (1988) confirmed this result in an analysis of census data of 100 species of British land birds living in small to moderate sized populations on 16 islands. Stanley (1987) found that among marine bivalve mollusks living along the Pacific rim during Pleistocene times, abundant clades survived the longest. Yet success (longevity) and dominance (abundance and impact on biota) are qualitatively different phenomena and can be uncoupled and opposed in particular cases. A clade can enjoy enormous success by virtue of chronological persistence, but also be so rare as to have negligible influence on the ecosystem in which it lives. Examples include many "living fossils", such as onychophorans and coelacanths, which tend to be local and rare. At the opposite extreme, a clade can be so abundant for a while as to exert a profound and lasting impact on the rest of life, yet be short-lived. The most striking examples of all time may include the extinct clades of social mammals. (We hope and trust that this assemblage will not soon include man.)

A clade is likely to increase its longevity not only if its populations are larger but also if it contains many species, representing an array of adaptive types, and occupies a broad geographical range; because the clade then has "spread its bets" across many independent gene pools, among which one or more may possess traits that insure survival when the remainder of the clade succumbs. This relation is likely to exist as a broad correlation, yet it too can be uncoupled in particular cases. A tardigrade genus such as *Echiniscoides*, for example, might be worldwide and ancient yet monotypic and scarce; a species swarm of cichlid fishes can be middle-aged and extremely diverse, yet limited to a single African lake; and so on through all of the combinations of the population-level traits pertaining to abundance and diversity.

In Table 3.1 I have used the population-level traits to categorize the four animal (and human) assemblages possessing the most complex social organizations. With the exception of *Homo*, none of these assemblages is monophyletic, which is good for purposes of analysis, since each thus embraces multiple evolutionary experiments. The higher social insects include at least twelve independent lines, and the nonhuman social mammals are at least that polyphyletic. The colonial invertebrates

Table 3.1 Evolutionary and ecological traits of the four major groups of organisms with the most complex modes of social organization.

Group	Traits				
	Success (clade longevity)	Dominance (abundance, impact)	Speciosity (species numbers)	Adaptive radiation	Breadth of geographic range
Colonial invertebrates	±	+	−	−	+
Eusocial insects (terminates, ants, eusocial bees, eusocial wasps)	+	+	−	±	+
Nonhuman social mammals	−	±	−	−	+
Man (*Homo*)	?	+	−	+	+

+ traits strongly developed; − weakly developed; ± with many clades in both categories; ? intermediate. Human beings are judged to have undergone adaptive radiation, but by culture rather than speciation.

are distributed as independent clades across two phyla, the Cnidaria and Ectoprocta (Bryozoa).

What emerges clearly from this subjective comparison is that highly social organisms tend to be ecologically dominant and geographically widespread, but not speciose. Their record of longevity is decidedly mixed, with eusocial insects showing remarkable persistence since their origins, and colonial invertebrates and nonhuman social mammals less so.

The ecological dominance of social animals is striking. In terra firma rainforest near Manaus, eusocial insects make up about 80% of the insect biomass and a third of the entire animal biomass (Fittkau and Klinge 1973). In some samples from the Peruvian rainforest canopy, ants compose 70% or more of the individual insects (T. L. Erwin and J. E. Tobin, pers. comm.). The prevalence of ants in particular holds equally well in arid habitats. It also extends into the cold temperate zones, where for example ants remain the dominant insects in southern Finland. Wood and Sands (1978), Brian (1983), Hölldobler and Wilson (1990), Wilson (1990) and others have reviewed evidence documenting the impact of ants and termites on soil structure and composition, seed dispersal, plant distribution, and the biology of myriads of species of insects and other animals.

There can be little doubt that this overwhelming presence is due to social organization. The following traits arising from social life have been identified by experiments and field observations as conveying competitive superiority over otherwise similar nonsocial insects (Wilson 1990): (1) series-parallel operations, by which workers change occupations from moment to moment in a way that keeps larval care, nest construction, and other sequential tasks continuously well attended; (2) high levels of altruism, conferring combat advantage over enemies with a reduced cost of inclusive fitness in case of injury or death; (3) ability of the colony as a whole to control nest sites and food sources and bequeath them to later generations; and (4) use of social homeostatic devices, such as nest design, water cooling, and heating from massed bodies, to control microclimate and permit longer occupancy of nest sites near stable resources. With these advantages ants and termites occupy center stage in the terrestrial environment. They have pushed out solitary insects from the generally most favorable nest sites. The solitary forms occupy the most distant twigs, the very moist or dry or excessively crumbling pieces of wood, the surface of leaves, in other words, the more remote and transient microniches. They are also either very small, or fast moving, or cleverly camouflaged, or heavily armored. At the risk of oversimplification, the picture I see is one of social insects at the ecological center, solitary insects at the periphery.

Marine zoologists have independently forged a similar conception of the dominance of colonial invertebrates (Jackson 1985, Vermeij 1987). By Early Cambrian times sponge-like colonial Archaeocyatha formed reefs, to be succeeded in this role by bryozoans, sponges and corals, and then noncolonial worms, foraminiferans, and brachiopods during the Paleozoic; next sponges and algae and then rudist pelecypods in the Mesozoic; and finally corals, sponges, and coralline algae during the Cenozoic. A gradual increase in differentiation of zooids and in integration of the colonies occurred during the Mesozoic and Cenozoic in the corals and cheilostome bryozoans, possibly enhancing the competitive ability of these prominent colonial forms.

Colonial invertebrates displace solitary invertebrates in stable hard-bottom communities around the world. Being larger than solitary animals, they can overgrow them and are less likely to be pulled up the turbulence or eaten by predators. In contrast, solitary immobile animals tend to prevail in less stable environments, including the rocky intertidal zone, unconsolidated bottoms, and small or transient substrata such as shells and seaweeds. Like solitary insects, solitary benthic invertebrates fill in the smaller and more ephemeral niches where their competitively superior social counterparts cannot take hold.

That *Homo sapiens* has also achieved great biomass and competitive superiority by means of social organization is too clear to warrant further discussion.

Remarkably, sociality promotes dominance but not diversity. The numbers of species of the most highly social forms is very small compared to related solitary forms in the current world biota. The number of described species of eusocial insects is as follows: termites, 2200; ants, 8800; eusocial bees, roughly 1000; eusocial wasps, 800. The total number of described insect species, social and solitary combined, is about 750 000. In other words, 2% of the known insect species of the world have appropriated more than half of the biomass. The story is similar for the colonial invertebrates: among 240 000 multicellular invertebrates outside the insects, fewer than 5000 are sponges, 6000 are corals, and 4000 are bryozoans. The highly social nonhuman mammals are even less speciose, comprising the lion, 3 species of canids, 2 elephants, and only about 35 bandforming primates. Being close to the top of the food chain, most of the mammals are also less than dominant. An exception is formed by the elephants, which are capable of building dense populations that destroy extensive areas of woodland and alter the soil through physical disturbance and the deposit of large amounts of excrement.

The apparent reason for lessened biodiversity of the most social animals in size. For the most part these animals are very large. A coral head is essentially one organism, a superorganism. The same is true of an ant or termite colony. Both are orders of magnitude larger than single organisms belonging to solitary species. It is a well-established principle that within a monophyletic assemblage, such as the beetles, birds, and mammals, the numbers of species belonging to different body sizes are highest near (but not precisely at) the lower extreme and taper off gradually toward the upper extreme (Dial and Marzluff 1988, May 1988). The generally accepted explanation, which is also plausible for animal societies (in other words, superorganisms) is the larger the organism the larger the spatial niche and the fewer the species that can coexist stably.

3.4
References

Brian, M. V. (1983) *Social insects: ecology and behavioural biology*. New York: Chapman and Hall.

Bush, G. L., Case, S. M., Wilson, A. C., Patton, J. L. (1977) Rapid speciation and chromosomal evolution in mammals. *Proc. Natl. Acad. Sci. USA* **74**, 3942–3946.

Dial, K. P., Marzluff, J. M. (1988) Are the smallest organisms the most diverse? *Ecology* **69**, 1620–1624.

Fittkau, E. J., Klinge, H. (1973) On the biomass and trophic structure of the central Amazonian rain forest. *Biotropica* **5**, 2–15.

Hölldobler, B., Wilson, E. O. (1990) *The ants*. Cambridge, MA: Belknap Press of Harvard Univ. Press.

Jackson, J. B. C. (1985) Distribution and ecology of clonal and aclonal benthic invertebrates. In: Jackson, J. B. C., Buss, L. W., Cook, R. E. (Eds.), *Population biology and evolution of clonal organisms*. New Haven, CT: Yale University Press, 297–355.

Magurran, A. E. (1988). *Ecological diversity and its measurement*. Princeton, NJ: Princeton Univ. Press.

Marzluff, J. M., Dial, K. P. (1990) Does social organization influence diversification? *Am. Midl. Nat.*, in press (cited with permission of authors).

May, R. M. (1988) How many species are there on Earth? *Science* **241**, 1441–1449.

Melnick, D. J. (1987) The genetic consequences of primate social organization: a review of macaques, baboons, and vervet monkeys. *Genetica* **73**, 117–135.

Melnick, D. J. (1988) *Why are we social? – Ann. Rep.* Stanford, CA: Center for Advanced Study in the Behavioral Sciences.

Michener, C. D., Grimaldi, D. A. (1988). A *Trigona* from Late Cretaceous amber of New Jersey (Hymenoptera: Apidae: Meliponinae). *Am. Mus. Nov.* **2917**, 1–10.

NEVO, E., BEILES, A., BEN-SHLOMO, R. (**1984**) The evolutionary significance of genetic diversity: ecological, demographic and life history correlates. In: MANI, G. S. (Ed.), Evolutionary dynamics of genetics diversity. *Lecture Notes in Biomathematics* **53**, 13–213.

PIMM, S. L., JONES, H. L., DIAMOND, J. (**1988**). On the risk of extinction. *Am. Nat.* **132**, 757–785.

STANLEY, S. M. (**1987**). Periodic mass extinctions of the Earth's species. *Bull. Am. Acad. Arts Sci.* **40**, 29–48.

VERMEIJ, G. J. (**1987**). *Evolution and escalation: an ecological history of life.* Princeton, NJ: Princeton Univ. Press.

WILSON, E. O. (**1975**) *Sociobiology: the new synthesis.* Cambridge, MA: Belknap Press of Harvard Univ. Press.

WILSON, E. O. (**1985**) Invasion and extinction in the West Indian ant fauna: evidence from the Dominican amber. *Science* **229**, 265–267.

WILSON, E. O. (**1990**) *Success and dominance in ecosystems: the case of the social insects.* Oldendorf-Luhe, Germany: Ecology Institute.

WOOD, T. G., SANDS, W. A. (**1978**). The role of termites in ecosystems. In: BRIAN, W. V. (Ed.), *Production ecology of ants and termites.* New York: Cambridge Univ. Press, 245–292.

4

The Theory of Biological Evolution:
Historical and Philosophical Aspects

Franz M. Wuketits

> *There is grandeur in this view of life ...*
> Charles Darwin

4.1
Introduction

Evolution, "the most exciting and the most portentous natural truth that science has ever discovered" (Gould 1997, p. X), has fascinated many naturalists and some philosophers for more than 200 years. While nowadays nobody should seriously deny that evolution, i.e., the transformation of species through the ages, is a plain fact heavily documented by data from practically all fields of biological research, there is – as the creationist movement shows – still resistance against evolutionary theory and many people feel taken aback when confronted with the ancestry of their own species. Indeed, we humans came up from the ape, and reconstructing the paths of our evolutionary history is one of the most fascinating and challenging adventures of modern science (see Chapter 6). It is true that biological concepts and theories differ from that of other natural sciences in a particular sense. As Mayr (2000, p. 80) writes: "Many biological ideas proposed during the past 150 years stood in stark conflict with what everybody assumed to be true. The acceptance of these ideas required an ideological revolution". However, it seems that this "ideological revolution" is still an unfinished task.

This contribution is a brief review of historical and philosophical aspects of evolutionary thinking in biology. It outlines the major steps in the history of an idea that has changed our worldview more dramatically than any other idea in the natural sciences (including Einstein's theories and quantum physics). In a way, evolutionary thinking requires its own philosophy and stands in contrast to any *idealistic* conceptions that have predominated Western tradition many centuries. Darwin's theory – which is of course one of the focuses in the present article – meant no less than a shift from *static* (*essentialist*) to *dynamic* thinking. This might be regarded as the very meaning of what has been frequently called "Darwinian Revolution", for Darwin showed that what is *real* in nature is not the "type", but "variation". The implications of this conception have been serious and far reaching, going beyond

Handbook of Evolution, Vol. 2: The Evolution of Living Systems (Including Hominids)
Edited by Franz M. Wuketits and Francisco J. Ayala
Copyright © 2005 Wiley-VCH Verlag GmbH & Co. KGaA, Weinheim
ISBN: 3-527-30838-5

the narrower domain of biological science. Besides, and philosophically no less notable, Darwin's theory – evolutionary theory in general – expresses a *materialist* view based on the (philosophical) assumption that the world is composed only of material things and not influenced by any supranatural principles (Bunge 1981).

This chapter aims at giving a brief presentation of the philosophical presuppositions of evolutionary theory as seen in its historical development. I argue that the history of evolutionary thinking can be understood as a "battle" against some old and venerable philosophical conceptions. I of course do not intend to give a complete history of evolutionary theory. This would require another *Handbook* volume. Anyway, there exist many comprehensive, full-length expositions of this topic (some of them I have used here for reference). I just want to highlight the most remarkable ideas and figures of this adventurous intellectual journey.

However, there are certainly more historically and philosophically relevant aspects that are not treated in this chapter; the reader will find them discussed in other contributions to this *Handbook*. Thus, for instance, the epistemological and ethical implications of evolutionary thinking are presented in the first volume.

4.2
"Forerunners" of Darwin

Darwin was not the founder of evolutionary thinking. He rather established a particular evolutionary *theory* that could – and, basically, still can – explain the diversity of life on Earth and its history (see Section 4.5). The idea that "natural things" are changeable in one way or another goes back to ancient times, finding its expression in many concepts; some of them – from our today's point of view – quite amusing. Let us just think of the idea of *generatio spontanea* ("spontaneous generation") according to which many plants and animals can emerge and develop spontaneously from some nonliving material or from other organisms. Thus, for instance, Aristotle (384–322 BC) believed that many insects come from putrefying earth or vegetable matter – a belief that was widespread among naturalists and philosophers also in later centuries (e.g., see Mayr 1982) and, of course, must not be confused with evolutionary thinking in a strict sense, for *evolution* always implies the change of *species* over comparatively long periods of time. However, many naturalists and philosophers before Darwin speculated about this change and attained at what can be called "evolutionary thinking" in a broader sense. But for methodological reasons, we have to be cautious in this context. [One should keep in mind that up to the 19th century there was no clear distinction between "naturalists" and "philosophers". The term *philosophy* was frequently used in the sense of *methodology* or denoted a comprehensive view, a synthesis of facts and ideas in science. A good example is Lamarck's *Philosophie zoologique* (see Section 4.3)].

Let us consider, for example, the speculations of Benoit de Maillet (1656–1738), also known as Telliamed (de Maillet spelled backwards) (see Zimmermann 1953; Haber 1959; Junker and Hoßfeld 2001). The French Consul-General in Egypt traveled widely in the Mediterranean regions, had good knowledge of history and natural history, and did some original research work in geology. He believed that the antecedents of humans were marine organisms (*hommes marins*) and that dogs descended from seals. According to de Maillet, actually all terrestrial animals descended from marine creatures. This sounds as if he had anticipated a "modern truth". However, all things considered, de Maillet's views were rather fantastic and inspired by the idea of spontaneous generation. He did believe in transformation, but his "transformationism" does not resemble evolutionary thinking in a strict sense.

We should not "hunt" for precursors of Darwin. Indeed, many researchers in pre-Darwinian days assumed – sometimes very cleverly – that we live in a changeable universe. But this does not at all mean that they *anticipated* Darwin's theory or that they were already concerned with Darwin's problems. "The efforts of pre-Darwinian naturalists", says Bowler (1984, p. 21), "are interesting not because they tried to solve Darwin's own problems and failed but because they found so many different ways to explore the implications of a changing universe". Anyway, what cannot be neglected is that many researchers before Darwin speculated about the variability of species and their common descent.

A good example is the philosopher Denis Diderot (1713–1784), one of the leading figures of the French Enlightenment, editor of the famous *Encyclopédie*, and author of numerous philosophical essays, novels, and dramas. He wrote:

> Imagine ... that the order which strikes you has always subsisted; but allow me to believe that it is not at all so; and that, if we went back to the birth of time and of things, and if we perceived matter stirring and chaos disentangling itself, we should encounter a multitude of shapeless beings for a few well-organized ... I may ask you, for instance, who has told you ... that in the very beginning of animal formation, some were not without a head and others without feet? I can assure you that some had no stomach, and others no intestines ... that all the monsters were successively annihilated; all the imperfect combinations of matter have disappeared; only those remained in whose mechanism there was no imperfect contradiction and which could exist by themselves and perpetuate themselves (Quoted after Crocker 1959, p. 120).

Diderot was also very impressed by the facts of comparative anatomy and pondered that this discipline could, maybe, help us to understand the origin of animals. However, he hesitated to go further in his speculations and at the end even resigned because there was, as he said, God who created all living beings and the idea that species are changeable was contradicted by the observations. Thus, Diderot after all did not transgress, so to speak, from pre-evolutionary conceptions.

His contemporary, Georges L. L. Buffon (1707–1788), one of the most eminent naturalists of the 18th century and author of the monumental *Histoire Naturelle*, took a more decisive step towards evolutionary thinking. Although not truly an evolutionist, he might be called the "father of evolutionism" (Mayr 1982) for his ideas were more explicit and better grounded on biological facts. He considered data from comparative anatomy, biogeography, and the variation of plants and animals under domestication. A donkey, he argued, is nothing but a "degenerated" horse and, more generally, he concluded:

> If one species had been produced by another, if, for example, the ass species came from the horse, the result could have been brought about only slowly and by gradations. There would therefore be between the horse and the ass a large number of intermediate animals. Why, then, do we not today see the representatives, the descendants, of these intermediate species? Why is it that only the two extremes remain? (Quoted after Lovejoy 1959, p. 99)

Taking this passage seriously, we have to admit that Buffon has already argued in terms of evolution. However, we read other passages in his heavy volumes, we must recognize his objections against evolutionary changes. Like Diderot, he did not see empirical evidence for such changes and stated that since Aristotle, nobody had observed the emergence of a new species. Also, he missed the "intermediate animals" between extant species like the horse and the donkey. Why, then, should Buffon be called the "father of evolutionism"? Because he posed questions and raised ideas that marked, historically and philosophically, the outset of modern evolutionary thinking. [The terms "evolutionism" and "evolutionary thinking" are used in a general way and indicate the belief in a change of organisms (and other natural objects as well as human social and cultural systems). However, they do not necessarily imply any statement concerning the mode and causes of such a change, and are therefore not to be confused with evolutionary *theory* (see Section 4.3).]

Several other naturalists and philosophers of the 18th and 19th century are also to be regarded as advocates of a kind of evolutionary thinking, although only very few of them reached Buffon's scrutiny and the rigor of his arguments. They were usually guided by typological thinking, which was one of the most serious obstacles to evolutionary theory (see Section 4.4). Thus, the zoologist Charles Bonnet (1720–1793) insinuated that "perhaps there will be a continued progress ... of all species towards a higher perfection, such as all degrees of the scale will be continually changing in a constant and determined order" [quoted after (Glass 1959, p. 168)]. Bonnet strongly supported the idea of the *scala naturae* or the chain of being which had predominated views of nature since ancient Greek philosophy (see Lovejoy 1936). The "chain" reflected the assumption that all natural objects are arranged in a linear way – from inorganic compounds to objects of ever-higher complexity (plants and animals) and, finally, to human beings – and that this arrangement represents the divine plan of creation (Figure 4.1). It was, of course, a static model of nature; however, Bonnet's achievement was to think that the chain of being "had unfolded step by step through time to give a progression of life from the simplest forms at the bottom of the chain to the most complex at the top" (Bowler 1984, p. 62). Hence, in a way, Bonnet anticipated the idea of evolution for he contributed to the temporalization of the chain of being (Lovejoy 1936), but he remained an advocate of idealistic philosophy and cannot be regarded a "true evolutionist". Evolution in a strict sense does not simply mean a step-by-step progression of living forms; the core of evolutionary theory is *common descent*.

Idée d'une échelle
des etats naturels

L'HOMME	PLANTES
Orang-Outang	Lychens
Singe	Moisissures
QUADRUPÈDES	Champignons, Agariez
Écureuil volant	Truffes
Chauve-souris	Coraux et Coralloïdes
Autruche	Lithophytes
OISEAUX	Amianthe
Oiseaux aquatiques	Talcs, Gyse, Sélénites
Oiseaux amphibies	Ardoises
Poissons volans	**PIERRES**
POISSONS	Pierres figurées
Poissons rampans	Crystallisations
Anguilles	**SELS**
Serpens d'eau	Vitriols
SERPENS	**MÉTAUX**
Limaces	**DEMI-MÉTAUX**
Limaçons	**SOUFRES**
COQUILLAGES	Bitumes
Vers à tuyau	**TERRES**
Teignea	Terre pâte
INSECTES	**EAU**
Gallinsectes	**AIR**
Ténia, ou Solitaire	**FEU**
Polypes	Matières plus subtiles
Orties de mer	
Sensitive	

Figure 4.1 Bonnet's *scala naturae* (chain of being) from inorganic compounds to human beings.

Typically, naturalists and philosophers in the 18th century searched for the "prototype" of organisms and, actually, of everything. (See also the notion of archetype to be briefly discussed below.) In this sense, most of them were "idealists". The French philosopher Jean B. Robinet (1735–1820) is another example. He claimed that nature has varied a single prototype, "a germ which tends naturally to develop itself" [quoted after Crocker (1959, p. 134)]. Such statements did imply a kind of evolutionary thinking, but they did not really anticipate Darwin's views and their originators should not *a posteriori* be classified as Darwin's forerunners in a narrower sense.

Things are different in the case of Charles Lyell (1797–1875), whose work inspired Darwin in a very direct manner and who urged Darwin to publish his theory. He contributed much to the expansion of the time scale (see Section 4.4) and was one of the founders of *historical geology*, a "codifier of modernity" (Gould 1988, p. 4), for he strongly influenced Darwin's thinking and reasoning. Darwin took the first volume of Lyell's *Principles of Geology* (1830) with him when he embarked on the expedition on the *Beagle* and read it quite enthusiastically (Manier 1978). [His teacher, the theologian and botanist John Steven Henslow (1796–1861) had advised him to read the book but not to believe it!] Lyell's central idea was that the Earth's surface is produced by physical, chemical, and biological processes through long periods of time. His *uniformitarianism* implied that (1) everything on our planet (mountains, lakes, oceans, land masses) changes over time and (2) past events can be explained by the same causes we observe around us today. This principle of *actual causes* was a remarkable methodological achievement in the earth sciences and important for the foundation of evolutionary theory.

To summarize briefly at this point, we can state that, when Darwin entered the scene, the idea that species are variable was not new. Darwin does not stand at the beginning of evolutionism, but he did take the last and most decisive step in a 150-year-old tradition. He was quite aware of this. It is interesting to see whom he regarded as his "forerunners". In a historical sketch "of the progress of opinion on the origin of species" (Darwin 1859 [1958, pp. 17–25]) he mentioned about two dozen researchers – naturalists and philosophers – "who believed that species undergo modification and that the existing forms of life are the descendants by true generation of pre-existing forms" (Darwin 1859 [1958, p. 17]). In the first place he mentioned – not surprisingly – Buffon; in the second – more explicitly – Lamarck (see Section 4.3). Among the other names we find, to give just a few examples, Thomas H. Huxley (see Section 4.4), Herbert Spencer (see Section 4.5), the Prussian-Estonian anatomist and embryologist Karl E. von Baer (1792–1876), the German geologist Leopold von Buch (1774–1853), the British zoologist Robert E. Grant (1793–1874), and the British anatomist and paleontologist Richard Owen (1804–1892).

Owen deserves particular attention. He coined the word "dinosaur" (Desmond 1979), described *Archaeopteryx*, the first known fossil bird, and published a considerable number of anatomical and paleontological books. Darwin was astonished to learn from correspondence between Owen and the editor of the *London Review* "that Professor Owen claimed to have promulgated the theory of natural selection before I had done so; and I expressed my surprise and satisfaction at this announcement" (Darwin 1859 [1958, p. 22]). Owen had studied and worked at the zoological material from Darwin's voyage, but finally became Darwin's opponent. The issue that divided the two eminent naturalists was the place of humans in nature – Owen claimed that the human brain is essentially different from that of the apes and that humans are to be regarded as superior beings. He simply disliked human ape ancestry. Owen also formulated the important notion of *homology* as contrasted to mere *analogy*. He noticed that, while analogies are just cases of similarities between unrelated organs (like the "wings" of birds and insects), "homologies ... are cases in which the same basic structure is used by various species for quite different purposes" (Bowler 1984, p. 132). (Well-known examples are the limbs of vertebrates.) However, Owen was driven by an idealistic notion of *archetype* (see Desmond 1982) and gave an ahistorical, a comparative definition of homology as sameness, whereas Darwin later conceived of homology – in a historical sense – as an indicator of common descent (see Müller 2003). Owen searched for a "ground plan" on which the diversity of living forms is modeled and thus retreated to a kind of romanticism. In his studies on the vertebrate skeleton he looked for an "idealized vision of the simplest vertebrate form, an imaginary creature that had the essence of the type without any of the specialized modifications required by every real animal" (Bowler 1984, p. 132).

In other words, Owen was a "victim" of typological thinking, convinced that all living animal species are varieties of an "ideal form" (archetype) designed by the Creator. This belief appeared to be one of the obstacles to modern evolutionism (see Section 4.4.). However, we must always keep in mind that scientific ideas do not develop in a vacuum, but that they, some way or other, reflect the zeitgeist, and are not independent of cultural and social constraints. Human intellectual history in general and the history of scientific ideas in particular is not a linear process, but rather resembles a "zigzag" pattern. "Romanticism was a reaction to the 18th century clockwork cosmos, and, in early 19th century England, to the horrors of industrialization and its utilitarian supporters" (Desmond 1982). Owen and many other naturalists of the 18th and 19th century came very close to the idea of evolution, but their conceptions still remained embedded in idealism.

It is certainly true that historical contingencies influence the development of science and that our understanding of science closely interacts with our understanding of nature (Hull 1988), and *vice versa*.

4.3
What is an Evolutionary *Theory* and Who was the First Evolutionary *Theorist?*

Speculating about the possibility of changes in nature, about the possible changes of plant and animal species through longer periods of time, does not necessarily imply – or lead to – an evolutionary theory. Any evolutionary theory in a strict sense has to pose and to answer at least three sets of questions (see Wuketits 1988, 1989, 2000a, 2003):

(1) Are species changeable? Are the extant species descendants of other, earlier species?
(2) What are the modes of evolution? How does evolution occur?
(3) What are the mechanisms, the "motors" of evolutionary change?

The first question marked the very beginning of evolutionary thinking and is no longer relevant. The statement that species *are* changeable does not yet imply a *theory* of evolutionary change. Such a theory has to answer the second and third question, and has always to include at least two sets of statements (see Ayala 1985):

(1) Statements concerning the degree of relationship and evolutionary history of particular species or "classes" of species.
(2) Statements concerning the mechanisms by which evolutionary changes occur.

The first set of statements also includes propositions about the tempo and modes of evolutionary changes (see Chapter 1).

In this sense the French naturalist Jean-B. de Lamarck (1744–1829), who also coined the word "biology", was the first evolutionary theorist (e.g., see Chapman 1873, Dobzhansky et al. 1977; Mayr 1982; Wuketits 1988; Young 1992; Oeser 1996; Junker and Hoßfeld 2001).

Lamarck was a very productive researcher, and contributed data and ideas to different scientific disciplines. He published a three-volume flora of France, a comprehensive system of invertebrate animals, a two-volume treatment on matter and energy, and many other book-length works and articles concerning important scientific questions of his time. He was appointed, at the age of 50, to a professorship of zoology devoted to invertebrates, the "inferior animals" including insects, spiders, snails, worms, and all the other innumerable creatures that most people usually do not tend to esteem. We have to keep in mind here the zeitgeist of the 18th century, the spirit of the Enlightenment, ideas and ideals regarding progress and improvement of the *conditio humana* – the role of invertebrates was, to say the least, an insignificant one. Yet Lamarck's accomplishments in evolutionary thinking were mainly due to his intensive study of these animals and to the application of the *analytical method*. It was therefore completely wrong to say that Lamarck's theory is lacking empirical evidence (Dürken 1924) and to think that Lamarck was just a "speculating forerunner" of Darwin.

Lamarck was the first naturalist who not only speculated about the possibility of evolution, but explicitly stated that complex organisms have evolved from simpler ones (Figures 4.2 and 4.3). Also, most importantly, he tried to *explain* evolution, i.e., to find a mechanism for evolutionary change. He argued that there is an inherent tendency towards greater complexity and that a major "driving force" of evolution is the *inheritance of acquired characteristics*. This is what is usually understood by *Lamarckian evolution* or *Lamarckism* – a term which was not coined by Lamarck – and that could never be seriously tested. It was therefore unwarranted to claim, as some authors at the beginning of the 20th century did, that the empirical evidence

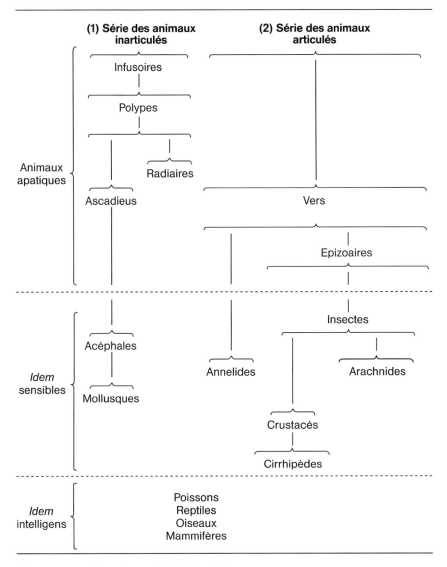

Figure 4.2 Lamarckian scheme of animal evolution.

in favor of Lamarckism is enormous (see Wagner 1908). In fact, the idea that (individually) acquired characteristics are inherited in a strict sense and that they, in the long run, cause evolutionary change has after all proved invalid (e.g., also see Zimmermann 1969), and the attempt to reanimate Lamarckism – neo-Lamarckism (see Table 4.4) – failed.

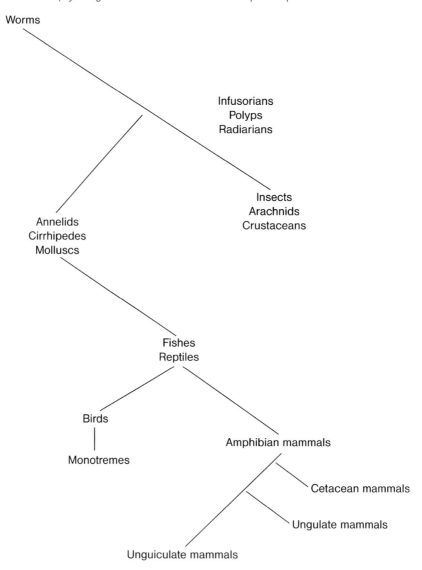

Figure 4.3 Lamarck's branching of animal groups.

Nevertheless, Lamarck's theory remains the first complete evolutionary theory. It is, by the way, a remarkable historical coincidence that his main methodological work containing the theory, his *Philosophie zoologique* (*Zoological Philosophy*), was published exactly 50 years before Darwin's *Origin of Species* which was – and remains to be – *the* book by which evolutionary thinking was finally established in science.

Why had 50 years to pass? Why did the scientific community – except for a few "adventurers" and "radicals" – ignore Lamarck's theory or not give it credit as an *evolutionary* theory?

4.4
Obstacles to Evolutionary Thinking

The more general question to be posed is: "Why, after all, did it take so long in the history of Western thinking to establish evolutionism, not to speak of an evolutionary theory?". As I have already mentioned, the emergence of evolutionary thinking can be regarded as an intellectual battle against philosophical conceptions, some of them strongly inspired by theology. Evolutionists had to fight against different dogmas and firmly established belief systems. (Considering the creationists movement these days, we must recognize that the battle is even now not yet won; see also Chapter 2).

Maybe the main obstacle to evolutionary thinking was – and sometimes still is – the plain fact that evolution does not fit our common sense (Mayr 1982). In our everyday life we do not observe the transmutation of species, but see instead that, for example, the progeny of horses is again horses, the elephants again procreate elephants, and so on and so forth. And it is, at first glance, indeed difficult to believe that dragonflies, woodpeckers, and elephants may have anything in common. Evolution connects all animal and plant species; however, it is usually a rather slow process – the development of new species, genera, families, etc., takes time. This is the point: time had to be discovered and, despite some earlier speculations – here again Buffon is to be mentioned – was not discovered before the 19th century (e.g., see Gould 1988). What is meant is the expansion of the time scale. The Earth was, as some naturalists (including Buffon) were tempted to think, rather old, thousands or even millions of years old (see also Taylor 1963). On the other hand, there were attempts to calculate the date of Creation and in the 17th century archbishop James Ussher, for instance, came to the conclusion that the world was created in the year 4004 BC. From today's point of view, this seems ridiculous, but one has to keep in mind that for many centuries the only accepted source for such calculations and conclusions was the Old Testament, and is was taken for granted that the Creation story of Genesis reveals the ultimate truth about the origin of the Earth and its inhabitants.

This story was the second obstacle to evolutionary thinking. Its influence was so pervasive that, as Mayr (2000, p. 81) reminds us, in the middle of the 19th century still "virtually all leading scientists and philosophers were Christian men". Thus, Lamarck's views were almost totally neglected or regarded as religiously offensive (see also Ruse 1982). Natural theologians, above all the English clergyman William Paley (1743–1805), had argued that the organization of living beings offered evidence for an omnipotent and omniscient designer, and the "intelligent-design argument" became, after all, a challenge to Charles Darwin (see Greene 1963; Ayala 2004). It is interesting to note that a philosopher in the second half of the 19th century who even claimed to establish a "cosmic philosophy" *based* on the theory of evolution could still declare that:

> ... it becomes desirable ... to inquire whether on the new [evolutionary] view there is any ground for assuming, as was necessarily assumed on the old view, that the divine Power works by methods analogous to human methods.

> The question we have to answer is not whether there exists a God. ... [O]ur Cosmic Philosophy is based upon the affirmation of God's existence, and not upon the denial of it, like irreligious Atheism or upon the ignoring of it, like non-religious Positivism. The question which we have now to answer concerns the existence of a limited personal God, who is possessed of a quasi-human consciousness, from which quasi-human volitions have originated the laws of nature, and to whose quasi-human contrivance are due the manifold harmonies observed in the universe (Fiske 1875, p. 377).

As can be seen, the author does see some particular problems with God in the light of evolutionary thinking, but finds no reason to doubt the existence of the Creator. In fact, Fiske's intention was to spiritualize evolutionary theory and to give evolution a transcendental direction. [Although nowadays almost unknown, John Fiske (1842–1901), historian and philosopher, was at his time an important popularizer of evolutionary thinking in the United States. His work *Outlines of Cosmic Philosophy* was well received in the United States and elsewhere.]

Generally, many scholars in the 19th century tried to connect evolutionism with theological ideas or at least to leave room for God in evolutionary thinking.

The third obstacle to the very idea of evolution was *typological thinking* or *essentialism* which is deeply rooted in Western philosophy and still quite influential – in philosophy and elsewhere. It is mainly a heritage of Plato's conceptions according to which, to put it briefly, only the "ideas" or "essences" are real, while the observed objects are just their representations; ideas or essences are unchangeable and give the observable natural kinds their "fixed" character. This point of view is the opposite to evolutionary thinking. It emphasizes the world's stability and invariance. For Plato and his successors the variety of natural objects is constrained, so to speak, by a limited number of *types*, each one forming a class separated from other such classes. Actually, variety is an illusion, only the types are real. Thus, evolution is not possible at all.

The eminent French educator, zoologist and founder of paleontology, George Cuvier (1769–1832), argued that animals possess so many diverse characters that they cannot be arranged in a single linear system. Thus, he was on the right track – the track towards evolution, that is – but since he was a "typologist", he could not embrace the idea of evolution. Just the opposite, he used his position and influence in science to defeat Lamarck (see Section 4.3), in particular, and evolutionism, in general. He arranged animals into four large groups (types) (Table 4.1), each of them representing one of the basic patterns of animal organization and subdivided into classes, orders, etc. These groups (types) he considered unchangeable and also "argued against the possibility of one species transforming itself into another" (Bowler 1984, p. 114). One might wonder how a paleontologist, aware of fossils, could reach such a conclusion. While (correctly) recognizing fossils as remnants of extinct animals, Cuvier (falsely) assumed that fossil species did not change and evolve into recent ones. Sure, as we know today, many species simply died out and did not become ancestors of any other kind of organisms. However, for Cuvier, extinction meant a proof for the stability of species and he believed "that each

Table 4.1 Cuvier's four types of animal organization.

Vertebrata (vertebrates)	Creatures possessing a backbone: mammals, birds, reptiles, and fish (in the 19th century the amphibians were usually included with the reptiles)
Mollusca (mollusks)	Creatures with no backbone but sometimes an external shell (oysters, clams, etc.)
Articulata (articulates)	Creatures with articulated or segmented bodies, like insects, spiders, worms, etc.
Radiata (radiates)	Creatures with a radial or circular body plan, e.g., sea urchin

species is a functional whole that cannot be disturbed by significant variation" (Bowler, 1984, p. 116), but only definitely destroyed by catastrophes. Therefore, his views have been characterized as *catastrophism* and were contrasted by Lyell's uniformitarianism (see Section 4.2).

Late in his career Cuvier was involved in a debate with his compatriot Etienne Geoffroy Saint-Hilaire (1772–1844), a brilliant and well-known naturalist, whom we have also to consider as a forerunner of Darwin. Geoffroy postulated the "unity of composition", i.e., a single consistent structural plan underlying the anatomy of all animals, but he conceded that one animal form might derive from another within its type and in a way he even anticipated the idea that evolutionary change could occur through sudden "mutations". Cuvier opposed this and his debate with Geoffroy is among the best known academic controversies. As he did with Lamarck's ideas, he demolished Geoffroy's theory – and the struggle for evolutionism was lost once again.

One more word on the meaning of fossils. From a philosophical or methodological point of view it is interesting to note that fossils initially did not serve as a proof for evolutionary change. Fossils had been known already in ancient times, but were continuously interpreted as *ludus naturae* ("plays of nature"), results of the Creator's mood or products of the Deluge (e.g., see Hölder 1960). That means that within the old *paradigms* (typological, statical thinking, belief in creation) they could not be properly understood. Only the paradigm of evolution made a correct interpretation of fossils possible. Hence, it is wrong to assume that – as is still often implied in textbooks – fossils offered evidence for evolution. Of course, the fossil record demonstrates the variation of species and helps us to reconstruct the modes of evolution. But, first of all, evolution is to be accepted because, as we can learn from Cuvier's arguments, fossils alone do not impose the idea of the transformation of species. Cuvier, in fact, uncovered the fossil record, but since he was an antievolutionist, an advocate of the old paradigm, he thought that fossils – especially the gap one sees in the record – speak against evolution (Ruse 1982).

The fourth obstacle to evolutionary thinking was *teleology*, the belief that there is universal purpose in nature, that there are final goals and a kind of "conscious actor". The problem of purpose plays still an important role in biology and the question "What for?" is a typically biological question. Whether one takes, for example, the neck of a giraffe, the eye of an octopus, the wings of birds, or the hands of humans – they all serve particular purposes. This is also true to physiological processes such as, e.g., respiration or digestion, and behavioral traits, and teleological explanations have been postulated as a special type of explanations in biology.

When biologists nowadays refer to purpose, they do not have in mind a universal or cosmic teleology. They do not believe in any "higher purpose", but argue that teleologically organized systems are based on programs that are result of evolution by natural selection. When, for example, a biologist states that the large ears of the African elephant serve the purpose of cooling, then he or she is not implying any kind of conscious goal-intended processes or a Creator designing ears and other structures for particular purposes. Rather, the biologist would argue that, under particular constraints, this type of ears has been favored by natural selection. In order to avoid a confusion with the old belief in teleology, the term *teleonomy* is now frequently used (e.g., see Mayr 1988, 1997). It refers to the fact that natural selection "breeds", as it were, structures, functions, and behavioral traits which aid the survival of an organism. Hence, the problem of purpose is no longer obscured by metaphysical speculations. It is fully compatible with the theory of evolution by natural selection and teleological (teleonomic) explanations do not contradict causal accounts (e.g., see also Ayala 1999).

However, up to the 19th century the interpretation of purpose hampered evolutionary considerations or even made them impossible. Darwin shocked (in the literal sense of the word) many of his contemporaries mostly because he explained the diversity of life without any reference to cosmic teleology and instead introduced a rather gloomy mechanism, i.e., natural selection. "Thus", he concluded, "from the war of nature, from famine and death, the most exalted object we are capable of conceiving, ... the production of the higher animals, directly follows" (Darwin 1859 [1958, p. 450]). Nevertheless, in some aspects Darwin was still influenced by teleological thinking and was tempted to use teleological phrases, so that he is regarded as a teleologist by some historians of science (Himmelfarb 1968; Lennox 1993) (see Section 4.5).

Closely related to the belief in a universal or cosmic teleology is the fifth obstacle to evolutionism, i.e., the belief in a special status of human beings in nature. Thomas H. Huxley (1825–1895), "Darwin's bulldog" and "devil's disciple", as he was called, and one of the chief spokesmen for evolution in the 19th century, relentlessly defended the view that humans are closely related to other living beings, and offered ample anatomical and paleontological evidence for the common ancestry of humans and apes. An engaged science educator, he popularized this view quite successfully (see also Weiss 2004). Yet, in his classical *Man's Place in Nature*, he concluded:

> Our reverence for the nobility of manhood will not be lessened by the knowledge, that Man is, in substance and in structure, one with the brutes; for, he alone possesses the marvelous endowment of intelligible and rational speech, whereby, in the secular period of his existence, he has slowly accumulated and organized the experience which is almost wholly lost with

the cessation of every individual life in other animals; so he now stands raised upon it as on a mountain top, far above the level of his humble fellows, and transfigured from his grosser nature by reflecting, here and there, a ray from the infinite source of truth (Huxley 1863 [2001, p. 114]).

Thus, Huxley did justice to the belief that several aspects of human activities are not strictly zoological (see di Gregorio 1984). Anyway, his (evolutionary) view of humans was a heresy. We have to keep in mind the long-standing (religiously informed) conviction that humans are in the center of the world and that they are essentially separated from all other creatures living on the Earth. Up to now philosophers have frequently characterized humans as fundamentally different from other living beings and until recently some biologists have also acknowledged the human species as something special. However, as Ruse (1986, p. 104) puts it: "If you take Darwin seriously ... then the special status of *Homo sapiens* has gone for ever. Any powers we have are no more than those brought through the crucible of the evolutionary struggle and consequent reproductive success". Having in mind that such statements still meet with fierce resistance in philosophy and the human sciences, it should not be difficult to imagine that scholars in the 19th century protested against those who were trying to place humans in the animal kingdom.

Paradoxically, nobody seemed to protest, when – a century before Darwin and Huxley – the Swedish naturalist and father of biological systematics and taxonomy, Carl von Linné (Linnaeus) (1707–1778), put humans together with chimpanzees and gorillas, and portrayed the apes as very humanlike creatures (Figure 4.4).

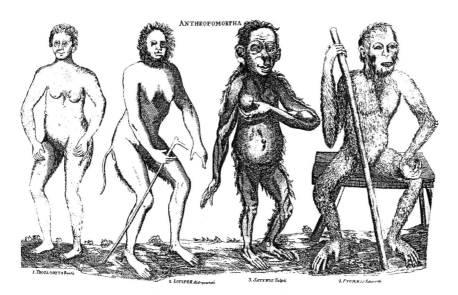

Figure 4.4 von Linné put humans together with chimpanzees and gorillas, and portrayed the apes as very humanlike creatures (reproduced from Zimmermann 1953).

However, Linné was not an evolutionist and therefore obviously not considered a dangerous person. He just tried to reveal the Divine order of nature and, modest as he was, "believed that he was privileged to see the outline of the Creator's plan and his efforts to represent it would become the basis of a new biology" (Bowler 1984, p. 64). [In a way, he even did not underestimate himself for he worked out some fundamental principles of biological classification that proofed to be very useful and are still valid. He introduced *binominal nomenclature*, a method of naming organisms based on Aristotelian logic. According to this method the scientific name of any plant or animal consists of two (Latin) words, the first designating the genus, the second emphasizing the species and its uniqueness, e.g., *Canis lupus* (wolf) or *Homo sapiens* (modern humans). Also, compared to earlier attempts to systematize living beings, Linné's systematics was, because of its coherence, a remarkable advance (see also Oeser, 1996).] Thus, Linné advocated a typological species concept and regarded each species as a fixed, discrete entity different from other species (e.g., see also Price 1996).

Finally, a general methodological aspect should be mentioned in this context. Evolutionary theory is a *historical theory* and it seems that for this reason it still causes some confusion even among scientists, at least among those who believe that anything and everything has to be *experimentally* tested. Indeed, evolutionists do rely on some experiments, e.g., the *mutation experiments* in *Drosophila* (e.g., see Ayala and Valentine 1979), but the major paths of evolution are not amenable to experiments. It is for this reason that Mayr (1997, 2002, 2004) uses the structure of evolutionary theory – and of evolutionary *biology* as the empirical study of evolution – to demonstrate the *autonomy* of biology among the natural sciences (e.g., see also Vollmer 1995, Wuketits 2004). Neither in physics (except for disciplines like cosmology) nor in chemistry does the historical aspect plays such an important role as it does in biology. Evolutionary theory (biology) has a special status among the natural sciences because:

(1) It deals, to a large extent, with (historically) unique phenomena, such as the emergence of vertebrates, the extinction of the dinosaurs, the origin of birds, or the evolutionary history of humans.
(2) The events to be studied cannot be experimentally tested (we cannot experiment on the emergence of vertebrates or the extinction of dinosaurs).
(3) The evolutionist tries to answer the question *why?* (see also Simpson 1980) or *what happened?* (e.g., to the dinosaurs).

Evolutionary theory (biology) relies on the method of *historical narrative*: "Just as in much of theory formation, the scientist starts with a conjecture and thoroughly tests it for its validity, so in evolutionary biology the scientist constructs a historical narrative, which is then to be tested for its explanatory value" (Mayr 2004, p. 22). What, from this point of view, can be said with regard to most pre-Darwinian thinkers is that they made a lot of interesting observations and quite often even posed the right questions, but that they were unable to construct the proper historical narratives. This brings us conveniently to Darwin.

4.5
Charles Darwin (1809–1882): The Theory of Natural Selection

In her introductory remarks to Darwin's *Autobiography*, Noral Barlow, his grand-daughter, states that:

> Darwin's whole trend of thought was against facile speculation, yet theories flowed freely through his mind ready for the essential tests of observation and experiment. He took twenty years of combined theorizing and fact-finding to prepare his case for evolution in the face of a predominantly antagonistic world. He had to convince himself by accumulated evidence before he could convince others, and his doubts are as freely expressed as his convictions. His books lie like stepping-stones to future knowledge. Dogmatic fixity was wholly alien to his central idea (Darwin 1958, p. 13).

This is a very sympathetic characterization of Darwin's attitude to his own work. In fact, Darwin had first to convince himself – of the variability of species for initially he, like most of his contemporaries, believed that the Earth and its inhabitants were created by God. One has always to keep in mind that he studied theology and was prepared for Holy Orders in the Church of England. He was strongly influenced by natural theology and Paley's interpretation of design (see Section 4.4), which he later dismissed. However, he never studied natural sciences at a professional level. In a way, Darwin represented the typical amateur naturalist in 19th century England (Finkelstein 2000).

So much has been written about Charles Darwin and his theory that it seems rather superfluous to comment on the man and his ideas once again. On the other side, how could Darwin's theory be omitted in a *Handbook of Evolution*! Besides, some aspects of Darwin's work are still often misunderstood, and the general reader might wish for some information concerning Darwin's very achievements in science and the historical and philosophical background of his theory.

Theory? Indeed, Darwin established *five theories* (Mayr 1991; see also Table 4.2), and again (see Section 4.3) a distinction must be made between (1) the *fact* that

Table 4.2 Darwin's five theories.

Evolution	Species are not constant but variable
Common descent	All species are genealogically related and are to be traced back to a single unique origin in the history of the earth
Multiplication of species	One species splits into two or more new species
Gradualism	Evolution occurs gradually, with no major breaks or discontinuities
Natural selection	The very mechanism of evolution is natural selection; evolution occurs on the basis of the abundance of genetic variation in each generation

evolution has occurred, (2) the *pathways* of evolution, and (3) the *theory* of mechanisms that explain evolutionary change (Oldroyd 1986). Darwin's genuine theory was that of natural selection, while for the other four theories we can find several other advocates and authors. Lamarck, for one, was convinced that species are variable (see Section 4.3) and also espoused gradualism which expressed the old and venerable philosophical doctrine that nature does not produce any saltations (*natura non facit saltus*). From the point of view of many naturalists of the 19th century "the two great merits of [Darwin's] work are its bringing together in a condensed form the evidences in favor of the Evolution of Life, and its offering Natural Selection as a cause of Evolution" (Chapman 1873, pp. 16–17). These two "merits", however, can also be seen from our today's point of view.

At the beginning of his scientific career, Darwin believed in creation and was not aware that the transmutation of species had been discovered by Lamarck (see Section 4.3). When he eventually took notice of Lamarck's work, he was obviously not impressed. The same is true to his grandfather's work. Erasmus Darwin (1731–1802) was a colorful person, a universal genius, physician, naturalist, engineer, and poet – at one and the same time. He developed a kind of Lamarckian theory of evolution and was, in a way, a forerunner of his grandson who, when reading his *Zoonomia*, was first quite enthusiastic about it, "but on reading it a second time after an interval of ten or fifteen years ... [was] much disappointed, the proportion of speculation being so large to the facts given" (Darwin 1958, p. 49). No doubt, Charles and his grandfather were scientists of a completely different mentality.

Most important in Darwin's life was the *Beagle* voyage (1831–1836), which gave him many opportunities to study biological and geological phenomena. When he returned to England, not only his thoughts of a clerical life had vanished, but his belief in the transmutation of species had developed. For the next two decades he was busy studying the collected material and formulating his theory of natural selection. He proceeded very cautiously with this theory, hesitating to write it down and publish it. "In June 1842", he remembered in his *Autobiography*, "I allowed myself the satisfaction of writing a very brief abstract of my theory in pencil in 35 pages" (Darwin 1958, p. 120). However, only 17 years later he published the results of his research. Probably, it would have taken him even longer to come to a reasonable – satisfactory for him – end, but then what Ruse (1982, p. 3) describes as follows happened:

> On June 18 1858, Charles Darwin sat down after breakfast in his study to go through his mail. One package came from the Far East; it was sent by a young naturalist and collector, Alfred Russell Wallace. Wallace and Darwin had never met but for a year or two now they had been corresponding, exchanging views and queries about the biological world. Wallace had written a short paper, mailing it to Darwin and hoping that the older man might think it worthy of publication. As Darwin had started to read, his heart started to sink. For twenty years, he, Darwin, had been sitting on a secret: a theory and a mechanism that would explain in a scientific way the organisms we find around us and in the fossil record: a theory of *evolution*. His friends

had long argued that he publish, lest someone else forestall him. Now the worst happened, for Wallace had hit on the same ideas, and the paper he had just sent to Darwin was as if Darwin himself had written it. Even the language was the same!"

In the next meeting of the Linnean Society (in July 1858) Wallace's paper and extracts from Darwin's unpublished writings were read, and then Darwin rushed to finish and publish his *On the Origin of Species by Means of Natural Selection*. This impressive work is perhaps not the best example for literary quality in the Victorian age, although some are inclined to characterize Darwin as "a superb writer, in the great Victorian tradition" (Gribbin and White 1995, p. vii). However that might be, the book contains a vast number of facts from different disciplines including geology, biogeography, anatomy, and embryology, and is probably best labeled as *One Long Argument* (Mayr 1991). That is what Darwin intended: to give, based on a body of evidence, convincing arguments for evolution by natural selection.

The importance of Alfred Russel Wallace (1823–1913) in the history of evolutionary thinking is undisputed. He developed a theory of evolution through natural selection independently of Darwin. (By the way, he also coined the term *Darwinism*.) Why, then, has he never been such a celebrity like his compatriot? I do think that Wallace has been underestimated, but this is not the place to go into biographical and historical detail. What seems to be true is that evolutionary thinking, in general, and the theory of natural selection, in particular, have been mainly promoted through Darwin's work. In addition, Darwin was more consequent than Wallace. He applied the theory of natural selection also to the evolution of human psychic and mental capacities. Wallace, however, influenced by spiritualism, believed that some supranatural or, at least, non-biological agency must be responsible for the emergence of "higher" phenomena such as the human mind.

How did Darwin arrive at his theory of natural selection? Table 4.3 presents his most important observations and his major conclusions.

While reflecting about a possible mechanism of evolutionary change, Darwin read – "for amusement", as he remembers in his *Autobiography* (1958, p. 120) – the *Essay on the Principle of Population* (1798) by the sociologist Thomas R. Malthus

Table 4.3 Central elements of Darwin's theory of natural selection.

Observations	Potentially unlimited reproduction
	Limited resources
	Uniqueness of the individual
	Differential reproduction
Conclusions	Competition among the individuals of one and the same species (struggle for existence)
	Natural selection and survival of the fittest
	Variation of species over many generations (evolution)

(1766–1834). Malthus argued that the resources are limited, but that human populations can increase exponentially; thus, the consequences would be hunger and disease. Darwin applied this argument to nature, he "saw that the potentiality for exponential population growth is quite universal in the living world … [so that] it follows that only a part of the progeny survive, and the rest are eliminated by death" (Dobzhansky et al. 1977, p. 97). Thus, he produced the theory of natural selection which replaced the "intelligent-design argument" (see Section 4.4). Using his own words, we can state that "there seems to be no more design in the variability of organic beings and in the action of natural selection, than in the course which the wind blows" (Darwin 1958, p. 87). [In the *Origin of Species* Darwin definitely dispensed with religious arguments, and thus irritated theologians and theologically oriented philosophers. Only a few of them did not really feel threatened by him (see Gregory 1991).] What counts in the living world is reproductive success and those individuals that are most successful in this aspect will be favored by natural selection. This is the very meaning of Darwin's formula *survival of the fittest* which he adopted from the philosopher Herbert Spencer (1820–1903). [Spencer was also the first who used the term "evolution" in a modern sense. Although in many aspects he was a Lamarckian, he shared important evolutionary ideas with Darwin and was also inspired by Malthus' *Essay*. He expanded evolutionary thinking to all natural phenomena, and to human social and political systems, and thus tried to establish a kind of "universal evolutionism" (for a review, see Gay 1999).]

In this context we should also remember Darwin's reference to *domestic breeding* which helped him to understand the meaning of variation. Darwin himself bred and raised pigeons, and was well acquainted with the literature on domestication. Fifteen years before the publication of the *Origin* he already drew the analogy between natural selection and an omniscient "super breeder" who selects not for the benefit of the human breeders, but only for the benefit of the organism itself. However, he did not simply apply *artificial selection* to nature. Rather, as Richards (1998) explains, he treated domestic breeding as an "experiment", from which he could establish the law of natural selection.

From a philosophical/methodological point of view, three aspects of Darwin's theory are of particular interest: gradualism, the theory of adaptation and the general meaning of theory.

Regarding gradualism, Darwin followed, as already stated, the old philosophical assumption *natura non factit saltus[m]* and concluded: "As natural selection acts solely by accumulating slight, successive, favorable variations, it can produce no great or sudden modifications; it can act only by short and slow steps … 'Natura non facit saltum'" (Darwin 1859 [1958, p. 435]). He was, of course, influenced by the zeitgeist and the idea of slow, but constant, progress at different levels. This idea fits pretty well the notion of adaptation – the assumption that, through natural selection, organisms get better and better adapted to their respective environment(s).

Concerning his views of adaptation, Darwin has been frequently misunderstood – as *adaptationist*, that is. Adaptationism (e.g., as criticized by Gould and Lewontin 1984) is the conviction that any – anatomical, physiological, or behavioral trait – of an organism is *adapted* to a given environment and that selection works exclusively, as it were, from outside (environmental selection). Darwin, however, did not believe that living beings are simply "molded" by their environment(s) and was prepared to acknowledge the organism's role in evolution (see Wuketits 2000b). Thus, he critically stated:

> Naturalists continually refer to external conditions, such as climate, food, etc., as the only possible source of variation. In one limited sense ... this may be true; but it is preposterous to attribute to mere external conditions, the structure, for instance, of the woodpecker, with its feet, tail, beak, and tongue, so admirably adapted to catch insects under the bark of trees (Darwin 1859 [1958, p. 28]).

Organisms are active, dynamical systems and Darwin himself anticipated what more than hundred years later the philosopher Popper (1984) claimed: that living beings are actively searching for a "better world".

With respect to the meaning of theory in Darwin's work, we have to abandon the view that Darwin worked on the basis of *induction*, that he accumulated "facts" in order to, finally, – establish a theory. As Ghiselin (1969, p. 4) has already pointed out, Darwin's "entire scientific accomplishment must be attributed not to the collection of facts, but to the development of theory". This is not to say that Darwin disregarded empirical evidence – just the opposite is true as one can learn from his expositions in the *Origin* and his zoological writings, like the work on barnacles (see Stott 2003). "From my early youth", Darwin (1958, p. 141) stated, "I have the strongest desire to understand or explain whatever I observed, – that is, to group all facts under some general laws". Mere "facts" are not enough. What counts is, as we have seen in the case of fossils, a proper explanation, a useful theory. Erroneously, attempts to corroborate evolutionary theory were based, for some time, mainly on empirical arguments and not on the theory itself (Gutmann 1979). Empirical arguments do matter in science, but the sophistication of theories is no less important – I would say even more important than the "facts".

Finally, it is to be emphasized that, if we take Darwin and his theory of natural selection really seriously, we can recognize some most fundamental philosophical implications and, besides, we have take into account that his theory is meaningful in the context of a new humanism. As Mayr (2000, p. 83) says:

> He [Darwin] developed a set of new principles that influence the thinking of every person: the living world, through evolution, can be explained without recourse to supernaturalism; essentialism or typology is invalid, and we must adopt population thinking, in which all individuals are unique (vital for education and the refutation of racism); natural selection, applied in social groups, is indeed sufficient to account for the origin and maintenance of altruistic ethical systems; cosmic teleology, an intrinsic process leading life automatically to ever greater perfection, is fallacious, with all seemingly teleological phenomena explicable by purely material processes; and determinism is thus repudiated, which places our fate squarely in our own evolved hands.

These statements also assert the meaning of Darwinian evolutionary thinking for education and general culture. A similar point has been recently made by Kutschera and Niklas (2004) in a paper published on the occasion of Mayr's 100th birthday. Evolutionary thinking is one of the most important enterprises of human intellectual development, and must be part and parcel of higher education. It is indispensable for a modern, secular world view for it helps us to develop a deep understanding of our own nature, our origins, and our possible future. As Simpson (1963, p. 25) accurately states: "It is a characteristic of [the] world to which Darwin opened the door that unless *most* of us do enter it and live maturely and rationally in it, the future of mankind is dim, indeed – if there is any future".

4.6
Darwin and After Darwin

Like any other scientific theory, Darwin's theory of natural selection is not – and has never been – a body of dogmas. Therefore, it has undergone several expansions and changes (e.g., see Stebbins and Ayala 1985; Beurton 1994; Wieser 1994; Wuketits 2003; Kutschera and Niklas 2004). At the same time, there have always been some alternative theories, some of them contrasting, others more or less compatible with Darwin's views. It has to be stressed – again and again – that controversies within evolutionary thinking, particularly different views of modes and mechanisms of evolutionary change, do not impinge on the *fact* of evolution. Evolution is a very complex process and it is understandable that it requires a broad view of the phenomena in question. As in other fields of scientific research, controversies have proved quite fruitful in evolutionary biology for they urge the involved "combatants" to reassess – and critically reflect upon – their own arguments.

Table 4.4 gives a synopsis of evolutionary theories after Darwin. Darwin's own views were expanded in the light of new biological findings. Neo-Darwinism included classical genetics (Weismann 1904), the synthetic theory, or, according to Huxley (1942), *Modern Synthesis* included population genetics. The meaning of the study of populations increased in the 20th century and today evolutionary thinking without population biology is unimaginable. However, other biological disciplines have also gained growing importance in evolutionary studies, e.g., developmental biology (see Chapter 5). Darwin's theory of natural selection has remained basically undisputed, but – in his own spirit – it has been frequently (and successfully) argued that selective forces have to be conceptualized at different levels of organismic organization (e.g., see Gould 1982). The synthetic theory was a kind of long-term project carried out in the 1930s and 1940s by many biologists in several countries (see, for a review, Reif et al. 2000). It brought together results from various fields of research and can be regarded as an integrative view of biological evolution. Some of its tenets, e.g., gradualism, have been challenged by more recent findings and theoretical considerations, but basically it proved to be very solid and heuristically extremely useful.

Table 4.4 Evolutionary theories after Darwin and their main elements.

Neo-Lamarckism	Inheritance of acquired characteristics; "will" in organisms to change their own structures and functions ("psycho-Lamarckism")
Neo-Darwinism	Expansion of Darwin's theory; the cell as a level of selection; classical Darwinism plus (Mendelian) genetics
Mutation theory	Evolution occurs mainly through mutations
Synthetic theory	Expansion of Darwin's theory and neo-Darwinism; genetic recombination and mutations are the material basis, natural selection is the "driving force" of evolution; neo-Darwinism plus population genetics
Saltationism	Drastic, sudden changes ("saltations") happen in evolution, for example through "macromutations"
Punctuated equilibria	There are long periods without significant evolutionary changes interrupted by short episodes of rapid change
Neutrality theory	Molecules evolve under the influence of mutation (by chance) and are neutral to selective forces
Systems theory	Expansion of the synthetic theory; inner constraints ("internal selection") and environmental selection are mutually related

From the point of view of the philosophy of science, it is remarkable that the history of evolutionary theory in the 20th century was a process of *integrating* scientific, biological, disciplines. The synthetic theory mirrors, so to speak, the attempt to unify biological sciences (Smocovitis 1992). At the same time, its arguments are based on practically all fields of biological research – from comparative anatomy to population genetics, from paleontology to cell biology, from physiology to biogeography. Only in this way was it possible to establish a really *synthetic* theory. Mayr (1963) distinguishes between two classes of evolutionary theories: *monistic* and *pluralistic theories*. The former are based on the conviction that there is *one* particular mechanism responsible for evolutionary change (e.g., mutations, according to the mutation theory); the latter operate with several mechanisms and their interplay. Today there is agreement that evolution can be properly explained only if we consider the combination of some mechanisms or factors (mainly natural selection, genetic recombination, and mutations).

Finally, I should like to give some remarks on the philosophically relevant idea of progress in evolution or progressive evolution. As can be observed, many evolutionary theorists have recently abandoned the idea of *progress* (e.g., see Gould 1996; Wuketits 1997, 1998; Nitecki 1988), but it seems that it is not so easy to dismiss this idea completely from evolutionary thinking. Maybe, "it is", as Simpson (1951, p. 107) states, "impossible to think in terms of history without thinking of progress". This, then, would be a psychological problem and the idea of progress would reflect what Gould (1989) characterized as "the iconography of an expectation". Many people still tend to believe that evolution has a direction, i.e., it is "going somewhere", and that the appearance of humans was the inevitable result of all

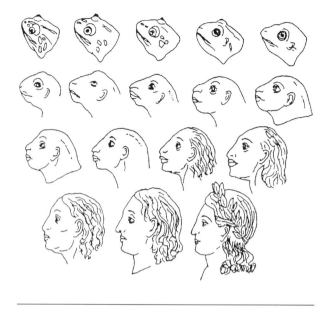

Figure 4.5 Ideas of progress – yesterday and today (reproduced from Gould 1989, Wuketits 1998).

previous evolutionary processes. Even some biologists (e.g., Bresch 1977) claim that evolution is goal-directed and in all its paths inevitable, that selection forces a particular direction, with *Homo sapiens* as the – preliminary – climax (Figure 4.5). Science is a process of changing ideas, but some ideas seem to persist, like "living fossils"; among them is the idea of evolutionary progress.

Earlier in this chapter (see Section 4.4) it was stated that teleological thinking and the belief in human superiority were obstacles to evolutionism. However, it seems that, after evolutionism was established, both ideas found new grounds. For all those who have been willing to accept evolution, but at the same time unwilling to renounce the conception of cosmic teleology – with humans still in the center of nature – evolution has become a new religion (see Midgley 1985). Now, more than 120 years after Darwin, it is time to disclaim this conception and to purge evolutionary thinking from metaphysical belief. A distinction must be made between *evolutionary theory* and *metaphysical interpretations of evolution*. The idea of progress is a metaphysical notion reflecting the hope of humans, but a mischievous understanding of biological evolution. What counts, is survival, and hence evolution is not going towards "perfection" – it is going nowhere (Ruse 1986).

What can be said in conclusion is that the world into which Darwin led us is far from being totally explored. Still, evolutionists have to fight against preconceptions and are at the same time constantly challenged by new findings ("facts") in different fields of research. It would be arrogant to maintain that all problems of evolution are already solved. However, it would be arrogant as well to assume that any of the "open problems" can find a satisfactory solution within the frame of some metaphysical belief systems disregarding Darwin's theory of natural selection and its more recent expansions.

4.7
References

AYALA, F. J. (1985) The theory of evolution. Recent successes and challenges. In: McMULLIN, E. (Ed.), *Evolution and Creation*. Notre Dame: University of Notre Dame Press, 59–90.

AYALA, F. J. (1999) Adaptation and novelty: teleological explanations in evolutionary biology. *Hist. Philos. Life Sci.* 21, 3–33.

AYALA, F. J. (2004) In Willam Paley's shadow: Darwin's explanation of design. *Ludus Vitalis* 12 (1), 53–66.

AYALA, F. J., VALENTINE, J. (1979) *Evolving: The Theory and Processes of Organic Evolution*. Menlo Park: Benjamin/Cummings.

BEURTON, P. (1994) Historische und systematische Probleme der Entwicklung des Darwinismus. *Jahrb. Geschichte Theor. Biol.* 1, 93–211.

BOWLER, P. J. (1984) *Evolution. The History of an Idea*. Berkeley: University of California Press.

BRESCH, C. (1977) *Zwischenstufe Leben: Evolution ohne Ziel?* Munich: Piper.

BUNGE, M. (1981) *Scientific Materialism*. Dordrecht: D. Reidel.

CHAPMAN, H. C. (**1873**) *Evolution of Life*. Philadelphia: Lippincott.

CROCKER, L. G. (**1959**) Diderot and Eighteenth century French transformism. In: GLASS, B., TEMKIN, O., STRAUS, W. L. (Eds.), *Forerunners of Darwin 1745–1859*. Baltimore: Johns Hopkins Press, 114–143.

DARWIN, CH. (**1859** [**1958**]) *On the Origin of Species*. New York: New American Library.

DARWIN, CH. (**1958**) *The Autobiography of Charles Darwin 1809–1882* (Ed.: BARLOW, N.). New York: Norton Library.

DESMOND, A. (**1979**) Designing the dinosaur: Richard Owen's Response to Robert Edmont Grant. *ISIS* **70**, 224–234.

DESMOND, A. (**1982**) *Archetypes and Ancestors: Palaeontology in Victorian London 1850–1875*. London: Blond & Briggs.

DI GREGORIO, M. A. (**1984**) *T. H. Huxley's Place in Natural Science*. New Haven: Yale University Press.

DOBZHANSKY, T., AYALA, F. J., STEBBINS, G. L., VALENTINE, J. (**1977**) *Evolution*. San Francisco: Freeman.

DÜRKEN, B. (**1924**) *Allgemeine Abstammungslehre. Zugleich eine gemeinverständliche Kritik des Darwinismus und des Lamarckismus*. Berlin: Borntraeger.

FINKELSTEIN, G. (**2000**) Why Darwin was English. *Endeavour* **24** (2), 76–78.

FISKE, J. (**1875**) *Outlines of Cosmic Philosophy, Based on the Doctrine of Evolution*. Vol. II. Boston: Osgood.

GAY, H. (**1999**) Explaining the universe: Herbert Spencer's attempt to synthesize political and evolutionary ideas. *Endeavour* **23** (2), 56–59.

GHISELIN, M. T. (**1969**) *The Triumph of the Darwinian Method*. Berkeley: University of California Press.

GLASS, B. (**1959**) Heredity and variation in the eighteenth century concept of the species. In: GLASS, B., TEMKIN, O., STRAUS, W. L. (Eds.), *Forerunners of Darwin 1745–1859*). Baltimore: Johns Hopkins Press, 144–172.

GOULD, S. J. (**1982**) Darwinism and the expansion of evolutionary theory. *Science* **216**, 380–387.

GOULD, S. J. (**1988**) *Time's Arrow, Time's Cycle: Myth and Metaphor in the Discovery of Geological Time*. London: Penguin Books.

GOULD, S. J. (**1989**) *Wonderful Life: The Burgess Shale and the Nature of History*. New York: Norton.

GOULD, S. J. (**1996**) *Full House: The Spread of Excellence from Plato to Darwin*. New York: Three Rivers Press.

GOULD, S. J. (**1997**) *Dinosaur in a Haystack: Reflections in Natural History*. London: Penguin.

GOULD, S. J., LEWONTIN, R. C. (**1984**) The spandrels of San Marco and the panglossian paradigm: a critique of the adaptationist programme. In: SOBER, E. (Ed.), *Conceptual Issues in Evolutionary Biology*. Cambridge: MIT Press, 252–270.

GREGORY, F. (**1991**) Darwin and the German theologians. In: WOODWARD, W., COHEN, R. S. (Eds.), *World Views and Scientific Discipline Formation*. Dordrecht: Kluwer, 269–278.

GREENE, J. C. (**1963**) *Darwin and the Modern World View*. New York: New American Library.

GRIBBIN, J., WHITE, M. (**1995**) *Darwin: A Life in Science*. London: Simon & Schuster.

GUTMANN, W. F. (**1979**) Entwickelt sich ein neues Evolutionsverständnis? Das Analogie-Denken Darwins und die physikalistische Evolutionstheorie. *Biol. Rdsch.* **17**, 84–99.

HABER, F. C. (**1959**) Fossils and the idea of a process of time in natural history. In: GLASS, B., TEMKIN, O., STRAUS, W. L. (Eds.), *Forerunners of Darwin 1745–1859*. Baltimore: Johns Hopkins Press, 222–261.

HIMMELFARB, G. (**1968**) *Darwin and the Darwinian Revolution*. New York: W. W. Norton.

HÖLDER, H. (**1960**) *Geologie und Paläontologie in Texten und ihrer Geschichte*. Freiburg: Alber.

HULL, D. L. (**1988**) *Science as a Process: An Evolutionary Account of the Social and Conceptual Development of Science*. Chicago: University of Chicago Press.

HUXLEY, J. (**1942**) *Evolution: The Modern Synthesis*. London: Allen & Unwin.

HUXLEY, T. H. (**1863 [2001]**) *Man's Place in Nature*. New York: Modern Library.

JUNKER, T., HOSSFELD, U. (**2001**) *Die Entdeckung der Evolution. Eine revolutionäre Theorie und ihre Geschichte*. Darmstadt: Wissenschaftliche Buchgesellschaft.

KUTSCHERA, U., NIKLAS, K. J. (**2004**) The modern theory of biological evolution: an expanded synthesis. *Naturwissenschaften* **91**, 255–276.

LENNOX, J. G. (**1993**) Darwin *was* a teleologist. *Biol. Philos.* **8**, 409–421.

LOVEJOY, A. O. (**1936**) *The Great Chain of Being. A Study of the History of an Idea*. Cambridge: Harvard University Press.

LOVEJOY, A. O. (**1959**) Buffon and the Problem of Species. In: GLASS, B., TEMKIN, O., STRAUS, W. L. (Eds.), *Forerunners of Darwin 1745–1859*. Baltimore: Johns Hopkins Press, 84–113.

MANIER, E. D. (**1978**) *The Young Darwin and His Cultural Circle*. Dordrecht: Reidel.

MAYR, E. (**1963**) *Animal Species and Evolution*. Cambridge: Harvard University Press.

MAYR, E. (**1982**) *The Growth of Biological Thought. Diversity, Evolution, and Inheritance*. Cambridge: Harvard University Press.

MAYR, E. (**1988**) *Toward a New Philosophy of Biology*. Cambridge: Harvard University Press.

MAYR, E. (**1991**) *One Long Argument*. Cambridge: Harvard University Press.

MAYR, E. (**1997**) *This is Biology: The Science of the Living World*. Cambridge: Harvard University Press.

MAYR, E. (**2000**) Darwin's Influence on Modern Thought. *Sci. Am.* **283** (1), 78–83.

MAYR, E. (**2002**) Die Autonomie der Biologie. *Naturw. Rdsch.* **55**, 23–29.

MAYR, E. (**2004**) The Autonomy of Biology. *Ludus Vitalis* **12**, 15–27.

MIDGLEY, M. (**1985**) *Evolution as a Religion: Strange Hopes and Stranger Fears*. London: Methuen.

MÜLLER, G. B. (**2003**) Homology: The Evolution of Morphological Organization. In: MÜLLER, G. B., NEWMAN, S. A. (Eds.), *Origination of Organismal Form: Beyond the Gene in Developmental and Evolutionary Biology*. Cambridge: MIT Press, 51–69.

NITECKI, M. H. (Ed.) (**1988**) *Evolutionary Progress*. Chicago: University of Chicago Press.

OESER, E. (**1996**) *System, Klassifikation, Evolution. Historische Analyse und Rekonstruktion der wissenschaftstheoretischen Grundlagen der Biologie.* Vienna: Braumüller.

OLDROYD, D. R. (**1986**) Charles Darwin's theory of evolution: a review of our present understanding. *Biol. Philos.* **1**, 133–168.

POPPER, K. R. (**1984**) *Auf der Suche nach einer besseren Welt: Vorträge und Aufsätze aus Dreißig Jahren.* Munich: Piper.

PRICE, P. W. (**1996**) *Biological Evolution.* Philadelphia: Saunders College Publishing.

REIF, W.-E., JUNKER, T., HOSSFELD, U. (**2000**) The synthetic theory of evolution: general problems and the German contribution to the synthesis. *Theor. Biosci.* **119**, 41–91.

RICHARDS, R. A. (**1998**) Darwin, domestic breeding and artificial selection. *Endeavour* **22** (3), 106–109.

RUSE, M. (**1982**) *Darwinism Defended: A Guide to the Evolution Controversies.* London: Addison-Wesley.

RUSE, M. (**1986**) *Taking Darwin Seriously: A Naturalistic Approach to Philosophy.* Oxford: Basil Blackwell.

SIMPSON, G. G. (**1951**) *The Meaning of Evolution.* New York: New American Library.

SIMPSON, G. G. (**1963**) *This View of Life: The World of an Evolutionist.* New York: Harcourt, Brace & World.

SIMPSON, G. G. (**1980**) *Why and How: Some Problems and Methods in Historical Biology.* Oxford: Pergamon Press.

SMOCOVITIS, V. B. (**1992**) Unifying biology: the evolutionary synthesis and evolutionary biology. *J. Hist. Biol.* **25**, 1–65.

STEBBINS, G. L., AYALA, F. J. (**1985**) The evolution of Darwinism. *Sci. Am.* **53** (1), 54–64.

STOTT, R. (**2003**) *Darwin and the Barnacle: The Story of one Tiny Creature and History's Most Spectacular Scientific Breakthrough.* London: Faber & Faber.

TAYLOR, F. S. (**1963**) *A Short History of Science and Scientific Thought.* New York: Norton.

VOLLMER, G. (**1995**) *Biophilosophie.* Stuttgart: Reclam.

WAGNER, A. (**1908**) *Geschichte des Lamarckismus. Als Einführung in die psychobiologische Bewegung der Gegenwart.* Stuttgart: Franck.

WEISMANN, A. (**1904**) *Vorträge über Deszendenztheorie.* Jena: Fischer.

WEISS, K. M. (**2004**) Thomas Henry Huxley (1825–1895) puts us in our place. *J. Exp. Zool.* **302B**, 196–206.

WIESER, W. (Ed.) (**1994**) *Die Evolution der Evolutionstheorie. Von Darwin zu DNA.* Heidelberg: Spektrum Akademischer Verlag.

WUKETITS, F. M. (**1988**) *Evolutionstheorien. Historische Voraussetzungen, Positionen, Kritik.* Darmstadt: Wissenschaftliche Buchgesellschaft.

WUKETITS, F. M. (**1989**) *Grundriß der Evolutionstheorie.* Darmstadt: Wissenschaftliche Buchgesellschaft.

WUKETITS, F. M. (**1997**) The philosophy of evolution and the myth of progress. *Ludus Vitalis* **5** (2), 5–17.

WUKETITS, F. M. (**1998**) *Naturkatastrophe Mensch: Evolution ohne Fortschritt.* Düsseldorf: Patmos.

WUKETITS, F. M. (**2000a**) *Evolution. Die Entwicklung des Lebens.* Munich: Beck.

WUKETITS, F. M. (**2000b**) The organism's place in evolution: Darwin's views and contemporary organismic theories. In: PETERS, D. S., WEINGARTEN, M. (Eds.), *Organisms, Genes and Evolution: Evolutionary Theory at the Crossroads.* Stuttgart: Steiner, 83–91.

WUKETITS, F. M. (**2003**) Evolution heute: Kontroversen, Paradigmen, Perspektiven. In: FASTERDING, M. (Ed.), *Aufbruch der Wissenschaft. Forschung am Anfang des dritten Jahrtausends.* Gelsenkirchen: Edition Archaea, 37–62.

WUKETITS, F. M. (**2004**) *This is Biology*: Ernst Mayr and the autonomy of biology as a science. *Ludus Vitalis* **12** (1), 149–160.

YOUNG, D. (**1992**) *The Discovery of Evolution.* London: Natural History Museum.

ZIMMERMANN, W. (**1953**) *Evolution. Die Geschichte ihrer Probleme und Erkenntnisse.* Freiburg: Alber.

ZIMMERMANN, W. (**1967**) *Vererbung „erworbener Eigenschaften" und Auslese.* Stuttgart: Fischer.

5
Evolutionary Developmental Biology

Gerd B. Müller

5.1
Introduction

The interrelation between proximate and ultimate causation is the central problem of any explanatory theory of organismal form. During certain periods of biological research these causalities were treated discretely, separating evolution from embryonic development, but during more synthetic times attention was paid to their systemic linkage. We have again entered such a phase. Whereas embryology and development had played an important role in the first inceptions of evolutionary theory, they were conspicuously absent from the neo-Darwinian orthodoxy prevailing through the second half of the 20th century. No formalized principles of development could be introduced into the central, population–genetic framework of neo-Darwinism and experimental embryology was much occupied with deciphering the very basic mechanisms of development. As a consequence, evolutionary biology and developmental biology were pursued as largely independent disciplines, even after both fields had reached molecular levels of analysis. The "black boxing" of development in evolutionary theory was often lamented, but until recently no significant alterations of the theory's major tenets were expected from adding "details" of development.

The past two decades have seen a remarkable revival of interest in the inter-relations between proximate and ultimate causation in biology. Development has returned with great vigor to the evolutionary arena. New research programs now focus explicitly on the interactions between developmental and evolutionary processes, a rapidly growing number of scientific publications and meetings address these topics, and funding agencies recognize their importance. This expansion of interest has led in rapid succession to the founding of new scientific journals, the creation of new divisions in professional societies, the specific dedication of academic positions, and other epistemological and institutional adjustments. A distinct field of study has emerged, which is often proposed to represent a new subdiscipline of the biosciences, called evolutionary developmental biology or, briefly, EvoDevo. Stimulated by the fresh agenda, development provides innovative

Handbook of Evolution, Vol. 2: The Evolution of Living Systems (Including Hominids)
Edited by Franz M. Wuketits and Francisco J. Ayala
Copyright © 2005 Wiley-VCH Verlag GmbH & Co. KGaA, Weinheim
ISBN: 3-527-30838-5

input for evolutionary biology and, at the same time, the evolutionary perspective has triggered new approaches to the study of development. EvoDevo is generally welcomed as a new research paradigm that might finally provide access to the long-sought rules relating genotype with phenotype.

In a short time span EvoDevo has attained the reputation of a coherent field of study or a unifying discipline in the making. With closer observation, however, it is apparent that rather diverse opinions exist about its conceptual origins, its aims, its empirical agendas, its theoretical significance, and its future potential. Very different expectations are associated with EvoDevo, depending on the points of view of the respective commentators. In addition, the inclusion of development imports a number of new (and old) problems into current evolutionary theory. Developmental data consist mostly of observations in single organisms, which cannot easily be integrated into a population-theoretical framework. However, development is also complex, nonlinear, stochastic, and self-organizing – properties that are not readily formalized. Most significantly, development cannot be reduced to the mechanisms of gene activity, but includes a multitude of nonprogrammed factors that interact with gene activation – the old problem of epigenesis, which many evolutionary theorists were happy to abandon, has returned.

An intensive debate has arisen over whether, and to what degree, evolutionary orthodoxy is challenged by the EvoDevo approach. It has been argued that EvoDevo could extend or even supersede neo-Darwinism, whereas others categorically deny any such possibility or necessity, or even reinforce Dawkins's (1976, p. 62) belief that "the details of the embryonic developmental process, interesting as they may be, are irrelevant to evolutionary considerations" (e.g., Wallace 1986). Some see in EvoDevo a straightforward extension of the evolutionary genetics agenda; others see in it an explicitly organismal and systemic approach that will enable us to solve long-standing macroevolutionary issues. Some regard EvoDevo as perfectly compatible with a strictly selectionist theory of evolution; others claim it represents a strong departure from it. However, beyond these debates, and beyond uniting core subjects of evolutionary and developmental biology, EvoDevo is becoming a conceptual and organizational nucleus for a comprehensive integration of research areas in organismal biology, including domains of genetics, ecology, paleontology, anthropology, behavior, cognition, and other fields. This integrative capacity, the range of positions, the multitude of theoretical implications, and the historical as well as philosophical aspects of EvoDevo have already generated a number of metatheoretical reflections (e.g., Robert 2004; Amundson 2005; Laubichler and Maienschein 2005) which indicate the importance attributed to the new discipline.

This chapter provides an overview of the historical roots and the present status of EvoDevo, including its key concepts, its empirical research programs, its theoretical consequences, and its future prospects. In going through these issues I will identify several conceptual and empirical domains within EvoDevo, and I will examine in which ways these differ from other programs. I will also explore whether EvoDevo can solve problems that are not solved by earlier approaches, whether it is heuristically fruitful and generates new research initiatives, and whether it can lead to an improved conceptual framework for the understanding of organismal evolution.

5.2
Historical Roots

Developmental and evolutionary biology are historically interwoven in many ways. After "parallels" between embryonic stages and the "scale of beings" had been much discussed in pre-Darwinian times, the foundation of a scientific theory of evolution was also linked to embryological arguments. Not only did Darwin assign an entire chapter of the *Origin of Species* to embryology and asserted (in a letter to Asa Gray) that it provided "by far the strongest single class of facts in favor of a change of form", but his first sketches of a phylogenetic tree possibly were also inspired by tree-like renderings of embryological differences between species – based on work of the contemporary embryologist von Baer (Richards 1992). Even the very term "evolution", so notably absent in Darwin's early publications, had a (preformationist) developmental meaning before it came to be widely applied to phylogenetic "transmutation" (Bowler 1975).

After evolution had become a distinct subject of biological investigation and before it was wholesale "geneticized", many scholars were interested in the relation of phylogenetic patterns with ontogenetic development. In the beginning much of this interest focused on the taxonomical uses of embryonic characters. Francis Balfour, William Brooks, Karl Gegenbaur, Alexandre Kowalevsky, Fritz Müller, and many others successfully applied the comparative method to embryology, and obtained new data that helped discern hitherto unknown relationships among taxa. Kowalevsky's finding that ascidians are related to the vertebrates because their larvae possess a notochord, gill slits, neural folds, and other shared traits is one of the great successes of this method (Kowalevsky 1866, 1871).

These comparative endeavors were mostly descriptive and phenomenological, but a number of more mechanistically oriented and theoretically grounded programs also emerged. One sprang from the joining of the concept of recapitulation with a mechanism for effecting developmental change. Recapitulation, a widespread notion in late 18th century "Naturphilosophie", rooted in the Greek analogistic tradition, was elaborated primarily by Ernst Haeckel into a mechanistic concept of morphological evolution (Haeckel 1866) by uniting it with developmental timing as a key mechanism for embryonic change. Under Haeckel's patronage this approach assumed programmatic and even ideological status. Recapitulation remained the only thinkable way by which ontogeny and phylogeny would be tied together well into the 20th century. The rise of experimental embryology, on the one hand, and that of genetics, on the other, eventually stifled the – by then often exaggerated – recapitulationist claims. The new paradigm of genetic variation and differential inheritance eclipsed recapitulation as a general explanatory principle for the progression of organic life.

The fall of recapitulationism was widely viewed as a "disproof" of all recapitulatory events in developing embryos. Although the temporary presence of ancestral character states is an embryological fact that should be distinguished from a *theory* of recapitulation (Mayr 1997), the disdain for recapitulation is widespread among experimental embryologists and evolutionary biologists alike. Any mention of such links between development and evolution has raised, and still raises, strong aversions (Wilkins 2002). Thus, it was generally overlooked that Haeckel's concept – although incorrect in many details – contained a mechanism for evolutionary change, i.e., the modification of development through heterochrony, a point that was only resurrected in the late 1970s (Gould 1977).

A second major movement related to the ontogeny–phylogeny question, taking place in the first half of the 20th century, was the study of environmental influences on embryogenesis and the maintenance of so-induced effects in subsequent generations. Most of these widespread endeavors were carried out in a neo-Lamarckian belief and the experiments were designed to test for the possibility of an inheritance of acquired characters. An extensive amount of data was generated by ingenious modifications of external parameters in the development of insects (Jollos 1934), amphibians (Kammerer 1923), and other taxa (Kammerer 1925; Hämmerling 1929). Entire institutions, such as the Vivarium Institute in Vienna (1902–1945), devoted their efforts to the study of the environment–development–evolution interaction. The conclusiveness of the results was debated heatedly (e.g., McBride versus opponents in *Nature* during the 1920s). Eventually the neo-Lamarckian interpretations lost credibility. However, these early experimental attempts to combine environmental modification with breeding regimes represent a body of evidence that merits attention independently from their Lamarckian interpretations. Recently the importance of "enduring modifications" and "epigenetic inheritance" have been reconsidered (Rubin 1990; Jablonka and Lamb 1995), and EvoDevo has actively readdressed the issue of environmental influences on development and evolution (Gilbert 2001; Gilbert and Bolker 2003; Hall et al. 2003).

A third initiative arose with early attempts to include the genetics of development into evolutionary theory, based on theoretical considerations, breeding studies, and quantitative genetics. A number of concepts were proposed that related to the development–evolution problem, such as reaction norms (Woltereck 1909), rate genes (Goldschmidt 1940), assimilation (Waddington 1956), and the whole field of epigenetics (in the Waddingtonian sense). These developments took place before the rise of DNA genetics and in the absence of molecular tools for genetic analysis. However, the calls for a more prominent role for these mechanisms in evolutionary theory, such as expressed by Goldschmidt and Schmalhausen, and later by Waddington, went largely unheard. Attention concentrated on transmission genetics and quantitative genetics, whereas developmental genetics, and developmental biology for that matter, was left aside.

These off-the-mainstream initiatives all addressed facets of the ontogeny–phylogeny or development–evolution interface and thus kept the connections between the fields alive even during prolonged periods of their largely separate study in the 20th century. Apart from certain conceptual traces, not much of these traditions has survived in modern EvoDevo and none of them can be considered its immediate forerunner or the key initiator of its sudden rise roughly 20 years ago. Two developments were more directly responsible. One stimulus was the increasing awareness of explanatory deficits in the prevailing paradigm of evolution. Neo-Darwinism worked well for the population genetic phenomena it concentrated on, but in the late 1970s and early 1980s concern accumulated about its difficulty to account for the characteristics of phenotypic evolution. Such considerations included biased variation (Alberch 1982; Maynard Smith et al. 1985), rapid changes of form (Eldredge and Gould 1972), the occurrence of nonadaptive traits (Gould and Lewontin 1979), and the origination of higher-level organization such as homology and body plans (Riedl 1978). Most of the criticisms attributed the explanatory deficits of neo-Darwinism to the neglect of the generative processes that relate between genotype and phenotype, and to the exclusion of developmental theory from the evolutionary synthesis (Hamburger 1980; Reid 1985b).

As a response to this situation a number of scientific meetings (such as those held in Dahlem in 1981, Sussex in 1982, Plzen in 1984, Columbia in 1985, and Woods Hole in 1985) and books (Gould 1977; Bonner 1982; Goodwin et al. 1983; Raff and Kaufman 1983) began to concentrate on the intersections between development and evolution. Empirical research initiatives took up the theme (e.g., Katz et al. 1981; Alberch and Gale 1983, 1985; Raff et al. 1984; Müller 1986, 1989), using classical techniques of comparative and experimental embryology at first, and later, increasingly, the methodologies of molecular biology. This new agenda, which aimed at delineating the role of developmental processes in organismal evolution, was initially called "ontophyletics" (Katz et al. 1981; Katz 1983) or "evolutionary embryology" (Müller 1991) until "evolutionary developmental biology" (Hall 1992; Wake 1996) became the generally accepted label.

A second initiating factor of modern EvoDevo was the rise of molecular developmental genetics, which brought about the cloning of regulatory genes and techniques for the visualization of their activation sites in the developing embryo. This created a new way of comparing the development of different taxa and led to the discovery of unexpected similarities in gene regulation among distantly related species (McGinnis et al. 1984). The similarities were found to extend to the spatial and temporal sequences of early gene expression in anatomically different embryos such as insects and mammals (Duboule and Dollé 1989; Graham et al. 1989). In contrast to earlier beliefs, which viewed the diverse ways in which different animals develop as a result of an equally diverse genetic apparatus, it became increasingly clear that relatively few genetic regulators are implicated in the embryonic foundations of all animal body plans (Akam 1989; Holland 1992; Holland et al. 1996). The search for commonalities and differences in gene expression patterns and gene regulation gained rapid momentum, and has led to an understanding of many molecular details of animal development (Carroll et al. 2001; Davidson 2001). Today, the evolution of developmental gene regulation and of gene regulatory networks has become a major theme in empirical EvoDevo research.

The general excitement over the successes of comparative developmental genetics has led to pronouncements that EvoDevo is synonymous with the comparative study of the developmental genetic machinery (Arthur 2002; Tautz 2002; Wilkins 2002). However, EvoDevo is more encompassing and explicitly reached beyond the genome from the very outset (Katz 1983). The deep questions that triggered the formation of this new field are predominantly situated in the realm of the phenotype (Love 2003, 2005). Organismal form, shape, morphological structure, and the generative mechanisms underlying their evolution represent the essential questions within EvoDevo. This objective predates the discoveries in developmental genetics, even though these have immensely boosted the research activity in the new field, providing new molecular insights and analytical tools. However, as pointed out above, the modern discipline of EvoDevo is also not a direct continuation of any one of the earlier approaches. Rather, its explosive rise in the mid-1980s was possible because of the continuous presence of the historical questions onto which a new, mechanistic, multilevel analysis of genetic and developmental processes was superimposed.

5.3
Agenda and Concepts

In modern EvoDevo a general agenda can be distinguished from individual concepts that address particular issues of the development–evolution relationship. The general agenda rests on the theoretical frameworks of developmental biology and evolutionary biology. It postulates that a causal–mechanistic interaction exists between the processes of individual development and the processes of evolutionary change (Wagner et al. 2000). The nature of these interactions and their consequences for organismal evolution represent the central research goal. Hence, EvoDevo consists of two interconnected parts – one addressing evolution's influence on development and the other development's influence on evolution. This reciprocal interrelationship constitutes a genuinely dialectical and systemic research agenda, with particular questions arising from both aspects. Although distinctions between the two principal subagendas (evo-devo and devo-evo) have been discussed (Hall 2000; Gilbert 2003), only the systemic combination of both makes EvoDevo a coherent discipline (Müller 2005).

A number of specific concepts can be distinguished within the general agenda. Some of these were formulated well before the EvoDevo era and are only of historical interest, or are now revived and elaborated based on fresh evidence. Others were prompted by the new research focus and the necessity to define principles of the mechanistic realization of the development–evolution linkage. These concepts can be grouped according to the specific questions they address (Arthur 2002; Müller 2005). Three major categories of concepts are distinguished. The first concerns the evolution of development, relating to issues such as the origin of developmental systems, the evolution of the developmental repertoire, and the evolutionary modification of developmental processes. A second set of concepts concerns the effects that the properties of developmental systems have on phenotypic evolution, including the patterns of phenotypic variation, the origin of phenotypic novelty, and the fixation of morphological motifs, i.e., the organization of the phenotype. A third group concerns the interrelations between environment and development, and their evolutionary consequences, such as polyphenism and plasticity. The following subsections briefly characterized some of the major concepts of each category.

5.3.1
The Origin of Developmental Systems

Relatively little is known about how the first systems of development arose, but clearly this problem is related to the origins of multicellularity and the evolution of life cycles. John Bonner, one of the most influential individuals in triggering the EvoDevo revolution, has concentrated on this issue, and has proposed a number of concepts on the relations between organism size, internal complexity, reproductive success, and life cycle selection (Bonner 1965, 1988). Many of these concepts are based on the study of extant colonial or aggregating unicellular organisms such as cellular slime molds. In a similar vein, Buss (1987) suggested that development resulted from competition among cells in early multicellular aggregates to become the ones capable of propagating the next generation. Development would thus have been a consequence of the transition between the cell as the unit of selection and the multicellular individual as the unit of selection.

A strikingly different conceptualization concentrates on the physical origins of developmental systems (Forgacs and Newman 2005). Given that single-cell organisms that existed before the emergence of multicellularity possessed certain physical properties, such as liquid-like viscoelasticity, adhesiveness, and chemical excitability, proto-metazoan aggregates must have had an inherent capacity to self-organize spatial patterns. Development would have arisen at the point when certain cells achieved organizational control over other cells, e.g., by releasing a diffusible chemical substance, and this capacity would have resulted in cell aggregates consisting of nonuniformly distributed cell states. In conjunction with differential adhesion (Steinberg 1963) and other generic physical mechanisms (Newman 1992, 1994), such systems could produce an array of "generic forms", whose shape and size would have been much influenced by the physicochemical conditions of the environment. Because of the strong environmental influence, it is assumed that in early forms of development the close correlation between genotype and phenotype observed in modern organisms would not have existed yet. Rather, the genotype–phenotype relation might have been one-to-many during what has been called a "pre-Mendelian phase" of evolution (Newman and Müller 2000). Only subsequent selectional fixation and genetic routinization would have resulted in robust forms of development and the faithful, Mendelian kind of inheritance seen in advanced taxa.

5.3.2
Evolution of the Developmental Repertoire

As developmental systems evolve, the mechanisms and components of such systems change and diversify. Concepts relating to these topics concern primarily the genetic aspects of developmental evolution, such as gene duplication, especially of regulatory genes (McGinnis and Krumlauf 1992; Holland 1999), and the evolution of gene regulatory networks (Davidson et al. 1995; Wray and Lowe 2000). The genetic redundancy generated by such mechanisms can be exploited through the acquisition of new functions for these genes – a process referred to as recruitment (Keys et al. 1999) or cooption (True and Carroll 2002). Present summaries on the evolution of developmental pathways rely almost exclusively on genetics (Carroll et al. 2001; Wilkins 2002), but the epigenetic mechanisms controlling gene activation also evolve, including the repertoire of mechanisms of cell and tissue interaction and embryonic induction, which has led to a number of conceptualizations of the evolutionary roles of these mechanisms (Løvtrup 1974; Hall 1983; Edelman 1988; Newman and Müller 2000).

Modularity is a concept that targets specifically the evolution of developmental repertoires (Schlosser and Wagner 2004; Callebaut and Rasskin-Gutman 2005). It recognizes that developmental systems are decomposable into components that operate according to their own, intrinsically determined principles. Such modules can be characterized as integrated structural or process units that depend on input from other components and, in turn, influence other components by their outputs, represented, for example, by gene signaling pathways or inductive interaction networks. The evolutionary function of developmental modules would be their phenotypic selectability. A selectable developmental module can consist of a set of genes, their products, and their developmental interactions, including the resulting character complex and the functional effect of that complex. The genes affecting the modular character complex would be characterized by a high degree of internal integration and a low degree of external connectivity, i.e., pleiotropic connections would be largely within-module. Modularity constitutes a principle connecting the genetic and epigenetic components of evolving developmental repertoires and has the potential to assume a central role in the EvoDevo framework (von Dassow and Munro 1999).

5.3.3
The Evolutionary Modification of Developmental Processes

This is the topic of a host of EvoDevo concepts. Into this category fall all mechanisms that have the capacity to modulate quantitatively or qualitatively the established processes that compose a developmental system, e.g., mutations affecting the regulation of developmental pathways. The most prominent concept here is heterochrony, i.e., evolutionary changes in the relative timing and rates of developmental processes. This classical idea has been revived by the books of Gould (Gould 1977) and Raff and Kaufman (Raff and Kaufman 1983), and has since been elaborated into a comprehensive explanatory framework (McKinney and McNamara 1991; Parichy et al. 1992; McNamara 1997). Different forms and mechanisms of heterochrony were shown to be associated with different life history strategies and produce different phenotypic results (Hall 1984; Raff and Wray 1989). Heterochrony has been documented in most groups of organisms, and its analysis is taken to molecular and genetic levels (Parks et al. 1988; Wray and McClay 1989; Kim et al. 2000). Mutations that directly affect developmental timing have been demonstrated in animals (Ruvkun and Giusto 1989) and plants (Dudley and Poethig 1991). A number of genetic mechanisms affecting developmental timing have been experimentally tested (Dollé et al. 1993; Zakany et al. 1997). There is no doubt that heterochrony based on genetic change represents a powerful mode for altering morphological characters and body plans (Duboule 1994). However, a major difficulty remains the distinction between those heterochronic phenomena that are simply a consequence of any change to development and those cases in which heterochrony of a particular process represents the causal mechanism for the evolutionary modification of a trait.

5.3.4
The Patterns of Phenotypic Variation

The question as to what extent the properties of developmental systems influence the variational and directional dynamics of phenotypic evolution is primarily addressed by the concept of developmental constraint. This was one of the key themes that spurred modern EvoDevo (Alberch 1982) and a consensus definition of developmental constraint was soon reached: "a bias on the production of variant phenotypes, or a limitation on phenotypic variability, caused by the structure, character, composition, or dynamics of the developmental system" (Maynard Smith et al. 1985). A rich variety of case studies provided empirical evidence for such constraints being effective in evolution, with data coming from comparative morphology (e.g., Wake 1982; Bell 1987; Vogl and Rienesel 1991; Caldwell 1994), comparative and experimental embryology (e.g., Alberch and Gale 1983, 1985; Müller 1989; Webb 1989; Streicher and Müller 1992), plant biology (e.g., Donoghue and Ree 2000), and quantitative genetics (e.g., Cheverud 1984; Rasmussen 1987; Wagner 1988). Whereas early conceptualizations concentrated on the limitational aspects of developmental constraints, later treatments also emphasized the heightened potentialities for change in a particular aspect of the phenotypic character space of a lineage that can arise from such limitations (Arthur 2001). Hence, counterintuitively, developmental constraints can also contribute to evolvability, i.e., the facilitation of variation. A taxon's capacity to generate heritable phenotypic variation or innovation can also depend on mechanisms that reduce (Kirschner and Gerhart 1998) or overcome (Müller and Wagner 1991) constraints, an issue that is controversially debated (Eberhard 2001; Wagner and Müller 2002).

5.3.5
The Origin of Phenotypic Novelty

The issue of evolutionary innovation and phenotypic novelty has recently come much to the fore, because it is perceived as one of the areas of evolutionary biology to which EvoDevo could make a major contribution (Müller and Wagner 1991; Wagner 2000). This contention is based on the observation that neo-Darwinian theory concentrates on variation and adaptation, and is, itself, adapted to the task of explaining this spectrum of evolutionary phenomena. In contrast, novelties are phenotypic changes in which no or very limited variation has preceded their origination and which are not seen to arise directly from a process of adaptation.

Thus novelties represent a distinct class of phenotypic change and the mechanisms for generating innovation should be distinguished from those generating variation (Müller and Newman 2003b; Müller and Wagner 2003). While earlier conceptions concerning innovation have dealt with function shift (Mayr 1960), macro-mutation (Goldschmidt 1940), and symbiosis (Margulis and Fester 1991), EvoDevo approaches concentrate on the roles of development.

One specific proposal is epigenetic causation (Müller 1990; Newman and Müller 2000). Here, the central issue is a distinction between general selectional trends and the specificity of phenotypic response conferred by the developmental system. Selection acting on general organismal features, such as shape, proportion, function, or behavior, can elicit epigenetic byproducts that arise from the generic properties of developing cell and tissue systems, e.g., following changes in blastema size or mechanical load. New structural elements, such as skeletal parts, can arise through this mode, without having been selected for, as a side-effect of the evolutionary modification of developmental parameters (Müller 1990). Epigenetic mechanisms could have had a significant role in the origination of body parts and organismal form (Müller and Newman 2003a). This approach can also be formulated as part of the developmental plasticity concept discussed below.

5.3.6
The Organization of the Phenotype

Although they seem to be themes of the past, the evolution of homology and of body plans represents a key issue of EvoDevo (Raff 1996). Such higher-level organizational phenomena had been difficult to explain from a strictly population genetic perspective. Although the connections between development and homology are rooted deep in the history of biological ideas (Laubichler 2000), only sporadic attempts were made towards a developmental explanation of homology. EvoDevo has revived the issue (Hall 1995). Many of the new ideas were triggered by the surprising conservation of the gene regulatory apparatus underlying the development of very diverse organisms. This led to gene-based definitions of homology (Holland et al. 1996; Abouheif 1997), although others have pointed out the shortcomings of such reasoning (Bolker and Raff 1996; Minelli 1997). While the most notoriously conserved developmental control genes, the homeobox genes, exhibit nonhomologous expression domains in comparative maps of vertebrate and invertebrate embryos, the reverse also applies: homologous structures can be specified by nonhomologous genes (Wray 1999). Several developmental concepts of homology concentrate on commonalities of developmental pathways (Wagner 1989; Wagner 1996) or on the modularity of developmental systems (Minelli 1997; Gilbert and Bolker 2001). Recently, it was proposed that the evolution of homology consists of three stages, in two of which (the generative stage and the integrative stage) development plays a decisive role, whereas the third, the autonomized stage, is characterized by achieving independence from the underlying generative mechanisms. According to this position, homology is a consequence of structural organization that maintains the identity of building elements despite variation in their molecular, developmental, and genetic makeup (Müller and Newman 1999; Müller 2003).

5.3.7
The Environment–Development–Evolution Interaction

This is yet another conceptual domain of EvoDevo. Once thought of as crucial for understanding evolution, the field had been marginalized for several decades because of its seemingly Lamarckian connotations. However, new data on the genetic and environmental components of developmental phenomena such as phenocopy, polyphenism, and plasticity have revitalized the interest in their evolutionary roles, and led to proposals of an enlarged scope of EvoDevo research (Gilbert 2001; Hall et al. 2003). The central conceptual framework in this domain is plasticity (Pigliucci 2001; West-Eberhard 2003). It provides a unifying theoretical background for the quantitative genetic, developmental, and phenotypic responses to environmental influences in evolving populations. The concept of plasticity is tightly interconnected with that of reaction norm (Schlichting and Pigliucci 1998; Sarkar 2003), i.e., the range of variation and phenotypes that can result from a single genotype as a response to different environmental conditions. Developmental plasticity, the phenotypic realization of this responsiveness, is thought to represent in itself an adaptive trait of a taxon. However, developmental plasticity has a second aspect in referring to the evolutionary modification of developmental systems without any significant effect on the phenotypic outcome, a phenomenon frequently observed in species level comparison. This notion of developmental plasticity is distinct from the classical concept of plasticity, understood as environmentally mediated variation, in that it refers to the changes that occur during the evolution of ontogenies (Chipman 2002).

   Although this brief overview is far from exhaustive, it is evident that a rich variety of individual concepts make up the theoretical background of the EvoDevo discipline. None of them covers the entire field (more inclusive theories will be discussed in the section on integration), but all contribute to its present unification. It is evident that the conceptual focus of EvoDevo is on the problems of phenotypic evolution. This does not always coincide with the predominant topics in the empirical research carried out.

5.4
Research Programs and Methods

Empirical research in EvoDevo is largely driven by methodological aspects. Hence, the individual projects do not exactly match the conceptual categories, but rather represent separate research strategies. Several distinct "programs" can be distinguished, although the respective researchers engaging in these studies may not have defined their particular approach in such a way. The four principal programs distinguished in the following also reflect a methodological evolution of EvoDevo.

5.4.1
The Comparative Morphology Program

This program represents the classical approach, but it is by no means outdated. In fact, much of present developmental biology represents a comparative approach with molecular tools. However, many of the major contributions do not even come from the embryology of extant organisms, but, increasingly, from paleontology. The paleontological data can include direct embryological evidence, such as information from fossilized dinosaur eggs (Carpenter et al. 1996) or from early stages of invertebrate development (Bengtson and Zhao 1997). Foremost, paleontology provides extensive information about the developmental factors underlying phenotypic evolution, such as the occurrences of heterochrony, developmental constraint, and ontogenetic innovation in many different taxa (McNamara 1997; Vrba 2003). This is especially true for characters that represent "frozen" stages of development such as teeth or hard shells. This kind of evidence yields the phylogenetic morphospaces against which to test the developmental capacities of extant taxa. Increasingly, the relationships between variational data from fossils and the developmental processes responsible for their generation are explicitly addressed (Shubin et al. 1997). At the same time the morphological approach provides the data sets for what could be called the phylogenetic systematics program of EvoDevo. This work seeks to provide robust phylogenetic reference systems for the interpretation of the evolution of developmental mechanisms – an essential prerequisite for the EvoDevo synthesis (Mabee 2000).

5.4.2
The Epigenetic and Experimental Program

Following the traditions of experimental embryology a variety of strategies have been devised in order to specifically address evolutionary questions (Müller 1991). The aim is to probe the epigenetic properties of developmental systems with regard to their intrinsic capacities to generate evolutionarily relevant phenotypes. Perturbations of cell number, cell cycle, developmental timing, or inductive interactions were shown to produce both ancestral character states as well as phenocopies of derived states (e.g., Alberch and Gale 1983 1985; Müller 1989). Such experiments directly address the developmental and, hence, the evolutionary dissociabilities of temporal, spatial, and functional interactions in developmental systems. The results highlight the existence of constraints and threshold effects in developmental processes that contribute to the evolvability of a lineage, influencing phenotypic evolution and providing new heritable variation and novel character states. In particular, the experimental approach exposes mechanisms through which quantitative selectional processes may be transformed into qualitative phenotypic change.

The classical perturbation method is expanded by genetic and molecular tools such as gain-of-function and loss-of-function experiments that are increasingly used in developmental genetics. This is further complemented by experiments that attempt to redesign organismal structures with well-chosen mutations. This designer approach (Dworkin et al. 2001; Larsen 2003) tests the range of possible morphologies that can be achieved by a developmental system through small changes in cell behavior. While it does not automatically follow that new structures actually arose through similar mutations, such experiments indicate that even highly conserved phenotypes are not necessarily strongly constrained. Using mutations to compare the relative stability of characters, this approach further elucidates the nature of developmental constraints and will assist in revealing the genetic backgrounds that are required for stabilizing phenotypic innovations. Combining experimental studies of new forms generated by a developmental system with theoretical morphospace predictions (see below) could lead to further EvoDevo insights.

The epigenetic and experimental program intersects with the field of study focusing on developmental plasticity, especially as it relates to environmental influences in what has recently been termed ecological developmental biology or eco-devo (Gilbert 2001; Hall et al. 2003). The essence here is that the regulating factors of developmental processes (and their evolutionary modifiability) do not all reside within the embryo but depend substantially on the ecological context. Although these effects eventually feed into developmental–genetic pathways, the causality resides in an interplay between internal and external factors, such as diet, pH, humidity, temperature, photoperiod, seasonality, population density, predator presence, etc. Particular attention is paid to physiological and metabolic processes that mediate interactions between the environment and development, such as endocrine and hormone activity.

A number of experimental studies begin to unravel these mechanisms, focusing on hormonal effects in vertebrates and invertebrates (Davey 2003; Rose 2003), predator-induced polyphenisms in crustaceans and other taxa (Tollrian and Harvell 1999), nutrient regimes in sea squirts (Newlon et al. 2003), environmental regulation during insect development (Nijhout 1999), and many more (see Schlichting and Pigliucci 1998; Gilbert 2001; Hall et al. 2003)). Although these kinds of study have a long tradition, there is a new awareness that developmental plasticity and environmental induction have an important function in the origination of evolutionary novelty (West-Eberhard 2003), an opinion pioneered by Riuchi Matsuda (Matsuda 1987) whose work is receiving renewed attention (Hall et al. 2003). One of the reasons for the notion that this mechanism could be even more important than mutational change is the fact that environmental factors affect entire populations at once, whereas mutation initially affects only one individual. The importance of epigenetic parameters is also increasingly recognized in hominid evolution (Lovejoy et al. 1999).

5.4.3
The Evolutionary Developmental Genetics Program

This program, now often called EvoDevo as such, is "focused on the developmental genetic machinery that lies behind embryological phenotypes" (Arthur 2002). This EvoDevo is said to be "born on the day when Bill McGinnis developed the first

'Zooblot' that was incubated with a homeobox probe" (Tautz 2002). Henceforth, "which *Hox* gene turns on, where does it turn on, and when does it turn on" represented "the crux of molecular evo-devo questions" (Goodman and Coughlin 2000). The rapid cloning of an increasing number of homeobox genes and the concurrent development of *in situ* hybridization techniques that enable the precise mapping of their embryonic expression patterns has made this the most active area of empirical EvoDevo research. Two overlapping subprograms can be distinguished. One is the molecular body plans program, which aims for an understanding of the role of developmental control genes in the patterning of phylogenetically and anatomically diverse body plans such as arthropods (Akam 1994), vertebrates (Holland 1992), and other taxa (Carroll et al. 2001). This program has moved beyond the model organisms usually studied by developmental biologists and reveals interesting gene expression patterns that are associated with the development of body plan novelty (Lee et al. 2003). It has led to an impressive amount of information concerning the commonalities and differences in the deployment of regulatory genes during the establishment of basic body patterns, including evidence that the evolutionary modification of major body regions in vertebrates is associated with the axial shifts of *Hox* gene expression boundaries (Burke et al. 1995) or that the evolution of mesopodial limb elements is associated with shifts in *Hoxa-11/a-13* expression regions (Wagner and Chiu 2001). In insects, the differences in bristle patterns on different species of *Drosophila* is associated with a variation of *Ubx* expression (Stern 1998) and the evolution of butterfly eyespot patterns involves recruitment of a *hedgehog* regulatory circuit (Keys et al. 1999). A future goal must be to ascertain that observed shifts of gene expression regions were actually causal for the derived developmental and phenotypic condition (Wagner 2001).

A second subfield or emphasis of evolutionary developmental genetics could be called the developmental regulation program. Here, the empirical research concentrates on changes in the evolving architecture of the regulatory circuitry. Tremendous progress was achieved regarding the understanding of gene regulatory pathways and networks (Davidson 2001; Wilkins 2002). Increasingly complex gene interaction networks are unraveled and a kind of regulatory gene cladistics is emerging. The conclusions following from this program posit that the evolution of organismal form is much less a direct consequence of mutational genetic innovation, as believed earlier, but depends on a continuous and continuing process of shift, recruitment, and re-wiring of regulatory interactions in development. In parallel with the evolution of redundancy in the regulatory genome, evolution favors the generation of alternative genetic circuits which are subsequently coopted into new regulatory functions. The working out of the complex details of these processes are going to keep the program active for a long time to come, especially as it is moving into genomics and proteomics. Whole genome analyses using high-throughput molecular technology will provide a new level of comparing gene activation quantitatively across taxa.

5.4.4
The Theoretical Biology Program

Biometrics, multivariate statistics, theoretical and computational modeling, computer simulation, and various domains of bioinformatics have recently taken much interest in EvoDevo, and begin to address some of its questions. A central issue, in one form or another, is the elucidation of the rules governing the genotype–phenotype relationship, which requires computation because of the complexity of epigenesis and the avalanche of data that is being generated on the genotype side. Since developmental evolution resides in the modification of the dynamics of gene, cell, and tissue interactions, the precise topology, timing, and quantity of gene expression as related to changes of cell behavior and tissue properties becomes a target issue in those studies that focus on realistic processes. These requirements have led to the development of computational tools for the representation and quantification of gene expression in developing embryos (Streicher et al. 2000; Sharpe et al. 2002; Weninger and Mohun 2002), and new bioinformatic techniques for the analysis of such data are explored (Fontoura Costa et al. 2004). The aim is to understand the topological evolution of gene expression patterns in a given developmental system in order to determine the spatiotemporal modifications of gene activation that are associated with phenotypic variation. This kind of detailed knowledge can be used for the biomorphic modeling of concrete developmental systems, such as tooth development (Jernvall 2000; Jernvall et al. 2000) or limb development (Hentschel et al. 2004), illustrating how the differential activation of genes and gene products can affect morphogenesis and evolutionary variation or innovation.

A more abstract approach is the modeling of evolving developmental networks, although still based on realistic processes, such as regulatory circuits with positive and negative feedback control. Using such models it was recently demonstrated that evolution has a tendency to substitute emergent developmental networks by hierarchically organized networks (Salazar-Ciudad et al. 2001). This indicates that the routinized and genetically entrenched ontogenies of extant species, from which our knowledge of development is derived, constitute a highly evolved and stabilized condition, whereas greater flexibility and innovative potential may have existed in primitive systems. In a "pre-Mendelian" world, genes and form would not have been inherited in the same closely corresponding manner as in the genetically overdetermined "Mendelian" period that ensued (Newman and Müller 2000). Such scenarios, emerging in part from a modeling approach, call for a re-evaluation of the earlier concepts of canalization and assimilation (Waddington 1942).

A different kind of theoretical biology of EvoDevo is based on the morphospace approach. The class of generative morphospaces, in particular, provides a framework in which a set of given rules produces a range of possible patterns that can be compared with forms that did or did not appear in natural systems (Thomas and Reif 1993; Eble 2001; Rasskin-Gutman 2003). These models can be used to detect general developmental rules that underlie the patterns of phenotypic variation and they permit predictions about the generative capacities of a given developmental system. Morphospace modeling indicates that only a limited number of phenotypic solutions can be obtained from a given developmental system, even in the presence of ample genetic variation. However, these effects are not only limitational. Certain morphological solutions, for example, are more likely to arise than others, independent from the molecular and genetic circuitry associated with their generation, pointing to inherent properties of the developmental system involved.

Multivariate statistics is a theoretical tool that is increasingly applied, in particular as EvoDevo extends into paleontology and anthropology. Although research on human evolution had a traditional focus on quantifying the patterns of variation represented by the fossil record, including patterns of heterochrony, relatively little attention had been paid to the actual ontogenetic mechanisms that underlie the crucial phenotypic transformations in hominid evolution, with few noticeable exceptions (Lovejoy et al. 1999; Minugh-Purvis and McNamara 2001; Lieberman et al. 2002). As landmark-based approaches are gaining wide acceptance in anthropology (Bookstein 1991; Eble 2001), the analyses of ontogenetic shape trajectories receives heightened attention (Prossinger and Bookstein 2003; Cobb and O'Higgins 2004; Mitteroecker et al. 2004), paving the way for EvoDevo in anthropology. Multivariate shape analyses help not only quantify the evolutionary modifications that took place in hominid evolution, but are essential for defining their ontogenetic locations and, hence, assist the identification of the developmental pathways that are responsible for effectuating these changes. In addition, the multivariate approach links EvoDevo with quantitative genetics and the concepts of morphological integration (Cheverud 1982).

Finally, the theoretical biology program in EvoDevo includes the aspect of modeling, an area in which a lot of advancement will take place in the future. Modeling approaches are well under way both in evolutionary and in developmental biology, but have only recently begun to address explicitly the evolution–development interface. First models concentrate on molecular worlds (Fontana 2002) and the evolution of developmental regulatory networks (Salazar-Ciudad et al. 2001). Projects are under way that attempt to model the interaction between genes, cell behavior, and higher-order tissue organization under the perspective of their evolutionary modifiability (Minsuk, in preparation). These new approaches are characteristic consequences of the EvoDevo agenda. They will facilitate the theoretical integration of EvoDevo, improve its predictive scope, and sharpen the questions that biologists can ask effectively in their analysis of embryonic development. The rapid progress in molecular technology, computing technology, and visualization technology greatly assists the development of modeling methodologies (Gerson 2005), in particular as they apply to morphogenesis. At the same time expectations that mathematical modeling will soon be able to describe an orderly mapping between genotype and phenotype variation must be met with caution, given the extent of cryptic genetic variation and the possibility that the inheritance of transcription and translation might be extensively nonadditive (Gibson and Dworkin 2004).

5.5
Practical and Theoretical Integration

It is frequently pointed out that one of the hallmarks of EvoDevo is its integrative potential. Integration is an important trait of EvoDevo indeed, but in at least two different ways. One is practical. EvoDevo unites several fields of research by defining new common questions and methods, and by creating joint research initiatives, academic curricula, etc. As well as developmental and evolutionary biology (and genetics as part of both), it is foremost paleontology, ecology, and anthropology that become directly involved. Beyond that, EvoDevo forms connections with theoretical biology, theoretical chemistry, systems biology, and the philosophy of science. This is an interesting feat, akin to the effect the evolutionary synthesis had in the last century, and it indicates that all participating disciplines expect progress in their own domains from the joint approach. This functional integrative effect is undeniable and takes place with astonishing alacrity. There is, however, a second kind of integration that is usually associated with EvoDevo and that is actually expected to result from it. This integration is theoretical. It would be achieved by the accommodation of a theory of development into the standard, neo-Darwinian theory of evolution. Thus, the black box(es) of the genotype–phenotype connection would be replaced by formalized rules, and the explanatory capacity of evolutionary theory would be enlarged. The extent to which these expectations will be satisfied is not yet clear.

On the one hand, extant EvoDevo has already expanded the explanatory reach of evolutionary theory in as much as explanation was extended to phenomena that were not formerly addressed by the theory. Foremost of these are phenomena pertaining to higher levels of phenotypic organization, such as the establishment of standardized building units (homology), the arrangement of such units in lineage-specific sets (body plans), the directionality of change imposed by the developmental systems (constraint), the generation of new structural elements (novelty), the repeated generation of similar forms in independent taxa (homoplasy), and the redeployment of established sets of organization (modularity). All these themes are addressed by the empirical and conceptual approaches described above. In addition, EvoDevo has achieved progress in explaining how development itself evolves and how the control of developmental processes is mutually effected by genetic, epigenetic, and environmental factors. But will EvoDevo also lead to an expanded theoretical framework, sometimes termed the EvoDevo synthesis, as predicted and hoped for by many? And will such an integrative theory be able to bridge the formal gap that exists between developmental and evolutionary theory in neo-Darwinism?

It seems that EvoDevo has not yet made much progress concerning the formal integration with neo-Darwinian theory, but a number of theoretical frameworks that hold such a possibility can already be discerned at this point. These tend to concentrate either on the mechanistic, regulatory interactions between realistic molecular, developmental, and evolutionary processes or they take an abstract, quantitative statistical approach to relating phenotypic, developmental, and environmental variables.

An example of the first kind, much overlooked in recent considerations, is the "morphoregulator hypothesis" proposed in the mid-1980s (Edelman 1986, 1988). This model assumes that the cell is the unit of morphogenetic control and that cell collectives are the units of morphogenetic signaling that are modified in evolution. The model accounts for the epigenetic component of morphogenesis in terms of a place-dependent, topological response of cells to a suite of inductive signals, mediated by the activation of specific regulatory genes that modulate cell behaviors (proliferation, migration, death, etc.). In this mechanistic framework, morphoregulatory and historegulatory gene activation play an essential role, but mechanochemical components contribute at all levels to the dynamic regulatory loop reiterating between genes, cells, tissues, and organs, and back from the levels of higher organization to the level of molecular regulation. Due to these epigenetic properties no complete specification of organismal form by the genome is required. Genes, in this view, are regulators of dynamical systems that depend on local, topological control. The key evolutionary proposal is heterochronic: small, mutational changes affecting the response time of morphoregulatory genes for cell adhesion would lead to amplified, nonlinear changes in morphology. The model is in accordance with the standard neo-Darwinian paradigm but extends it by a feedback loop joining the morphogenetic level and the evolving regulatory genome.

The greatest potential for formalizations that could pave the way toward a unified theory possibly resides in a fusion of developmental plasticity calculations with quantitative genetics and the life history framework. Major works summarizing the pivotal role of developmental plasticity in connecting the environment with evolution have recently appeared (Pigliucci 2001; West-Eberhard 2003). Developmental plasticity can be defined as the potential of a developmental system based on a given genotype to express a series of morphogenetic trajectories that result in different phenotypes, depending on the environmental conditions to which the individual (or population) is exposed. The range of phenotypic variation that can be attained can be quantified and expressed as the "reaction norm" of a developmental system or phenotype. These data can be entered into formalisms relating genetic variation and its constraints with phenotypic variation and reproductive success via the middle ground of reaction norms and trade off structures – a framework readily provided by life history theory (Stearns 1992). The mechanism of developmental plasticity implies that no genetic change is required for an organism (or a population) to respond to environmental conditions. However, natural selection can act on the phenotypic variation thus obtained and lead to the fixation of adaptive phenotypes – a process known as assimilation (Waddington 1956). In certain threshold cases the plastic response can go beyond the "norm" and provide phenotypic innovations outside the reaction norm, due to new epigenetic interaction between parts of the developmental system that do not usually interact (Müller 1990; Müller and Wagner 1991). Plasticity theory thus holds the potential to provide the formal basis for a long awaited second synthesis in evolutionary theory, a synthesis that would include environmental, developmental, and evolutionary principles (Hall et al. 2003).

5.6
Prospects

EvoDevo has enormously stimulated biological research, both empirically and theoretically, and much has been achieved in a very short period of time. This success is based on conceptual and technological advances in the last quarter of a century. During that period EvoDevo has developed a distinct set of questions, concepts, methods, and empirical research programs. New kinds of data are generated and explanations are provided for biological problems that were out of reach for earlier explanatory frameworks. Academic structures adapt in response to these activities. EvoDevo thus bears all the signs of distinctiveness required of an independent discipline. The scientific scope of this discipline, as shown in the sections above, is far greater than evolutionary developmental genetics, which represents one of the subprograms of EvoDevo, but is not synonymous with it as sometimes claimed. Such claims are unfortunate because they reduce the capacities of the discipline to a single level of explanation, when integration of multiple levels is the main feature of EvoDevo.

Where will the field go from here? The four empirical programs will certainly be pursued for some time to come. More information will be gained from the comparative and biometric analyses of embryological, paleontological, and anthropological data, and more will be learned from the environment–development interaction and other epigenetic conditions such as the generic physical properties of embryonic materials. A host of new data will also come out of the developmental genetics program which will be extended to genomics and proteomics, increasingly using high-throughput molecular technologies and bioinformatic tools, in an approach now often called "systems biology". But with regard to conceptual advancement, maybe the greatest future potential of EvoDevo lies in the theoretical biology program. New methods of visualization, quantification, and mathematical analysis of developmental processes are being developed, and can be used in a comparative way. Computational modeling and simulation, in particular, will be able to address EvoDevo questions, and enhance the predictive capacities of evolutionary theory. These methods can assist the formal integration of developmental and evolutionary theory in defining the epigenetic rules of the genotype–phenotype relation. They are also going to help identify new biological questions for empirical study.

Innovation, modularity, and plasticity appear as major guiding themes of EvoDevo for the near future. These topics have in common that a generative component is required for their explanation, an element that was missing in the pre-EvoDevo frameworks. Here, the principles contained in the black box need to be entered, and EvoDevo goes beyond the central neo-Darwinian tenets of historical contingency and deterministic adaptation. The new tenets added by EvoDevo may be called "emergence" and "inherency". Emergence complements adaptation in that it introduces the nondeterministic factors responsible for the origination of innovations. Emergence is a consequence of interactive systems, such as development, in which the recombination of pre-existing components can lead to unpredicted results (Reid 1985a). Inherency complements contingency. Whereas historical contingency signifies the lawful dependence on conditions that involve a large component of chance, inherency is something that will always happen because the potentiality is immanent to the system and can actually only be inhibited (Newman and Müller 2004). The accommodation of emergence and inherency expands evolutionary theory by those generative principles that were missing in the traditional framework and that are now captured by Evolutionary Developmental Biology.

Acknowledgments

I thank Werner Callebaut, Scott Gilbert, Brian Hall, Manfred Laubichler, Stuart Newman, and Günter Wagner for sharing their thoughts on EvoDevo so generously with me over many years, although they might not agree with everything said herein. I am grateful to Werner Callebaut, Ellen Larsen, Sharon Minsuk, and Stuart Newman for their critique and comments on earlier drafts of this chapter. The staff and fellows of KLI are cordially thanked for their continued dedication to supporting the development of EvoDevo theory.

5.7
References

ABOUHEIF, E. (**1997**) Developmental genetics and homology: a hierarchical approach. *Trends Ecol. Evol.* **12**, 405–408.

AKAM, M. (**1989**) Hox and HOM: homologous gene clusters in insects and vertebrates. *Cell* **57**, 347–349.

AKAM, M. (**1994**) The evolving role of Hox genes in Arthropods. *Dev. Suppl.*, 209–215.

ALBERCH, P. (**1982**) Developmental constraints in evolutionary processes. In: BONNER, J. T. (Ed.), *Evolution and Development*. Berlin: Springer-Verlag, 313–332.

ALBERCH, P., GALE, E. A. (**1983**) Size dependence during the development of the amphibian foot. Colchicine-induced digital loss and reduction. *J. Embryol. Exp. Morphol.* **76**, 177–197.

ALBERCH, P., GALE, E. A. (**1985**) A developmental analysis of an evolutionary trend: digital reduction in amphibians. *Evolution* **39**, 8–23.

AMUNDSON, R. (**2005**) *The Changing Role of the Embryo in Evolutionary Thought*. Cambridge: Cambridge University Press.

ARTHUR, W. (**2001**) Developmental drive: an important determinant of the direction of phenotypic evolution. *Evol. Dev.* **3**, 271–278.

ARTHUR, W. (**2002**) The emerging conceptual framework of evolutionary developmental biology. *Nature* **415**, 757–764.

BELL, M. A. (**1987**) Interacting evolutionary constraints in pelvic reduction of threespine sticklebacks, *Gasterosteus aculeatus* (Pisces, Gasterosteidae). *Biol. J. Linnean Soc.* **31**, 347–382.

BENGTSON, S., ZHAO, Y. (**1997**) Fossilized metazoan embryos from the earliest Cambrian. *Science* **277**, 1645–1648.

BOLKER, J. A., RAFF, R. A. (**1996**) Developmental genetics and traditional homology. *BioEssays* **18**, 489–494.

BONNER, J. T. (**1965**) *Size and Cycle*. Princeton: Princeton University Press.

BONNER, J. T. (Ed.) (**1982**) *Evolution and Development*. Berlin: Springer Verlag.

BONNER, J. T. (**1988**) *The Evolution of Complexity by Means of Natural Selection*. Princeton: Princeton University Press.

BOOKSTEIN, F. L. (**1991**) *Morphometric Tools for Landmark Data: Geometry and Biology*. Cambridge: Cambridge University Press.

BOWLER, P. (**1975**) The changing meaning of "evolution". *J. History Ideas* **36**, 95–114.

BURKE, A. C., NELSON, C. E., MORGAN, B. A., TABIN, C. (**1995**) *Hox* genes and the evolution of vertebrate axial morphology. *Development* **121**, 333–346.

BUSS, L. W. (**1987**) *The Evolution of Individuality*. Princeton: Princeton University Press.

CALDWELL, M. W. (**1994**) Developmental constraints and limb evolution in Permian and extant lepidosauromorph diapsids. *J. Vertebr. Paleontol.* **14**, 459–471.

CALLEBAUT, W., RASSKIN-GUTMAN, D. (**2005**) *Modularity: Understanding the Development and Evolution of Complex Natural Systems*. Cambridge: MIT Press.

CARPENTER, K., HIRSCH, K. F., HORNER, J. R. (**1996**) *Dinosaur Eggs and Babies*. Cambridge: Cambridge University Press.

CARROLL, S. B., GRENIER, J. K., WEATHERBEE, S. D. (**2001**) *From DNA to Diversity: Molecular Genetics and the Evolution of Animal Design*. Malden: Blackwell Science.

CHEVERUD, J. M. (**1982**) Phenotypic, genetic, and environmental morphological integration in the cranium. *Evolution* **36**, 499–516.

CHEVERUD, J. M. (**1984**) Quantitative genetics and developmental constraints on evolution by selection. *J. Theor. Biol.* **110**, 155–171.

CHIPMAN, A. D. (**2002**) Variation, plasticity, and modularity in anuran development. *Zoology* **105**, 97–104.

COBB, S. N., O'HIGGINS, P. (**2004**) Hominins do not share a common postnatal facial ontogenetic shape trajectory. *J. Exp. Zool. B Mol. Dev. Evol.* **302**, 302–321.

DAVEY, K. G. (**2003**) Evolutionary aspects of thyroid hormone effects in invertebrates. In: HALL, B. K., PEARSON, B. J., MÜLLER, G. B. (Eds.), *Environment, Development, and Evolution* Cambridge: MIT Press, 279–295.

DAVIDSON, E. H. (**2001**) *Genomic Regulatory Systems: Development and Evolution*. San Diego: Academic Press.

DAVIDSON, E. H., PETERSON, K., CAMERON, R. A. (**1995**) Origin of the adult bilaterian body plans: evolution of developmental regulatory mechanisms. *Science* **270**, 1319–1325.

DAWKINS, R. (**1976**) *The Selfish Gene*. Oxford: Oxford University Press.

DOLLÉ, P., DIERICH, A., LeMEUR, M., SCHIMMANG, T., SCHUHBAUR, B., CHAMBON, P., DUBOULE, D. (**1993**) Disruption of the *Hoxd-13* gene induces localized heterochrony leading to mice with neotenic limbs. *Cell* **75**, 431–441.

Donoghue, M. J., Ree, R. H. (2000) Homoplasy and developmental constraint: a model and an example from plants. *Am. Zool.* 40, 759–769.

Duboule, D. (1994) Temporal colinearity and the phylotypic progression: a basis for the stability of a vertebrate bauplan and the evolution of morphologies through heterochrony. *Dev. Suppl.* 135–142.

Duboule, D., Dollé, P. (1989) The structural and functional organisation of the mouse *Hox* gene family resembles that of *Drosophila* homeotic genes. *EMBO J.* 8, 1497–1505.

Dudley, M., Poethig, R. S. (1991) The effect of a heterochronic mutation, Teopod2, on the cell lineage of the maize shoot. *Development* 111, 733–739.

Dworkin, I. M., Tanda, S., Larsen, E. (2001) Are entrenched characters developmentally constrained? Creating biramous limbs in an insect. *Evol. Dev.* 3, 424–431.

Eberhard, W. G. (2001) Multiple origins of a major novelty: moveable abdominal lobes in male sepsid flies (Diptera: Sepsidae), and the question of developmental constraints. *Evol. Dev.* 3, 206–222.

Eble, G. J. (2001) Multivariate approaches to development and evolution. In: Minugh-Purvis, N., McNamara, K. J. (Eds.), *Human Evolution through Developmental Change*. Baltimore: Johns Hopkins University Press, 51–78.

Edelman, G. M. (1986) Evolution and morphogenesis: the regulator hypothesis. In: Gustafson, J. P., Stebbins, G. L., Ayala, F. J. (Eds.), *Genetics, Development, and Evolution*. New York: Plenum Press, 1–27.

Edelman, G. M. (1988) *Topobiology*. New York: Basic Books.

Eldredge, N., Gould, S. J. (1972) Punctuated equilibria: an alternative to phyletic gradualism. In: Schopf, T. J. M. (Ed.), *Models in Paleobiology*. San Francisco: Freeman, Cooper, 82–115.

Fontana, W. (2002) Modelling 'evo-devo' with RNA. *BioEssays* 24, 1164–1177.

Fontoura Costa, L., Barbosa, M. S., Manoel, E. T., Streicher, J., Müller, G. B. (2004) Mathematical characterization of three-dimensional gene expression patterns. *Bioinformatics* 20, 1653–1662.

Forgacs, G., Newman, S. A. (2005) *Biological Physics of the Developing Embryo*. Cambridge: Cambridge University Press.

Gerson, E. (2005) The juncture of evolutionary and developmental biology. In: Laubichler, M. D., Maienschein, J. (Eds.), *From Embryology to Evo-Devo: A History of Embryology in the 20th Century*. Cambridge: MIT Press, in press.

Gibson, G., Dworkin, I. (2004) Uncovering cryptic genetic variation. *Nat. Rev. Genet.* 5, 681–90.

Gilbert, S. F. (2001) Ecological developmental biology: developmental biology meets the real world. *Dev. Biol.* 233, 1–32.

Gilbert, S. F. (2003) Evo-Devo, Devo-Evo, and Devgen-Popgen. *Biol. Phil.* 18, 347–352.

Gilbert, S. F., Bolker, J. A. (2001) Homologies of process and modular elements of embryonic construction. *J. Exp. Zool. B Mol. Dev. Evol.* 291, 1–12.

Gilbert, S. F., Bolker, J. A. (2003) Ecological developmental biology: preface to the symposium. *Evol. Dev.* 5, 3–8.

GOLDSCHMIDT, R. (1940) *The Material Basis of Evolution*. New Haven: Yale University Press.

GOODMAN, C. S., COUGHLIN, B. C. (2000) The evolution of evo-devo biology. *Proc. Natl Acad. Sci. USA* 97, 4424–5.

GOODWIN, B. C., HOLDER, N., WYLIE, C. C. (Eds.) (1983) *Development and Evolution*. Cambridge: Cambridge University Press.

GOULD, S. J. (1977) *Ontogeny and Phylogeny*. Cambridge: The Belknap Press of Harvard University Press.

GOULD, S. J., LEWONTIN, R. C. (1979) The spandrels of San Marco and the Panglossian paradigm: a critique of the adaptationist programme. *Proc. R. Soc. Lond. B* 205, 581–598.

GRAHAM, A., PAPALOPULU, N., KRUMLAUF, R. (1989) The murine and *Drosophila* homeobox gene complex have common features of organisation and expression. *Cell* 57, 367–378.

HAECKEL, E. (1866) *Generelle Morphologie der Organismen*. Berlin: Verlag von Georg Reimer.

HALL, B. K. (1983) Epigenetic control in development and evolution. In: GOODWIN, B. C., HOLDER, N., WYLIE, C. G. (Eds.), *Development and Evolution*. Cambridge: Cambridge University Press, 353–379.

HALL, B. K. (1984) Developmental processes underlying heterochrony as an evolutionary mechanism. *Can. J. Zool.* 62, 1–7.

HALL, B. K. (1992) *Evolutionary Developmental Biology*. London: Chapman & Hall.

HALL, B. K. (1995) Homology and embryonic development. *Evol. Biol.* 28, 1–37.

HALL, B. K. (2000) Evo-devo or devo-evo – does it matter? *Evol. Dev* 2, 177–178.

HALL, B. K., PEARSON, B. J., MÜLLER, G. B. (Eds.) (2003) *Environment, Development, and Evolution*. Cambridge: MIT Press.

HAMBURGER, V. (1980) Embryology and the modern synthesis in evolutionary theory. In: MAYR, E., PROVINE, W. B. (Eds.), *The Evolutionary Synthesis: Perspectives on the Unification of Biology*. Cambridge: Harvard University Press, 96–112.

HÄMMERLING, J. (1929) *Dauermodifikationen*. Berlin: Borntraeger.

HENTSCHEL, H. G., GLIMM, T., GLAZIER, J. A., NEWMAN, S. A. (2004) Dynamical mechanisms for skeletal pattern formation in the vertebrate limb. *Proc. R. Soc. Lond. B* 271, 1713–1722.

HOLLAND, L. Z., HOLLAND, P. W., HOLLAND, N. D. (1996) Revealing homologies between body parts of distantly related animals by *in situ* hybridization to developmental genes: Amphioxus versus vertebrates. In: FERRARIS, J. D., PALUMBI, S. R. (Eds.), *Molecular Zoology*. New York: Wiley-Liss, 267–295.

HOLLAND, P. W. (1992) Homeobox genes in vertebrate evolution. *BioEssays* 14, 267–273.

HOLLAND, P. W. (1999) Gene duplication: past, present, and future. *Semin. Cell Dev. Biol.* 10, 541–547.

JABLONKA, E., LAMB, M. J. (1995) *Epigenetic Inheritance and Evolution*. Oxford: Oxford University Press.

JERNVALL, J. (2000) Linking development with generation of novelty in mammalian teeth. *Proc. Natl Acad. Sci. USA* 97, 2641–2645.

JERNVALL, J., KERANEN, S. V., THESLEFF, I. (2000) Evolutionary modification of development in mammalian teeth: quantifying gene expression patterns and topography. *Proc. Natl Acad. Sci. USA* **97**, 14444–14448.

JOLLOS, V. (1934) Inherited changes produced by heat treatment in *Drosophila melanogaster. Genetica* **16**, 476–494.

KAMMERER, P. (1923) Breeding experiments on the inheritance of acquired characters. *Nature* **111**, 637–640.

KAMMERER, P. (1925) *Neuvererbung oder Vererbung erworbener Eigenschaften.* Stuttgart: Walter Seifert Verlag.

KATZ, M. J. (1983) Ontophyletics: studying evolution beyond the genome. *Perspect. Biol. Med.* **26**, 323–333.

KATZ, M. J., LASEK, R. J., KAISERMAN-ABRAMOF, I. R. (1981) Ontophyletics of the nervous system: eyeless mutants illustrate how ontogenetic buffer mechanisms channel evolution. *Proc. Natl Acad. Sci. USA* **78**, 397–401.

KEYS, D. N., LEWIS, D. L., SELEGUE, J. E., PEARSON, B. J., GOODRICH, L. V., JOHNSON, R. L., GATES, J., SCOTT, M. P., CARROLL, S. B. (1999) Recruitment of a hedgehog regulatory circuit in butterfly eyespot evolution. *Science* **283**, 532–534.

KIM, J., KERR, J. Q., MIN, G. S. (2000) Molecular heterochrony in the early development of *Drosophila. Proc. Natl Acad. Sci. USA* **97**, 212–216.

KIRSCHNER, M., GERHART, J. (1998) Evolvability. *Proc. Natl Acad. Sci. USA* **95**, 8420–8427.

KOWALEVSKY, A. (1866) Entwicklungsgeschichte der einfachen Ascidien. *Mém. Acad. Sci. St Petersbourg* **7**, 11–30.

KOWALEVSKY, A. (1871) Weitere Studien über die Entwicklung der einfachen Ascidien. *Arch. Mikrosk. Anat.* **13**, 181–204.

LARSEN, E. (2003) Genes, cell behavior, and the evolution of form. In: MÜLLER, G. B., NEWMAN, S. A. (Eds.), *Origination of Organismal Form.* Cambridge: MIT Press, 119–131.

LAUBICHLER, M. D. (2000) Homology in development and the development of the homology concept. *Am. Zool.* **40**, 777–788.

LAUBICHLER, M. D., MAIENSCHEIN, J. (Eds.) (2005) *From Embryology to Evo-Devo: A History of Embryology in the 20th Century.* Cambridge: MIT Press, in press.

LEE, P. N., CALLAERTS, P., DE COUET, H. G., MARTINDALE, M. Q. (2003) Cephalopod Hox genes and the origin of morphological novelties. *Nature* **424**, 1061–1065.

LIEBERMAN, D. E., MCBRATNEY, B. M., KROVITZ, G. (2002) The evolution and development of cranial form in Homosapiens. *Proc. Natl Acad. Sci. USA* **99**, 1134–1139.

LOVE, A. C. (2003) Evolutionary morphology, innovation, and the synthesis of evolutionary and developmental biology. *Biol. Phil.* **18**, 309–345.

LOVE, A. C. (2005) Morphological and paleontological perspectives for a history of Evo-Devo. In: LAUBICHLER, M. D., MAIENSCHEIN, J. (Eds.), *From Embryology to Evo-Devo: A History of Embryology in the 20th Century.* Cambridge: MIT Press, in press.

LOVEJOY, C. O., COHN, M. J., WHITE, T. D. (**1999**) Morphological analysis of the mammalian postcranium: a developmental perspective. *Proc. Natl. Acad. Sci. USA* **96**, 13247–13252.

LØVTRUP, S. (**1974**) *Epigenetics.* London: Wiley.

MABEE, P. M. (**2000**) Developmental data and phylogenetic systematics: evolution of the vertebrate limb. *Am. Zool.* **40**, 789–800.

MARGULIS, L., FESTER, R. (Eds.) (**1991**) *Symbiosis as a Source of Evolutionary Innovation.* Cambridge: MIT Press.

MATSUDA, R. (**1987**) *Animal Evolution in Changing Environments with Special Reference to Abnormal Metamorphosis.* New York: Wiley.

MAYNARD SMITH, J., BURIAN, R., KAUFFMAN, S., ALBERCH, P., CAMPBELL, J., GOODWIN, B., LANDE, R., RAUP, D., WOLPERT, L. (**1985**) Developmental constraints and evolution. *Q. Rev. Biol.* **60**, 265–287.

MAYR, E. (**1960**) The emergence of evolutionary novelties. In: TAX, S. (Ed.), *Evolution after Darwin.* Chicago: University of Chicago Press, 349–380.

MAYR, E. (**1997**) *This is Biology.* Cambridge: Harvard University Press.

MCGINNIS, W., GARBER, R. L., WIRZ, J., KUROIWA, A., GEHRING, W. J. (**1984**) A homologous protein-coding sequence in *Drosophila* homeotic genes and its conservation in other metazoans. *Cell* **37**, 403–408.

MCGINNIS, W., KRUMLAUF, R. (**1992**) Homeobox genes and axial patterning. *Cell* **68**, 283–302.

MCKINNEY, M. L., MCNAMARA, K. J. (**1991**) *Heterochrony.* New York: Plenum Press.

MCNAMARA, K. J. (**1997**) *Shapes of Time: The Evolution of Growth and Development.* Baltimore: Johns Hopkins University Press.

MINELLI, A. (**1997**) Molecules, developmental modules, and phenotypes: a combinatorial approach to homology. *Mol. Phylogenet. Evol.* **9**, 340–347.

MINUGH-PURVIS, N., MCNAMARA, K. J. (**2001**) *Human Evolution through Developmental Change.* Baltimore: Johns Hopkins University Press.

MITTEROECKER, P., GUNZ, P., BERNHARD, M., SCHAEFER, K., BOOKSTEIN, F. (**2004**) Comparison of cranial ontogenetic trajectories among hominoids. *J. Human Evol.* **46**, 679–697.

MÜLLER, G. B. (**1986**) Effects of skeletal change on muscle pattern formation. *Bibl. Anat.* **29**, 91–108.

MÜLLER, G. B. (**1989**) Ancestral patterns in bird limb development: a new look at Hampé's experiment. *J. Evol. Biol.* **2**, 31–47.

MÜLLER, G. B. (**1990**) Developmental mechanisms at the origin of morphological novelty: a side-effect hypothesis. In: NITECKI, M. H. (Ed.), *Evolutionary Innovations.* Chicago: University of Chicago Press, 99–130.

MÜLLER, G. B. (**1991**) Experimental strategies in evolutionary embryology. *Am. Zool.* **31**, 605–615.

MÜLLER, G. B. (**2003**) Homology: The evolution of morphological organization. In: MÜLLER, G. B., NEWMAN, S. A. (Eds.), *Origination of Organismal Form.* Cambridge: MIT Press, 51–69.

MÜLLER, G. B. (**2005**) Six memos for EvoDevo. In: LAUBICHLER, M. D., MAIENSCHEIN, J. (Eds.), *From Embryology to Evo-Devo: A History of Embryology in the 20th Century*. Cambridge: MIT Press, in press.

MÜLLER, G. B., NEWMAN, S. A. (**1999**) Generation, integration, autonomy: Three steps in the evolution of homology. In: CARDEW, G., BOCK, G. R. (Eds.), *Homology*. Chichester: Wiley, 65–73.

MÜLLER, G. B., NEWMAN, S. A. (Eds.) (**2003a**) *Origination of Organismal Form: Beyond the Gene in Development and Evolution*. Cambridge: MIT Press.

MÜLLER, G. B., NEWMAN, S. A. (**2003b**) Origination of organismal form: the forgotten cause in evolutionary theory. In: MÜLLER, G. B., NEWMAN, S. A. (Eds.), *Origination of Organismal Form*. Cambridge: MIT Press, 3–10.

MÜLLER, G. B., WAGNER, G. P. (**1991**) Novelty in evolution: restructuring the concept. *Annu. Rev. Ecol. Syst.* **22**, 229–256.

MÜLLER, G. B., WAGNER, G. P. (**2003**) Innovation. In: HALL, B. K., OLSON, W. (Eds.), *Keywords and Concepts in Evolutionary Developmental Biology*. Cambridge: Haward University Press, 218–227.

NEWLON, A. W., 3RD, YUND, P. O., STEWART-SAVAGE, J. (**2003**) Phenotypic plasticity of reproductive effort in a colonial ascidian, *Botryllus schlosseri*. *J. Exp. Zool. A Comp. Exp. Biol.* **297**, 180–188.

NEWMAN, S. A. (**1992**) Generic physical mechanisms of morphogenesis and pattern formation as determinants in the evolution of multicellular organization. In: MITTENTHAL, J. B. A. (Ed.), *Principles of Organization in Organisms, SFI Studies in the Sciences of Complexity*. Reading: Addison-Wesley, 241–267.

NEWMAN, S. A. (**1994**) Generic physical mechanisms of tissue morphogenesis: a common basis for development and evolution. *J. Evol. Biol.* **7**, 467–488.

NEWMAN, S. A., MÜLLER, G. B. (**2000**) Epigenetic mechanisms of character origination. *J. Exp. Zool. B Mol. Dev. Evol.* **288**, 304–317.

NEWMAN, S. A., MÜLLER, G. B. (**2004**) Inherency, interaction, and integration in the evolution of developmental mechanisms. In: REHMANN-SUTTER, C., NEUMANN-HELD, E. (Eds.), *Genes in Development. Rereading the Molecular Paradigm*. Durham: Duke University Press, in press.

NIJHOUT, H. F. (**1999**) Control mechanisms of polyphenic development in insects. *BioScience* **49**, 181–192.

PARICHY, D. M., SHAFFER, H. B., MANGEL, M. (**1992**) Heterochrony as a unifying theme in evolution and development. *Evolution* **46**, 1252–1254.

PARKS, A. L., PARR, B. A., CHIN, J.-E., LEAF, D. S., RAFF, R. A. (**1988**) Molecular analysis of heterochronic changes in the evolution of direct developing sea urchins. *J. Evol. Biol.* **1**, 27–44.

PIGLIUCCI, M. (**2001**) *Phenotypic Plasticity: Beyond Nature and Nurture*. Baltimore: Johns Hopkins University Press.

PROSSINGER, H., BOOKSTEIN, F. (**2003**) Statistical estimators of frontal sinus cross section ontogeny from very noisy data. *J. Morphol.* **257**, 1–8.

RAFF, R. (**1996**) *The Shape of Life*. Chicago: University of Chicago Press.

RAFF, R. A., KAUFMAN, T. C. (**1983**) *Embryos, Genes, and Evolution*. New York: MacMillan.

Raff, R. A., Wray, G. A. (1989) Heterochrony: developmental mechanisms and evolutionary results. *J. Evol. Biol.* **2**, 409–434.

Raff, R. A., Anstrom, J. A., Huffman, C. J., Leaf, D. S., Loo, J.-H., Showman, R. M., Wells, D. E. (1984) Origin of a gene regulatory mechanism in the evolution of echinoderms. *Nature* **310**, 312–314.

Rasmussen, N. (1987) A new model of developmental constraints as applied to the Drosophila system. *J. Theor. Biol.* **127**, 271–299.

Rasskin-Gutman, D. (2003) Boundary constraints for the emergence of form. In: Müller, G. B., Newman, S. A. (Eds.), *Origination of Organismal Form*. Cambridge: MIT Press, 305–322.

Reid, R. G. B. (1985a) Unpredicted factors of evolution. The theory of emergence revisited. *Riv. Biol.* **78**, 493–512.

Reid, R. G. B. (1985b) *Evolutionary Theory: The Unfinished Synthesis*. New York: Cornell University Press.

Richards, R. J. (1992) *The Meaning of Evolution*. Chicago: University of Chicago Press.

Riedl, R. (1978) *Order in Living Organisms*. Chichester: Wiley.

Robert, J. S. (2004) *Embryology, Epigenesis, and Evolution: Taking Development Seriously*. Cambridge: Cambridge University Press.

Rose, C. S. (2003) Thyroid hormone-mediated development in vertebrates: What makes frogs unique. In: Hall, B. K., Pearson, B. J., Müller, G. B. (Eds.), *Environment, Development, and Evolution*. Cambridge: MIT Press, 197–237.

Rubin, H. (1990) On the nature of enduring modifications induced in cells and organisms. *Am. J. Physiol.* **258**, L19–L24.

Ruvkun, G., Giusto, J. (1989) The *Caenorhabditis elegans* heterochronic gene *lin-14* encodes a nuclear protein that forms a temporal developmental switch. *Nature* **338**, 313–319.

Salazar-Ciudad, I., Newman, S. A., Sole, R. V. (2001) Phenotypic and dynamical transitions in model genetic networks. I. Emergence of patterns and genotype–phenotype relationships. *Evol. Dev.* **3**, 84–94.

Sarkar, S. (2003) Generalized norms of reaction for ecological developmental biology. *Evol. Dev.* **5**, 106–115.

Schlichting, C., Pigliucci, M. (1998) *Phenotypic Evolution: A Reaction Norm Perspective*. Sunderland: Sinauer.

Schlosser, G., Wagner, G. P. (Eds.) (2004) *Modularity in Development and Evolution*. Chicago: University of Chicago Press.

Sharpe, J., Ahlgren, U., Perry, P., Hill, B., Ross, A., Hecksher-Sorensen, J., Baldock, R., Davidson, D. (2002) Optical projection tomography as a tool for 3D microscopy and gene expression studies. *Science* **296**, 541–545.

Shubin, N., Tabin, C., Carroll, S. (1997) Fossils, genes and the evolution of animal limbs. *Nature* **388**, 639–648.

Stearns, S. C. (1992) *The Evolution of Life Histories*. Oxford: Oxford University Press.

Steinberg, M. S. (1963) Reconstruction of tissues by dissociated cells. *Science* **141**, 401–408.

STERN, D. L. (**1998**) A role of Ultrabithorax in morphological differences between *Drosophila* species. *Nature* **396**, 463–466.

STREICHER, J., DONAT, M. A., STRAUSS, B., SPÖRLE, R., SCHUGHART, K., MÜLLER, G. B. (**2000**) Computer based three-dimensional visualization of developmental gene expression. *Nat. Genet.* **25**, 147–152.

STREICHER, J., MÜLLER, G. B. (**1992**) Natural and experimental reduction of the avian fibula: developmental thresholds and evolutionary constraint. *J. Morphol.* **214**, 269–285.

TAUTZ, D. (**2002**) Evo-Devo – Evolution von Entwicklungsprozessen. *Laborjournal* **5**, 18–21.

THOMAS, R. D. K., REIF, W.-E. (**1993**) The skeleton space: a finite set of organic designs. *Evolution* **47**, 341–360.

TOLLRIAN, R., HARVELL, C. D. (Eds.) (**1999**) *The Ecology and Evolution of Inducible Defenses.* Princeton: Princeton University Press.

TRUE, J. R., CARROLL, S. B. (**2002**) Gene co-option in physiological and morphological evolution. *Annu. Rev. Cell Dev. Biol.* **18**, 53–80.

VOGL, C., RIENESEL, J. (**1991**) Testing for developmental constraints: carpal fusions in urodeles. *Evolution* **45**, 1516–1519.

VON DASSOW, G., MUNRO, E. (**1999**) Modularity in animal development and evolution: Elements of a conceptual framework for EvoDevo. *J. Exp. Zool. B Mol. Dev. Evol.* **285**, 307–325.

VRBA, E. S. (**2003**) Ecology, development, and evolution: perspectives from the fossil record. In: HALL, B. K., PEARSON, B. J., MÜLLER, G. B. (Eds.), *Environment, Development, and Evolution.* Cambridge: MIT Press, 85–105.

WADDINGTON, C. H. (**1942**) Canalization of development and the inheritance of acquired characters. *Nature* **150**, 563–565.

WADDINGTON, C. H. (**1956**) Genetic assimilation. *Adv. Genet.* **10**, 257–290.

WAGNER, G. P. (**1988**) The influence of variation and of developmental constraints on the rate of multivariate phenotypic evolution. *J. Evol. Biol.* **1**, 45–66.

WAGNER, G. P. (**1989**) The biological homology concept. *Annu. Rev. Ecol. Syst.* **20**, 51–69.

WAGNER, G. P. (**1996**) Homologues, natural kinds, and the evolution of modularity. *Am. Zool.* **36**, 36–43.

WAGNER, G. P. (**2000**) What is the promise of developmental evolution? Part I: why is developmental biology necessary to explain evolutionary innovations? *J. Exp. Zool.* **288**, 95–98.

WAGNER, G. P. (**2001**) What is the promise of developmental evolution? Part II: a causal explanation of evolutionary innovations may be impossible. *J. Exp. Zool.* **291**, 305–309.

WAGNER, G. P., CHIU, C. H. (**2001**) The tetrapod limb: a hypothesis on its origin. *J. Exp. Zool.* **291**, 226–240.

WAGNER, G. P., MÜLLER, G. B. (**2002**) Evolutionary innovations overcome ancestral constraints: a re-examination of character evolution in male sepsid flies (Diptera: Sepsidae). *Evol. Dev.* **4**, 1–6.

WAGNER, G. P., CHIU, C., LAUBICHLER, M. (2000) Developmental evolution as a mechanistic science: the inference from developmental mechanisms to evolutionary processes. *Am. Zool.* **40**, 819–831.

WAKE, D. B. (1982) Functional and developmental constraints and opportunities in the evolution of feeding systems in urodeles. In: MOSSAKOWSKI, D., ROTH, G. (Eds.), *Environmental Adaptation and Evolution.* Stuttgart: Gustav Fischer, 51–66.

WAKE, D. B. (1996) Evolutionary developmental biology – prospects for an evolutionary synthesis at the developmental level. *Mem. Calif. Acad. Sci.* **20**, 97–107.

WALLACE, B. (1986) Can embryologists contribute to an understanding of evolutionary mechanisms? In: BECHTEL, W. (Ed.), *Integrating Scientific Disciplines.* Dordrecht: Martinus Nijhoff, 149–163.

WEBB, J. F. (1989) Developmental constraints and evolution of the lateral line system in teleost fishes. In: COOMBS, S., GÄNER, P., MÜNZ, H. (Eds.), *The Mechanosensory Lateral Line: Neurobiology and Evolution.* New York: Springer, 79–97.

WENINGER, W. J., MOHUN, T. (2002) Phenotyping transgenic embryos: a rapid 3D screening method based on episcopic fluorescence image capturing. *Nat. Genet.* **30**, 59–65.

WEST-EBERHARD, M. J. (2003) *Developmental Plasticity and Evolution.* Oxford: Oxford University Press.

WILKINS, A. (2002) *The Evolution of Developmental Pathways.* Sunderland: Sinauer.

WOLTERECK, R. (1909) Weitere experimentelle Untersuchungen über Artveränderung, speziell über das Wesen quantitativer Artunterschiede bei Daphniden. *Verh. Deutsch. Zool. Gesell.* **1909**, 110–172.

WRAY, G. A. (1999) Evolutionary dissociations between homologous genes and homologous structures. In: BOCK, G. R., CARDEW, G. (Eds.), *Homology.* Chichester: Wiley, 189–203.

WRAY, G. A., LOWE, C. J. (2000) Developmental regulatory genes and echinoderm evolution. *Syst. Biol.* **49**, 151–174.

WRAY, G. A., McCLAY, D. R. (1989) Molecular heterochronies and heterotopies in early echinoid development. *Evolution* **43**, 803–813.

ZAKANY, J., GERARD, M., FAVIER, B., DUBOULE, D. (1997) Deletion of a *HoxD* enhancer induces transcriptional heterochrony leading to transposition of the sacrum. *EMBO J.* **16**, 4393–4402.

6
Human Biological Evolution

Winfried Henke

6.1
Summary

This chapter provides a comprehensive compilation of evidence, current ideas, and interpretation of human biological evolution. Paleoanthropological research aims to explain the process of hominization. The biological approach to understanding human evolution as a self-organizing process focuses on the structural and functional adaptations within the order Primates. During the past decades, significant advances in our biological research concepts allowed an increased understanding of the complexity of the origin of *Homo sapiens* from early primate ancestors. The issue is introduced by a short review of the main subjects involved in reconstruction of the straightforward evolutionary processes, which led to our uniqueness. Humans are primates, and thus it is the challenge of evolutionary research to explain their emergence exclusively through the mechanisms of natural selection. The main subjects of research being discussed are taxonomy, functional, constructional, and evolutionary morphology, taphonomy as well as paleoecology and paleogenetics.

Within a multidisciplinary and multifaceted approach, paleoanthropology tries to decipher the adaptive problems that have been important in human evolution. This chapter outlines biological human evolution from the emergence of primates to the origin of anatomically modern humans. The main discussions concentrate on evolutionary trends within nonhuman primates, the adaptive problems that the australopithecines faced living in tropical environments as large, ground-dwelling mammals, the evolution of habitual bipedality, the nutritional change from herbivory to carnivory, and the increase of neuronal complexity (encephalization) within the genus *Homo*.

Finally, the taxonomical problems within our genus and the Eurasian dispersal of an early *Homo* are discussed on the basis of the actual fossil record. At present, the most debated paleoanthropological problem is the origin of modern humans. The competing models, the 'recent African origin model' and the 'multiregional model', are discussed thoroughly. A special section debates the image and fate of the Neanderthals, also taking recent aDNA results into consideration.

Handbook of Evolution, Vol. 2: The Evolution of Living Systems (Including Hominids)
Edited by Franz M. Wuketits and Francisco J. Ayala
Copyright © 2005 Wiley-VCH Verlag GmbH & Co. KGaA, Weinheim
ISBN: 3-527-30838-5

6.2
Introduction to the Issue

Although Darwin did not mention the descent of man in his principal work published in 1859, his masterpiece left no doubt about the consequences for our self-understanding. Darwin focused on this hot issue for the first time in 1871 in his essential book *The Descent of Man and Selection in Relation to Sex*. T. H. Huxley published soon after, despite the bitter protests of traditional creationists, a comparative analysis *On Evidences as to Man's Place in Nature* in 1863. He gathered all contemporary arguments for the provoking thesis that human beings are just another kind of animal, but he professed simultaneously that man is not just an animal, but more than that. No evolutionist will deny that humanity has a special status, because there are obvious differences that set us apart, e.g., our capacity for communication and our complex social interaction, as well as civilization and technology. Thus, to characterize *Homo sapiens* by a paradox, we are merely 'another unique species' (Foley 1987). Uniqueness arises for all species through the mechanism of natural selection, and it is the challenge of the paleoanthropological sciences to give conclusive answers regarding the questions of our own origins, the most mysterious and complex process in all of biology (Campbell 1998, Foley 1995, Henke and Rothe 1994, 1998, Jones et al. 1992, Tattersall 1995, 1997). The major questions about human evolution still remain the same:

- When and where did the first ancestors of our own species evolve?
- What are the essential phylogenetic adaptations of the hominins?
  - What were the ecological niches of our different Plio- and Pleistocene ancestors?
  - What did our early hominin forerunners look like?
  - How did they behave?
- What are the special relationships within the hominin primates?
- Where and when did our own species evolve?
- How did we become a bipedal, large-brained, culturally dependent animal?

Evolutionary anthropology must explain not only upright walking, orthognathy, and loss of body hair, but it is also necessary to explain how evolution, based mainly on natural selection and not on a creative master plan, could bring about an organism that has 'culture as its nature' (Vogel 2000). A biological reconstruction of the path to humanity is based on two sources: fossils and living, 'recent' primates. Fossils, petrified remains of earlier animals (or plants), represent important evidence of phylogeny, but the biological remains do not constitute direct factual information concerning the course of evolution. The idea that fossils are capable of narrating their own history is therefore erroneous: well founded theories concerning our phylogenetic pathway are the necessary precondition for valid thought processes about hominization.

Although fossils provide the basic physical evidence and characteristics for our attempts at phylogenetic reconstruction, hominization can be understood only as a complex psychophysical adaptive process and as such can only be explained by a broad multifaceted multi- and interdisciplinary approach within the framework of primatology (Jones et al. 1992, Martin 1990). Phylogenetic research exploits the all-encompassing investigation of both extinct and extant forms of primates. Human phylogenetics research proves to be the interaction of numerous neighboring disciplines (Figure 6.1). Sciences of major importance include primatology, taxonomy, functional and evolutionary morphology, taphonomy, and evolutionary ecology. The revolution in molecular biology has overrun paleoanthropology and established new fields of research: paleogenetics (archaeometrics) and molecular population genetics, which have yielded much-popularized results on human evolution, especially concerning the phylogenetic relationships within the order Primates. The science of paleoanthropology is – as is anthropology itself – a comparative human biology that describes and interprets:

- communalities between nonhuman primates and humans
- the 'distinctiveness' of the human being

The understanding of evolution as a self-organizing problem-solving process shapes paleoanthropology as a discipline that integrates knowledge from biological and earth sciences as well as the social sciences, especially archeology and cultural anthropology (ethnology). Within an integrative and broadly conceived approach,

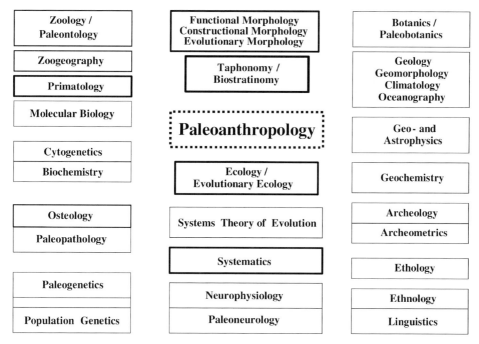

Figure 6.1 Disciplines that contribute to the reconstruction of the process of human evolution. Bold: subjects discussed in detail in this chapter (from Henke and Rothe 1994).

paleoanthropology tries to establish a valid phylogenetic model on the adaptive processes that made us human. As a result of sophisticated planned excavations there has actually been a dramatic increase in fossilized bones and archaeological material, which enables us to unravel our history and its spatial and chronological patterns. But just as important are methodological and technical improvements, as in the paleoecology and molecular biology. Although many special questions are still open, there is a consensus that human evolution followed the same principles as that of all other organisms. From an evolutionary standpoint, we are the product of straightforward evolutionary processes. The most fascinating aspect (and the most confusing, not only for laymen) is that we are the only recent species with self-consciousness. A pithy characterization of man is "Humans are animals who wonder intensely and endlessly about their origin" (anonymous).

Even this unique feature, which enables us to reflect about our origin, is explainable within the pattern of teleonomy and does not need a teleological theory (Pittendrigh 1958). Although Darwin's great idea transforms and illuminates our traditional view of humanity's place in the universe, we have to realize that the most fundamental and contentious question, that of *why* we evolved, is more than a biological problem.

In spite of vehement criticism from creationists and their intellectual allies, evolutionary biologists have an enormous amount of convincing evidence and concise arguments for our evolution by natural selection; as Dobzhansky put it: "Nothing in biology makes sense except in the light of evolution" (Freeman and Herron 1998, p. 51). But science has evidence without certainty. Paleoanthropologists just model, and there is no doubt that evolution is a fact or – somewhat less apodictically formulated – at least the evidence for it is plausible or even convincing by reason of the overwhelming details that have been gathered. And despite the fact that no alternative interpretation for this evidence has ever been offered, there are many controversies within the science of paleoanthropology. Not only do the scanty and scrappy fossils generate divergent interpretations, but different methodological approaches and their steady improvement also sometimes yield dramatic changes in our phylogenetic models. For this reason a paleoanthropologist "must constantly refine the research strategy, thoughtfully integrate the diversity of disciplines, carefully phrase questions, apply new methodologies, and implement targeted survey and excavation to significantly augment our knowledge of human evolution" (Johanson and Edgar 1996, p. 22).

The main subjects involved in the reconstruction of the human evolutionary process are characterized below.

Primatology is a vast field of biological and paleontological research on primates, an order of mammals. The broad catalogue of scientific approaches reaches from the fossil record of the primates that originated around 70–80 myr ago in the Paleocene to phylogenetic reconstruction according to all kinds of biological features. The primary objective of the interspecific morphological, physiological, cytogenetic, serological, molecular, biochemical, and behavioral comparison is to identify key characteristics in the earliest primates and investigate the fate of these features during the subsequent evolution of the group (Martin 1990, Szalay and Delson

1979). This kind of approach within a broad evolutionary context is essential for gaining knowledge about human evolution.

Long before Darwin's theory of a historical–genetic relationship among all organisms, the similarity between monkeys, apes, and humans had been pointed out by natural scientists from ancient Greece and the late Middle Ages. The first person to examine the anatomy of an ape was the Dutch physician Nicolaas Tulp (1641), who published a book on *Satyrus indicus*. Although he described the ape as an orangutan, there is evidence that it was a pygmy chimpanzee (bonobo). Somewhat later, the English anatomist Edward Tyson (1699) examined the common chimpanzee in comparison with a monkey and a human and placed the 'pygmie', as he called the autopsied ape, as an intermediate link between them. From this kind of pre-Darwinian phenomenological studies resulted the view that all superficial similarities are evidence for a homogeneous creative plan for the immutability of species. Darwin's theory refuted this traditional view by describing a mindless, purposeless, mechanical process he called natural selection. Darwin wrote in a notebook: "He who understands baboon would do more towards metaphysics than Locke" (Barrett et al. 1987, cited in Dennett 1995, p. 61).

T. H. Huxley (1863) was the first who presented comprehensive primatological arguments for a human–ape relationship. Somewhat later, Darwin (1871) pointed to Africa as the continent where the human lineage originated, a prognosis that proved to be correct. But primatology not only gives answers about where the cradle of mankind is situated. Comparison of the human with the nonhuman serves to enlighten the determination of what is specifically human. Thus, primatology can contribute essentially to our self-comprehension. Not only are the differences, i.e., the contrasting features, of anthropological interest, but also the similarities. A look at the broad spectrum of specializations within the order primates allows the recognition of evolutionary tendencies (optimization of the visual and tactile senses; enlargement of brain capacity, increase in body size, extension of the individual life span and developmental stages, reduction of the number of offspring, increase of parental investment, intensified behavioral flexibility, intensive build up and growth of traditions within a social group). In light of this view, humans are "intensified progressions of phylogenetic trends" (Vogel 1975). Within the pre-adaptations (predispositions) for human evolution there is, above all, a recognizable shift from emphasis on information by inheritance via the genes to a much greater emphasis on a tradigenetic evolution, that is, the passing on of learned/acquired abilities and individual experiences. The challenge of evolutionary biology is to explain, within Darwinian principles, the enormous intensification of cognitive/intellectual abilities, the conscious control of one's own behavior, and the increasingly complex tool manufacturing and tool usage.

An essential condition for technological evolution within the framework of culture was the development of symbolic language(s), combined with a system of information communication from one individual brain to another (tradigenetic information).

Furthermore, there are the problems concerning the evolution of personal and social responsibility and morality. These questions continuously lead to the sociobiological problem of whether the 'good' is also the human. If we ask how our animal heritage affects the way we live, we touch fundamental problems that were formerly only the demesne of existentialist philosophy and ethics – no wonder that primatology and especially the ethological field studies (e.g., Boesch 1996, Boesch-Achermann and Boesch 1994, Goodall 1986), as well as laboratory research on cognition and language (e.g., Premack 1976, Savage-Rumbough and Lewin 1995, Savage-Rumbough et al. 1978), have piqued tremendous public interest. The same is true concerning the gene called *FOXP2*, which is involved in the face and jaw movements necessary for speech. This is hopefully the first of many language genes to be discovered, and the geneticist Wolfgang Enard further estimates that this gene variant that permits language may have become widespread during the last 200 000 years (Enard et al. 2002).

The intensive anthropological focus on primates comes from the fact that primatological research allows us to test hypotheses by further observation and discoveries. Comparative research on living and fossil primates is therefore indispensable for the reconstruction of human evolution.

Taxonomy is the theoretical study of the laws, procedures, and principles of the arrangements of organisms. A taxon is thus defined as a group of organisms at any level of the hierarchy (e.g., species, genus, family). Although the listing and application of names (nomenclature) must follow established rules of the International Code of Zoological Nomenclature, the recognition of species and taxa, their integration and assignment in a classificatory hierarchy, is less clear-cut, engendering much discussion. The most controversial viewpoints concern the definition and recognition of species. The most common definition of a species is the biospecies (Mayr 1969, 1975), i.e., a group of actually or potentially interbreeding natural populations, which is reproductively isolated from other species. Because this definition is applicable only to contemporaneous living organisms, alternative definitions have been proposed (Simpson 1961, Wiley 1978). Paleontologists very often describe fossil species as morphospecies, based on morphological or anatomical similarity, or refer to temporally successive species in a single lineage, so-called paleospecies or chronospecies. The most accepted definition of a species is given by Wiley (1978, p. 18) as "a single lineage of ancestral descendant populations of organisms which maintains its identity from other such lineages and which has its own evolutionary tendencies and historical fate."

An evolutionary analysis of fossils is based on the simple principle of morphological (or molecular) similarity. But we know from many false pre-Darwinian classifications that there is a serious problem, because similarity is not equality.

On the one hand, one must describe similar features, which are based on common ancestry or a common ancestor, the homologs. The homology of characters is determined by the similarity of location, composition, and development. Although homologies can have a structural similarity to each other, this is not always true, due to functional differentiations. Take, for example, the bat wing and the forelimb of humans.

On the other hand, there can be a similarity between features of different animals without inheritance from a common ancestor. These kinds of features, which are similar by virtue of function, are called homoplasies. Within the homoplastic features are different categories such as parallelism, convergence, and analogy, whose difference is basically one of degree (Aiello and Dean 1990, contra Martin 1990).

Evolutionary analyses must be based on homologous features, because only these features provide evidence of genetic continuity. However, it is not sufficient to rely on a simple distinction between apparently homologous similarities and apparently analogous (convergent, parallel) similarities. A successful attempt needs a further separation of homologous similarities into initial ancestral features, which may be labeled 'primitive' or plesiomorphic (Hennig 1950, 1966), and derived (Mayr 1969) or apomorphic, features, respectively (Hennig 1950, 1966). The more remote homologies are called symplesiomorphic features, or ancestral features, and homologies shared by closely related taxa are called synapomorphic features (or shared ancestral features). Those features that are unique to a particular taxon are called autapomorphic features.

Evolutionary analyses aim to reconstruct the phylogenetic relationships and thus hierarchical grouping of taxa based on three different theoretical and methodological concepts:

- Evolutionary taxonomy (Simpson 1961, Mayr 1975).
- Phylogenetic systematics or cladistic (Hennig 1950, Wiley 1978, Ax 1984).
- Numerical taxonomy or phenetic (Sneath and Sokal 1973).

A main aspect of these different approaches is the fact that they are either preferentially based on grades, i.e., groups of animals similar in general levels of organization, or strictly on clades, i.e., groups of organisms with a common genetic origin.

Evolutionary systematics (taxonomy) is a grade-based concept. Grades are composed of independent lineages that may or may not be monophyletic. This kind of grouping is characterized by the inherent general level of organization or the share of a set of features. Simpson (1961) and Mayr (1969), the proponents of this approach, argue that a classification should not only be consistent with the evolutionary history of a group of organisms, but should also reflect other things about the organisms besides merely the branching pattern of a cladogram. Although the grade systematicists allow polyphyletic or paraphyletic taxa at higher levels of the hierarchy, they do not reject the criterion of monophyly completely, and phylogenetic systematics are strictly bound to a clade-based concept. The method of cladistic analysis is, despite much confusion on the diversity of viewpoints, a straightforward approach. The analysis of relationships relies most heavily on informative characters, but which are the most informative features for inferring evolutionary relationships?

Besides the just-mentioned criteria (homology, synapomorphy), phylogenetically informative traits must be independent and available in great numbers. Since every living or fossil species exhibits a large amount of traits, not all of them are useful for reconstructing evolutionary relationships. If a cladist has chosen features that are both shared and derived, the phylogenetic analysis proceeds by a series of logical steps:

- establishing a morphocline, a serial arrangement of the conditions of a trait such that each condition is logically derivable from its neighbors in the sequence,
- determining the direction of change in the morphocline,
- constructing a cladogram, i.e., a construct that indicates the order in which the species diverged; a cladogram shows which species are more closely related to each other,
- deriving a phylogeny,
- postulating an evolutionary scenario.

Finally, the concept of numerical taxonomy, which is based directly on the assessment of similarities between species, needs to be explained. This approach does not recognize the need to link classification to inferred phylogenetic relationships at all, and furthermore, does not require that taxa should be monophyletic.

It is obvious from this short review that much remains to be done to establish methods that will lead us to objective, reliable, and valid reconstructions of our ancestry (Henke and Rothe 1994, 1998, Johanson and Edgar 1996, Martin 1990, Wiesemüller et al. 2003).

Functional, constructional, and evolutionary morphology: Morphology is the science of the form and structure of animals. Within the concept of comparative biology, this discipline actually provides the main basis for our knowledge of the fossil trail to humanity. Although the traditional morphological approach was highly descriptive and – as some biologists proposed – would better be called 'morphography', the actual concept attempts especially to analyze those processes that enforce and determine the organismic form. The new morphological approach aims at the operational interactions between form and function. Fossils, mostly very scanty sources, provide us with little more than structural information. But because bone shapes are so closely related to function, they offer an opportunity for an understanding of extinct animals and our human ancestors, too. The importance of functional morphology/anatomy is eclipsed by its basic contribution to a deeper appreciation of biological adaptation (Bock and von Wahlert 1965). Although the analysis of the correlation between form (shape) and function is the subject of functional morphology, constructional morphology is defined by Schmidt-Kittler and Vogel (1991, p. 1) as

> the study of organisms and their parts as coherent systems and subsystems that obey statically and dynamically (at rest and in movement) physical principles and processes. The choice of this term also focuses attention on potentials and limitations set on these systems by ontogenetic growth and evolutionary history. It includes exploration of how mechanical and chemical forces and processes influence each other and how construction, behavior, and environment act and react on each other.

The phylogenetic interpretations of form–function complexes are the subject of evolutionary morphology. The simplified scheme in Figure 6.2 illustrates the hierarchy and relationships between the components of the organism and the

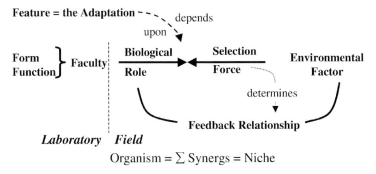

Figure 6.2 Simplified scheme of the hierarchy and relationships between components of the organism and the living space (after Bock and von Wahlert 1965).

environment, which are both pertinent to the understanding of biological adaptation. Because bone shapes are closely related to function (form follows function), paleoanthropologists have an opportunity to understand, how our forerunners mastered the mechanical problems connected with the acquisition of uprightness, orthognathy, and encephalization. Not only did the improvement of the methodological basis of the morphological sciences lead to an enormous increase in our paleoanthropological knowledge, but technological progress (e.g., computerized tomography, cinematography; Figure 6.3) is also responsible for our growing confidence in our conclusions about the nature of our evolutionary ancestors (Helmuth and Henke 1999, Henke and Rothe 1998).

Taphonomy is the study of the process that affects the remains of organisms from their death through to fossilization. It seeks to answer the question: How representative of life in the past is the fossil record? (Foley 1987, p. 82). The Russian paleontologist Efremov (1940, p. 85) conceived this new branch of paleontology (formerly biostratinomy). He described the concerns of this new science as the "study of the transition (in all its details) of animal remains from the biosphere into the lithosphere."

A fossil in its broadest definition is "any trace, impression, or remains of a once-living organism" (Shipman 1981, p. 5), and it is an obvious and trivial fact that "fossils are dead animals" (Hill 1975, p. 18). For this reason, as Shipman (1981, p. 5) emphasized, some traditional assumptions cannot be maintained any longer:

- that life environments are equivalent to death environments,
- that abundance in the fossil record reflects abundance in the original state,
- that species found in the same fossil assemblage reflect sympatry of those species in life,
- that the absence of a species from the fossil record reflects absence or rarity in the original animal community.

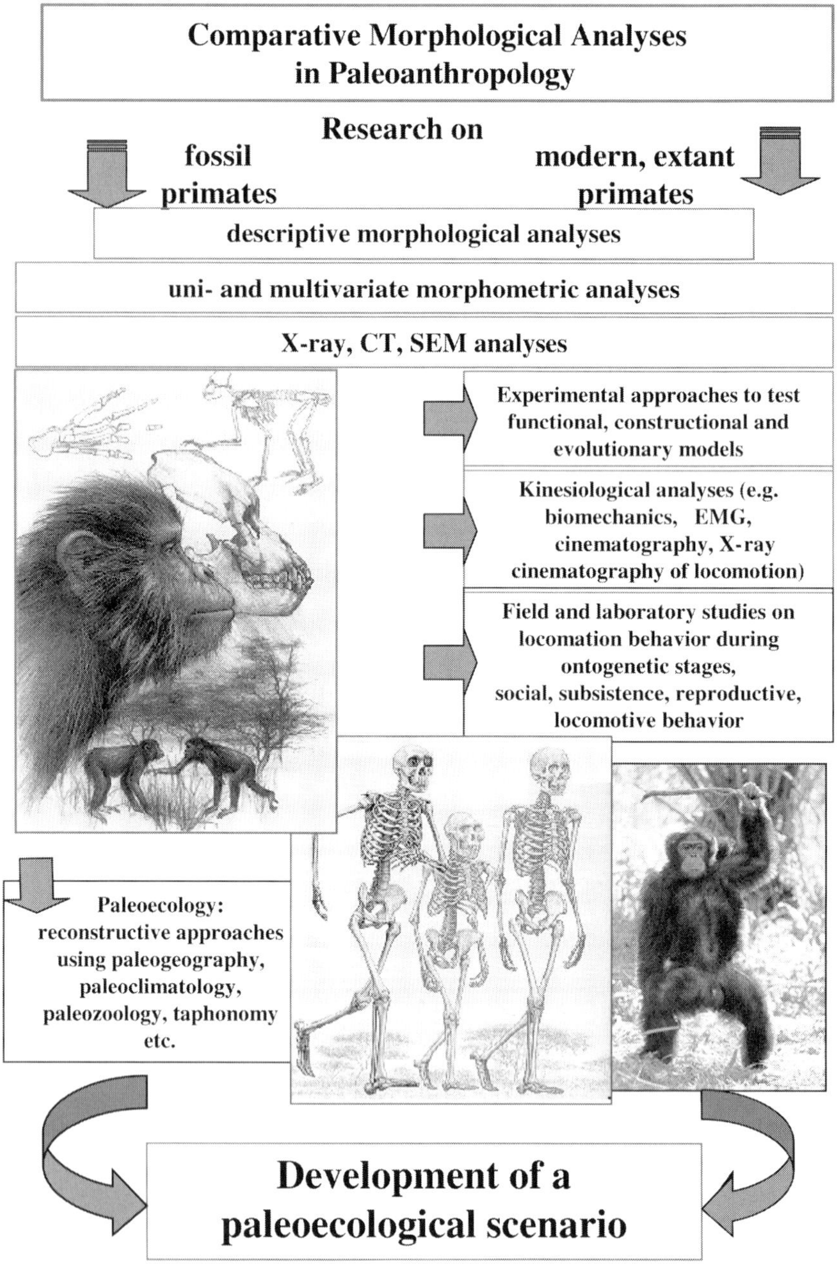

Figure 6.3 Approaches to solving the question of how bipedality evolved (after Henke and Rothe 1994).

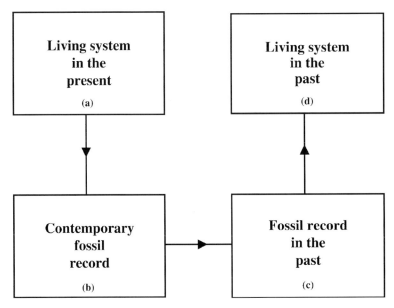

Figure 6.4 Linking the past and the present: the route of inference for knowledge about life in the past (d) depends on observations of the fossil record (c), but interpreting this record depends on understanding the process of transformation during fossilization (c and d). This in turn depends on observations of contemporary fossil formation (b) or on the way in which observable behavior in the present (a) would be visible in the fossil record (after Foley 1987, Figure 4.4, p. 82).

Binford (1977, 1981) referred to the principles that relate past to present as 'middle-range theory'. The route of inference for knowledge about life in the past depends on observations of the fossil record, but interpreting this record depends on understanding the process of transformation during fossilization. This in turn depends on observations of contemporary fossil formation or of the way in which observable behavior in the present would be visible in the fossil record (Figure 6.4).

Paleoanthropology deals, foremost, with bones and teeth of our ancient ancestors and – within a paleoecological approach – with, of course, all other species of the early hominin ecological community, competitors and predators as well as potential prey. The fossilized remains differ substantially from their nonfossilized counterparts. During the long geological times, the organic material of hard tissue is slowly impregnated with minerals present in the sediments.

Another interesting variant are 'behavioral fossils', e.g., the famous footprints from Laetoli. But these kind of fossils are extremely rare. The reconstruction of the 'life history of a fossil' deals with the problem of obtaining and arranging the historical data and of determining the processes that operated to produce the transformation. Shipman (1981, p. 12) formulated the first law of taphonomy: "The present is the key to the past." The second law stipulates: "The occurrence of the past event can be deduced only by demonstrating that its effects differed from those of other similar events."

The effects of postmortem events are caused by different destructive forces, e.g., predators and scavengers, hydraulic, subaerial, aeolian transport, manipulation by tools, and all kinds of decay by chemicals, roots, invertebrates, soil, and water. The most important aspect is that only a very small percentage of all bones 'survive'. In other words, the taphonomic history of a bone assemblage can be taken as a story of information loss. The preservation potential of a bone is a function of many biotic and abiotic environmental factors. Sophisticated approaches in paleotaphonomy, the branch of taphonomy that studies the characteristics of fossil assemblages and their enclosing sediments, refer to neotaphonomy, which studies the modern processes of death, decay, destruction, dispersal, and the concentration of skeletal remains. Experimental approaches in paleoanthropological taphonomy and ethnoarchaeology try to establish methods for a differential diagnosis of manmade cutting, chopping, and slicing marks from those that stem from other causes. A continued pursuit of taphonomic research and improvements in its theoretical and methodological bases are essential for developing our understanding of the past and for finally giving exact answers as to what the past was really like.

Evolutionary paleoecology is the branch of paleontology that deals with the function of ancient organisms and their relationships to their environments and to each other (Etter 1994). In particular, paleosynecology enables us to see assemblages of fossils as parts of former communities and ecosystems. The close correlations between communities and their environment make fossil community relics suitable tools for identifying ancient environments. Paleoecology enables us to trace changes in ecosystems through time. Because fossils cannot tell their own stories, taphonomy has been established as a successful tool for paleoenvironmental reconstructions (see above). A reliable taphonomic analysis is a prerequisite for any paleoecological analysis. Ecology is understood by Pianka (1983, p. 3) as "the study of the relations between organisms and the totality of the physical and biological factors affecting them or influenced by them."

To learn more about these interactions in paleoanthropology, one seeks to understand how the early hominins affected their surroundings and how, in turn, the surroundings affected them. The understanding of adaptations as the solution to the problems an animal faced leads paleoecologists to ask which problems the early hominins faced (Foley 1987, p. 90). Within this context, paleoecologists describe the broad spectrum of biological and ecological problems in early mankind and analyze the significance of tropicality, body size, terrestriality, seasonality, and interspecific competition within the framework of uniformitarian principles and the explicit use of middle-range theory. The central statement stipulates: "What will constitute human uniqueness is not any particular adaptations, but the combinations of them" (Figure 6.5). Pianka (1983, p. 3) is correct, when he claims that no one can master this enormous field of research alone, because there are direct or indirect interactions between almost all organisms in a given area, and great complexity couples these with a multifaceted physical environment, which makes ecology an exceedingly broad subject.

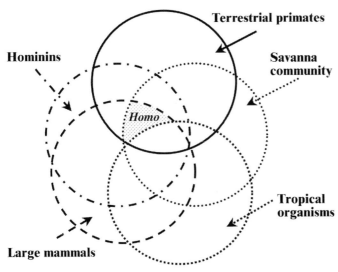

Figure 6.5 Venn diagram of human uniqueness and the human adaptive strategy (shaded area) as the interaction of the biological categories to which hominins belong (after Foley 1987, p. 91).

One fascinating output of paleoecological research is that there was not just a single lineage of hominins. There are good indications that the ecological niches of our Plio-Pleistocene and Pleistocene relatives were very different. But evolutionary ecological research on hominins is still in the beginning. Because paleoecological studies are beyond the expertise of individual paleontologists/anthropologists, they require teamwork involving ecologists, taxonomists, and others. A topic of special interest is paleobiogeography which is still in its infancy, but there is good hope that increased knowledge of the global distribution pattern of organisms with respect to the evolutionary and adaptational history of taxa, as well as to major climatic and oceanographic changes and plate tectonic will improve our picture of the distribution pattern of early mankind.

Paleogenetics, the study of human evolution, as well as the whole field of biological research, was dramatically revolutionized in the second half of the twentieth century by innovative techniques.

The first revolution in biology was induced by absolute dating techniques, such as radioisotope tests, the determination of paleomagnetism, electron spin resonance, and thermoluminescence. The rise of molecular biology, and, in particular, of molecular evolutionary studies (Zuckerkandl and Pauling 1962, Goodman 1962, Sarich and Wilson 1967) suggested that molecules could serve as an evolutionary clock. The hypothesis of a molecular evolutionary clock asserts that informational macromolecules (such as proteins and nucleic acids) evolve at rates that are constant through time and for different lineages. However, this general assumption has been criticized many times. Ayala (1997) referred to the 'vagaries of the molecular clock', alluding to the fact that these clocks sometimes behave in an erratic manner, which calls their use and the neutral theory of evolution into question.

In spite of much criticism on the molecular genetics data, a spate of new evidence was derived from analysis of mitochondrial DNA (mtDNA) in recent populations. MtDNA became a source of new perspectives concerning the evolutionary history of our species and the genetic relatedness of human populations (Wilson et al. 1985). By virtue of its ease of analysis, rapid rate of evolution, and strictly maternal and haploid mode of inheritance, analysis of mtDNA promised to unravel the genetic history of mankind (Stoneking and Cann 1989). Genealogical analysis based on the mtDNA types of people occupying five different geographic regions, published by Cann et al. (1987), was at first glance taken as proof of a recent African origin of modern humans (Brenner and Hanihara 1995, Clark and Willermet 1997). Until these initial studies, comparison of mtDNA had been particularly used for tracing the inheritance and migration patterns of human populations during the Holocene and even later. Cavalli-Sforza (1992) and Cavalli-Sforza et al. (1994) came up with an innovative new approach to bringing genetic, archaeological, and linguistic data together. A new subject, molecular population genetics, thus tries to bring such entities as population size, migration, selection, admixture, and gene flow, as well as demographical data into the focus of research. The second revolution in biology emerged after the realization that DNA may survive in ancient tissues. Because ancient DNA molecules are normally few and fragmented, and preserved soft tissue is rare, there was initially little hope of finding and analyzing ancient genetic material. The breakthrough in the study of DNA came with the application of polymerase chain reaction (PCR) technology in 1983, a method for copying any fragment of DNA (Mullis 1990). Sourcing ancient remains for traces of DNA was all but impossible until the development of this procedure. PCR starts with splitting the double helix of a sample of target DNA into its two single strands. Then enzymes build a new second strand from a bath of free-floating nucleic acid bases. By repetition of the process, a single molecule of DNA becomes two, then four, and so on. The successful recovery of ancient DNA from preserved hard tissues, bones and teeth, allow taxonomic and genealogical relationships to be analyzed (Herrmann and Hummel 1994, Hummel 2002) and yields many other exciting results. Nucleic acids and proteins trapped in ancient bones thus serve as time capsules of history. An aDNA study of the classical Neanderthal is the latest milestone in this kind of research (Krings et al. 1997), and there is ground for optimism that there will be much more highlights in the near future. But there are essential problems concerning the degree of degradation of the aDNA as well as contamination. The locational milieu of the samples under study is essential for their preservation, but until now there has been very little research on artificial diagenesis in paleontological material (Burger et al. 1999, Hummel 2002).

The fascinating catalogue of aDNA studies includes the ability to detect aDNA from archaeological and paleontological findings, as well as from medical and forensic specimens. Microbiologists have been successful in amplifying aDNA segments from the major histocompatibility complex (MHC), which regulates the immune system. The MHC alleles contain information that allows one to infer what diseases ancient populations were or were not resistant to (Ross 1992). Furthermore, there is a good chance in the near future to monitor rapid genetic processes such as recombination events, for example, in human parasites (e.g., *Yersinia pestis*, *Mycobacterium tuberculosis*). Besides the human biological research, there are other paleogenetics projects that focus on the evolution and relationships of domestic animals and plant seeds or compression fossils. For the past few years has been enormous progress in this field of research, which ranges from purely scientific to very practical applications (Burke et al. 1991, Ross 1992, Hummel 2002).

6.3
Human Biological Evolution: a Current Review

6.3.1
Emergence of Primates

The question as to which diagnostic features essentially define primates is rather difficult to answer. Many attempts at a definition are a compromise, because they suffer from the inclusion of likely plesiomorphic features of placental mammals and features that have arisen by convergent evolution in other mammal groups. Furthermore, Martin (1986, p. 8) complains that it is not useful to include trends – as is usually done – in any definition of primates. Unlike other mammals, primates lack any clear-cut diagnostic features. Interpretation of the skeletal features identifiable from the fossil record especially creates fundamental classification problems. On the other hand, there are fewer difficulties to decide on what anatomical and behavioral characteristics in living species define a primate. A preliminary catalogue of characteristics for the 'definition' of primates is given below (Table 6.1 and Figure 6.6).

Based on the Hennigian phylogenetic systematics, taxa have to be defined in terms of synapomorphies, or shared derived traits. Comparative anatomical analyses deliver good indications that the primates may be closely related to tree shrews (Order Scandentia), flying lemurs or colugos (Order Dermoptera), and bats (Chiroptera) (Figure 6.7). Actually there is great disagreement and debate about the correct phylogeny among these groups (Cartmill 1992, Schmitz and Zischler 2002, Schmitz et al. 2003).

Although fossil evidence from the earliest representatives of the order primates is missing, there is a broad consensus that the first primates evolved from insectivore-like, archaic terrestrial and nocturnal mammals some time in the latest part of the Cretaceous period (70–80 myr). Among the substantial fossils from the latest epoch of the Cretaceous or the earliest part of the Paleocene period is the mouse-sized *Purgatorius* from North American sites, the fossil that anatomically most closely resembles later primates. The dentition of this genus indicates a primarily insectivorous nutrition, but a trend from vertical shearing toward more transverse shearing and crushing is an indication of an omnivorous diet. Because the features *Purgatorius* shares with later primates are so general, its classification as a primate has been questioned by several experts. Still, Fleagle (1988, p. 271) emphasizes: "this genus is a placental mammal that could easily have given rise to

Table 6.1 Catalog of criteria for the definition of primates (from Martin 1986, Henke and Rothe 1994, 1998; numbers correspond to illustrations in Figure 6.6).

- Arboriality (arboricoly) in tropical and subtropical forest ecosystems (1).

- Hands and feet with grasping abilities at extremities, but essentially adapted for prehension rather than grappling arboreal supports (2–5).

- Opposable thumbs and opposable great toes (2, 3).

- Fingers and toes with flat nails, not bilaterally compressed claws (4, 5).

- Palms and soles have cutaneous ridges (dermatoglyphs) with optimized grasping and touch sensitivity functions (4, 5).

- Hind limb-dominated primate loco-motion, center of gravity is located near the hind limbs (6, 7).

- Diagonal gait, i.e., forefoot preceding hind foot on each side (7).

- Foot typically adapted for tarsi-fulcru-mation, with at least some degree of relative elongation of the distal segment of the calcaneus, commonly resulting in reverse alteration of the tarsus (calcaneo-navicular articulation) (8).

- Visual sense greatly emphasized; eyes relatively large and the orbits possess (at least) a postorbital bar; forward rotation of the eyes ensures a large degree of binocular overlap (10–13).

- Ipsilateral and contralateral fibers are approximately balanced in numbers on each side of the brain and organized in such a way that the contralateral half of the visual field is represented (13).

- Enlargement and medial approximation of the orbits is typically associated with ethmoid exposure in the orbital wall (17).

- The ventral floor of the well developed auditory bulla is formed predominantly by the petrosal (15).

- The olfactory system is unspecialized in most nocturnal forms and reduced in diurnal forms.

- The brain is typically moderately enlarged relative to the body size, in comparison to other living mammals (16–18).

- The brains of living primates always possess a true Sylvian sulcus (confluent with the rhinal sulcus) and a triradiate calcarine sulcus (18).

- Primate brains constitute a significantly larger portion of body weight at all stages of gestation than among other living mammals.

- Male primates are characterized by permanent precocial descent of the testes into a postpenial scrotum.

- Female primates are characterized by the absence of an urogenital sinus.

- In all primates, involvement of the yolk-sac in placentation is suppressed, at least during the latter half of gestation (22).

- Primates have long gestation periods relative to maternal body size, and produce small litters of precocious neonates.

- Fetal growth and postnatal growth are characteristically slow in relation to maternal size.

- Sexual maturity is attained late and life spans are correspondingly long relative to body size; primates are, in short, adapted for slow reproductive turnover.

- The dental formula exhibits a maximum of 2.1.3.3/2.1.3.3 (24).

- The size of the premaxilla is very limited, in association with the reduced size of incisors, which are arranged more transversely than longitudinally.

- The cheek teeth are typically relatively unspecialized, although cusps are generally low and rounded and the lower molars possess raised, enlarged talonids (25, 26).

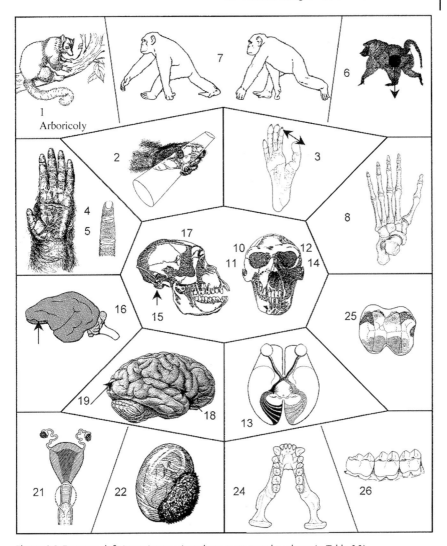

Figure 6.6 Features defining primates (numbers correspond to those in Table 6.1) (from Henke and Rothe 1998).

all later primates, and it is the only primate-like mammal from the earliest part of the Palaeocene."

Otherwise, Fleagle and Kay (1994a, p. 678) notice that "all known plesiadapiforms are either too specialized to be ancestral to any later primates or, in the case of *Purgatorius*, too generalized or poorly known to share any particular features with primates to the exclusion of many other orders. At this time, all that seems certain is that primates, plesiadapiforms, tree shrews, and bats seem to share a common ancestry relative to other orders, but the branching sequence among these groups, usually grouped in the superorder Archonta, is unsolved."

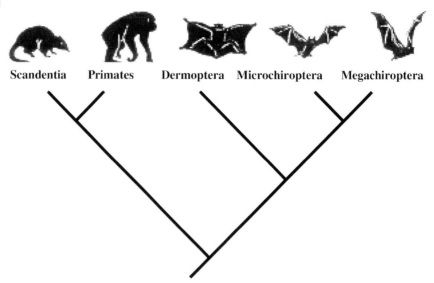

Figure 6.7 Archontan phylogeny – traditional hypothesis (adapted from Cartmill 1992).

The fact that the earliest primate fossils have been described from the New World exemplarily demonstrates that the fossil record is geographically unbalanced, which in turn yields many controversies. Although only little primate fossil material from Africa has been described before the Oligocene period, many authorities have no doubt, that Africa was the center of early primate evolution and that the fossil samples from the Paleo- and Eocene of North America as well as Europe reflect migrations from the south (Simons 1992). The principal sites of discovery of archaic primates in the Paleocene and Eocene indicate that the initial radiation of the plesiadapiforms occurred in the Northern Hemisphere. Fossils of the genus that lend its name, *Plesiadapis*, have been found in Colorado (U.S.A.) and France, suggesting that this group may indeed have originated in these regions. Because there are strong anatomical similarities between the archaic Primates (e.g., *Plesiadapis*) and the Eocene forms called adapids (e.g., *Nothartcus*) some authorities suggest a direct relationship between the infraorder Plesiadapiformes and the Eocene and more recent primates. The archaic primates exhibit many skeletal and dental specializations that show an evolutionary trend toward leaf- and fruit-eating, setting them apart from the primitive dental patterns of their insect-eating predecessors (Hartwig 2002, Ross and Kay 2004).

In the early Eocene the plesiadapiforms began to decline, while early modern-looking primates from the families Omomyidae (e.g., *Teilhardina*) – tiny nocturnal, preferably fruit-eating species that were probably ancestral to modern tarsiers – and Adapidae (e.g., *Cantius*) – diurnal folivores and frugivores, which were probably ancestral to the lemurs and lorises – started to increase in number. Although there are no indications that the first (eu-)primates out-competed and replaced the archaic group, it seems to be most likely that they took over their vacant ecological niches (Simons 1992, p. 202).

Although there are more than 40 genera of Eocene tarsier-like and lemur-like primates that share many dental and postcranial features with modern prosimians, none of these northern Eocene forms has characteristics of the suborder Anthropoidea, which must have arisen before the second half of the Eocene, very likely in Africa (Simons 1992). As yet, it is unclear whether anthropoids (monkeys, apes, and humans) are derived from adapids, from omomyids, or from some still unknown African taxon. Ross et al. (1998) prefer the phylogeny shown in Figure 6.8.

Whether the origin of the anthropoids was within the adapids or the omomyids makes no difference concerning the time span of their expected origin, ranging between 35 and 50 myr. But if neither one is ancestral to anthropoids, the roots may date back to approximately 60 myr, should it turn out that *Algeripithecus minutus* from Algeria is indeed a reliable candidate for an archaic ancestor of the anthropoids (Lewin 1993, p. 48). As can be seen in a review by Fleagle and Kay (1994a, p. 675), "there are proponents for an African, Asian, or, ultimately, North American or European origin for anthropoids" (Fleagle and Kay 1994b, Hartwig 2002, Ross and Kay 2004).

The earliest known relatives of tarsiers (*Afrotarsius*), lorises, and Old World anthropoids have been found in the ca. 31–40-myr-old Qatrani Formation at the Oasis El-Fayum near Cairo. There are many indications from these Egyptian and other fossil sites in Algeria and Morocco that a primate radiation took place in the late Eocene at the latest. A cranial fossil has been recovered, probably from late Eocene layers; it is classified as *Catopithecus* and exhibits, among other features, postorbital plates, a distinctive anthropoid feature. Further possible catarrhine primates of a somewhat younger age and dating in the earliest Oligocene are classified as genera *Oligopithecus* and *Qatrania*. Both of these show close similarities in their dental features with the later Oligocene El-Fayum anthropoids *Aegyptopithecus* (Figure 6.9) and *Propliopithecus*.

Besides the described genera, there are two additional anthropoid primates from El-Fayum of great interest: *Apidium* and *Parapithecus*. Together with *Qatrania* they form the family Parapithecidae. Although *Apidium* especially resembles recent Old World monkeys in many features of the teeth, *Aegyptopithecus* and *Propliopithecus* exhibit similarities with apes and are considered by some authorities to be the earliest members of the Hominoidea, although others regard them simply as primitive catarrhines (Simons 1992, p. 207).

Actually, all the El-Fayum species are thought to antedate the evolutionary divergence of the anthropoid stock into Old World monkeys and apes. *Aegyptopithecus* and *Propliopithecus* may even represent the basic condition prior to the split between Old World and New World anthropoids. *Aegyptopithecus*, a monkey-like frugivore with a body weight around 6 kg, as a generalized arboreal quadruped, best resembles the howler monkey, exhibiting neither arm swinging nor upright walking. The skeletal and dental material of *Aegyptopithecus*, *Propliopithecus*, and *Apidium* indicates significant sexual dimorphism, possibly as a correlate of some kind of polygynous social system.

As has been roughly sketched, there is a remarkable Oligocene collection of intermediate forms, which led to competing hypotheses concerning anthropoid

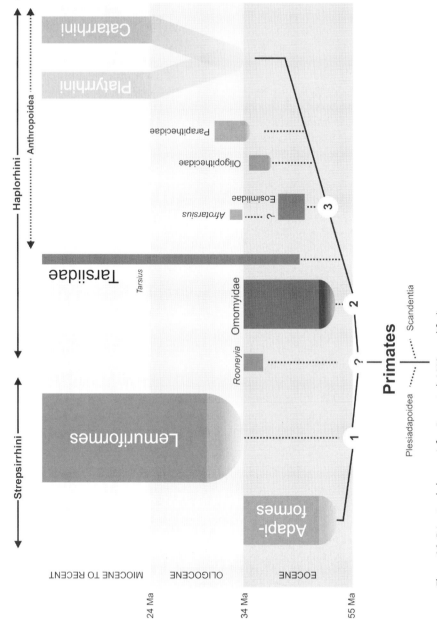

Figure 6.8 Primate phylogeny (after Ross et al. 1998, modified).

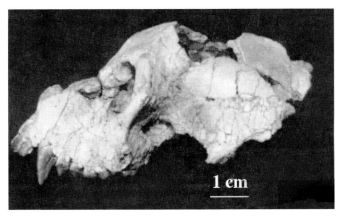

Figure 6.9 Skull of *Aegyptopithecus zeuxis*.

origins (anthropoids = primate taxon including monkeys, apes, and humans, which are separated from the prosimians). The phyletic issue is however much more complicated, especially if we take Far Eastern anthropoids into consideration (e.g., Holroyd and Ciochon 1994). As Fleagle and Kay (1994a, p. 694 f.) emphasize, "there are proponents for an African, Asian, or, ultimately, North American or European origin for anthropoids." Their final conclusion is that "Anthropoidea appears to be very bushy at its base; only when we can more clearly sort through these early branches will we have a clearer view of the roots of the group."

In terms of human evolution, the discussion of the origin of the Old World higher primates, the Catarrhini, is of first-rate interest. They are taxonomically more diverse today, but morphologically are more homogeneous than the New World sister taxon (Platyrrhini) and separated into only two superfamilies, the Old World monkeys (Cercopithecoidea) and the apes and humans (Hominoidea).

As mentioned above, the African parapithecines *Aegyptopithecus* and *Propliopithecus* could be Oligocene forerunners of the Miocene Hominoidea, while the genus *Victoricapithecus,* which is known by abundant fossils from deposits at Maboko Island, Kenya, is a key to Old World Monkey and Catarrhine origins (Benefit 1999). This middle Miocene monkey exhibits semiterrestrial adaptations that may hint of a nonarboreal lifestyle.

Although Old World monkeys are quite rare throughout the early Miocene, there is an intensive radiation from the Pliocene to the present. During the past 20 myr, the diversity of monkeys in Africa has increased as the diversity of hominoids has decreased (Andrews 1986). During this early period Cercopithecoids even lived in those parts of Africa, Europe, and Asia from which they are absent today (Fleagle 1988, p. 397). The well founded aspect of coevolution of Hominoidea and Cercopithecoidea yields more and more interest.

Fossil evidence place the cradle of the Hominoidea to the early Miocene of East Africa. The classification and phylogenetic relationships of the fossils are still under discussion. Early Miocene Hominoidea such as *Proconsul* (Figure 6.10), *Afropithecus,*

Figure 6.10 *Proconsul africanus* (drawn by Pavel Dvorský).

Kenyapithecus, Rangwapithecus, and *Turkanapithecus* are generalized as above-branch arboreal, quadrupedal, tailless, middle-sized to big forms that lived in very different habitats. Their postcranial features correspond highly not only with one another but also with much more ancient primates, such as the late Oligocene proplio-pithecids (e.g., *Aegyptopithecus*) and many of the recent cebids, especially the Atelinae, as far as this can be judged from the fossil material. Their postcranial configuration is described as ancient hominoid morphology. They are characterized by intensified mobility (degrees of freedom) of the extremities as well as a more developed grasping ability of the hands and the feet, an adaptation that was intensified during the hominoid evolution in the Miocene and the development of a suspensory locomotion.

The described changes in the morphological pattern of the extremities and the trunk skeleton are an essential (pre)adaptive step towards the evolution of upright-ness in recent Hominoidea and the expansion to new niches. None of the known Miocene hominoids show the morphological locomotion pattern of recent great apes or humans, which were obviously developed during terricoly of the Pliocene hominoids.

In 1934 Lewis described a new primate that he had unearthed in the Siwalik Hills of India as *Ramapithecus brevirostris* and considered this species the most manlike of the dryopithecines. Dryopithecines are the most widespread and widely discussed fossil primates, which have been subject to several revisions and innumerable comments (Szalay and Delson 1979). In his Ph.D. thesis, in 1937, Lewis described *Ramapithecus* as a hominid [new taxonomy: hominin] ancestral to *Australopithecus*, a sister taxon of *Homo*, but his study remained unpublished. In the 1960s Simons and Pilbeam (1965) began a revision of the Dryopithecinae (Pongidae, Anthropoidea) and concluded that the morphological evidence and functional conclusions drawn from the facial–dental complex were characteristic of later 'hominids' [traditional taxonomy]. Pilbeam (1972, p. 95) summarized:

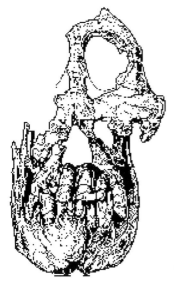

Figure 6.11 *Sivapithecus indicus* (GSP 15 000), frontal view.

"On paleontological evidence we can therefore place it on the line leading to *Australopithecus* and *Homo*, after this lineage diverged from the other hominoids."

Although only dentition was known from *Ramapithecus*, he concluded that it was more hominid- than pongid-like [traditional taxonomy] and for this reason he classified the *Ramapithecus* as a 'hominid'. But the issue of humans' Miocene–Pliocene ancestry was far from being closed. New fossil discoveries increased doubt about the hominin affinities of *Ramapithecus*. Especially the discovery of many additional new Miocene hominoid specimens of the genus *Sivapithecus* (Figure 6.11) in the late 1970s and reinterpretations of the postcranial data in Miocene hominoid evolution, particularly the more derived upper Miocene material from Potwar (India) and Rudabanya (Hungary) as opposed to the more primitive remains known from the lower Miocene deposits in Africa, changed many minds about the phylogenetic position of *Ramapithecus*.

Besides the morphological findings, which demonstrate that *Ramapithecus* has its nearest affinities with *Sivapithecus* and the recent orangutan (*Pongo*), the most important fact for a reconceptualization of the Miocene Hominoidea was the increasing acceptance or influence of biomolecular data for understanding the timing and relationships of hominoid cladogenesis (Ciochon and Corruccini 1982, 1983, Cronin et al. 1983, Goodman et al. 1983, Sarich 1983).

In the 1980s the scenario of hominoid evolution, as based on molecular data, changed from an early divergence of the gorillins/panins and the hominins to a very late divergence. The broad outline of human evolution has been painted in a very unexpected way, because the divergence of the lineages leading to recent African apes and humans occurred in the range of 5.5–8 myr BP (Figure 6.12; Henke and Rothe 1994, Lewin 1998). The fact that *Homo*, *Pan*, and *Gorilla* form a clade that split at a maximum of 8 myr BP indicates that all fossils older than this cannot be regarded as hominin (traditional taxonomy: hominid).

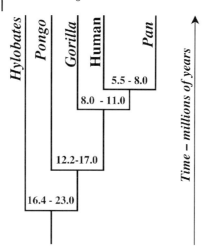

Figure 6.12 Phylogenetic tree: data from DNA hybridization experiments give a range of dates for the evolutionary origins of the hominoids.

During the last two decades it became increasingly evident from molecular as well as cytogenetic data that African apes and modern humans differ in < 2% of their genomes. This extremely close relationship is reflected in the a change of the traditional systematics (Figure 6.13; Bilsborough 1992, Henke and Rothe 1998).

In spite of the steady increase in knowledge concerning the possible appearance of the last common ancestor of apes and man, the relationships of the Miocene hominoids remain a matter of controversy. Although the features of the teeth and the cranial pattern of the Miocene genera *Proconsul* and *Dryopithecus* (generalized frugivors), *Afropithecus* and *Ouranopithecus* (specialized for tough, hard nutrition), and finally *Kenyapithecus*, *Rangwapithecus*, and *Oreopithecus* (folivors) differ intensively in relation to their food resources, the postcranial patterns are very similar in the early and middle Miocene and maybe even in the late Miocene. The diagram (Figure 6.14) of the ecology of fossil and recent Hominoidea by Begun and Kordos (1997) demonstrates the functional changes within the hominoids during phylogeny.

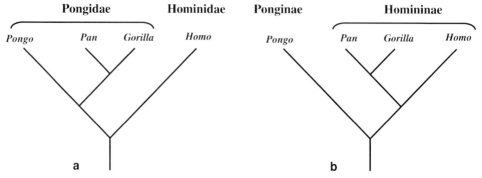

Figure 6.13 Phylogenetic relationship of the hominoids: (a) traditional systematic; (b) revised systematic.

FUNCTION

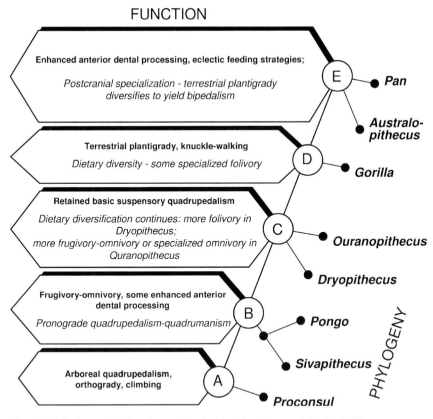

Enhanced anterior dental processing, eclectic feeding strategies;

Postcranial specialization - terrestrial plantigrady diversifies to yield bipedalism

E ● Pan

● Australo-pithecus

Terrestrial plantigrady, knuckle-walking

Dietary diversity - some specialized folivory

D

● Gorilla

Retained basic suspensory quadrupedalism

Dietary diversification continues: more folivory in Dryopithecus; more frugivory-omnivory or specialized omnivory in Quranopithecus

C

● Ouranopithecus

● Dryopithecus

Frugivory-omnivory, some enhanced anterior dental processing

Pronograde quadrupedalism-quadrumanism

B

● Pongo

● Sivapithecus

Arboreal quadrupedalism, orthogrady, climbing

A

● Proconsul

PHYLOGENY

Figure 6.14 Ecology of fossil and recent hominoids (after Begun and Kordos 1997).

In terms of the radiation of the nonhuman Hominoidea, different models, which can be described as African or Asian hypotheses, exist. None of them can be falsified, but doubtless the more ancient forms lived in the marginal areas of the distribution zones, while the more evolved forms of the superfamily were in the core regions. The most parsimonious cladogram, which was calculated by Begun et al. (1997) on the basis of 240 craniodental and postcranial features, mirrors the most probable relationships of Miocene–Pliocene Hominoidea (also Martin 1990, Henke and Rothe 1994). As can be seen in Figure 6.15, the recent chimpanzee (*Pan*) clusters with the fossil genus *Australopithecus*. The adelphotaxon of this clade is *Gorilla*, and the plesion *Dryopithecus* forms the adelphotaxon of *Gorilla* and the *Pan-Australopithecus* clade. Furthermore, it must be mentioned that the relationship of *Kenyapithecus* and *Proconsul* to the hylobatids is somewhat unexpected, because for a long time, *Kenyapithecus* was thought to be the last common ancestor of the recent African apes and man. One explanation could be the scarce fossil report, and for this reason we have to keep in mind that an increase in the fossil record may change the cladogram. But if we look for alternative models on the basis of current fossils, the phylogenetic relationships of *Gorilla*, *Pan*, and *Australopithecus* remain stable.

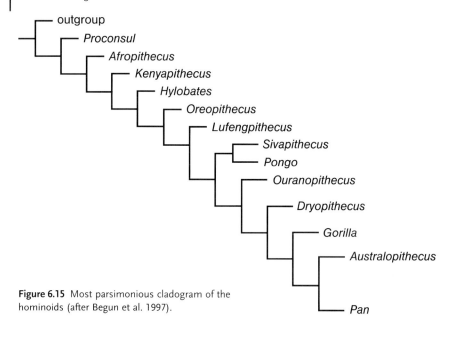

Figure 6.15 Most parsimonious cladogram of the hominoids (after Begun et al. 1997).

6.3.2
The First Hominins: *Ardipithecus, Australopithecus, Paranthropus*

6.3.2.1 Historical Outlines of Discovery and Fossil Evidence

Since humans and chimpanzees share more than 98.4% of their nDNA, paleo-geneticists are convinced that fossils older than 8 myr can no longer be considered hominin. The phylogenetic scenario of how recent apes and modern humans can be traced back in time through a sequence of separate species that converge at the most ancient common ancestor is one of the foremost problems in paleoanthropo-logy. There is no doubt that the common forerunner of the panins/gorillins and the Hominins lived in Africa. But the scenarios concerning the putative candidate change frequently. Those models that describe our ancient forerunners as an imperfect version and mixture of recent apes and humans are obsolete, because every species – including the extinct ones – are successful in their own right (Foley 1987, 1995, Johanson and Edgar 1996). Therefore we have to assume that a common hominid ancestor was a generalized ape and that ancient hominins were more ape-like in appearance.

It took more than 50 years after Darwin's prognosis that Africa could be the cradle of mankind before Dart (1925), a young anatomist from the University of Witwatersrand, discovered and proposed a possible 'ape-man' that fits the role of common ancestor. The type specimen from Taung (Rep. South Africa) was an infant facial skeleton and a brain cast, which was discovered in 1924 by limestone workers in dolomite cave deposits. This first piece of evidence to demonstrate that fossil hominins existed beside our own genus *Homo* was harshly criticized by the paleoanthropological authorities of that time. In their opinion, the infant fossil was not convincing proof of the existence of an early hominin taxon – in addition to the fact that the scientific establishment regarded the research of a paleoanthropological newcomer such as Dart as presumptuous.

Dart's hypothesis stated that *Australopithecus* was an ancient forerunner of humans, but which should be excluded from the genus *Homo* due to its small cranial capacity. In addition, it did not belong to the recent African apes (Panini) due to undoubtedly hominin features. The intermediate status made the Taung fossil a 'problem child' until Broom (1937) discovered a cranium in Sterkfontein, which has been nicknamed Mrs. Ples (Sts 5) and was first classified as *Australopithecus transvaalensis* (renamed *Plesianthropus* soon afterwards). Additional fossils from Makapansgat (Transvaal, Rep. South Africa) demonstrated the similarity of the taxa, which is why all of them are now included in the same species *Australopithecus africanus*. For a very long time the taxon *Australopithecus* included all those fossil hominins of the Pliocene and early Pleistocene that were not considered *Homo*.

That another, mostly contemporary and sympatric, taxon in the Pliocene and early Pleistocene existed became evident by fossils that had been described as much more 'robust' in comparison to the 'gracile' forms from Taung, Sterkfontein, and Makapansgat (Figure 6.16). The term 'gracile' means 'slender' and has been used as an antonym to 'robust', a characterization that was later strongly criticized (Grine 1993; Table 6.2). The split into gracile and robust forms is still pertinent, and what the functional relevance of the morphological differences may be is explained later.

The robust specimens came from Kromdraai and Swartkrans (Rep. South Africa) and were classified by Broom as *Paranthropus robustus* and *Paranthropus crassidens*, respectively. Although the phylogenetic relationships of the *Australopithecus* and *Paranthropus* taxa are still under discussion, which can be seen from the use of the trivial name australopithecines as well as from the terms *Australopithecus sensu lato* and *sensu stricto*, there is certainty that the differences between the samples cannot be explained by sexual dimorphism within a single species, since the two forms are obviously adapted to different ecological niches and demonstrate different dietary adaptations. Furthermore, the single-species explanation has been simply falsified by the chronological order: *Australopithecus* is the older taxon, and *Paranthropus* the somewhat younger hominin. Although there are still arguments against a generic differentiation, I consider this solution the most uncontroversial classification. Strong arguments for the split into *Australopithecus* and *Paranthropus* are given by fossils from Kromdraai and Swartkrans and additional fossils from Drimolen (Keyser 2000). Especially are the so-called hyper-robust australopithecine specimens from Plio–Pleistocene sites in eastern Africa a strong argument for generic separation.

The first primate fossil from eastern Africa, which was later classified as *Praeanthropus africanus* and *Meganthropus africanus*, was found in 1936 in Garusi (Laetoli, Tanzania) by the German ethnologist Kohl-Larson. It was more than 33 years after this first discovery that a hominin fossil was found in the Rift Valley by the Leakeys, who were still sounding out this area in 1935 without success – a skull

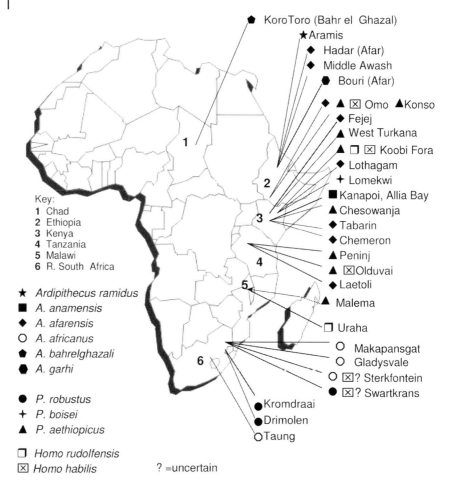

Figure 6.16 Locations of the main Plio-Pleistocene sites in Africa (redrawn from Henke and Rothe 1999).

with a 'nutcracker face', the so-called 'dear boy' (O.H. 5). In 1959 L. Leakey attributed this massively built australopithecine cranium from Bed I of Olduvai Gorge to the taxon *Zinjanthropus boisei*.

Robinson (1954) compared this fossil to those from South Africa and developed the hypothesis that the gracile and robust forms represented separate phylogenetic lines. Since the late 1950s, East Africa had come more and more into the focus of paleoanthropological research, and sites other than Olduvai still yielded fossils from *Parantropus boisei*, e.g., Peninj (Tanzania), Omo and Konso-Gardula (Ethiopia), Illert, Koobi Fora, and Chesowanja (Kenya).

The fact that gracile hominins of Plio–Pleistocene age existed in east Africa became evident for the first time in 1964 by the discovery of a skull from West Turkana, labeled KNM-ER 1470, which was classified as *Homo habilis*. Although

Table 6.2 Species of the Genera *Ardipithecus*, *Australopithecus*, and *Paranthropus*.

Taxa	Synonyma (excerpt)
Ardipithecus ramidus ramidus White et al., 1995	
Ardipithecus ramidus kadabba Haile-Selassie, 2001	
Australopithecus garhi Asfaw et al., 1999	
Australopithecus anamensis M. G. Leakey et al., 1995	
Australopithecus bahrelghazali Brunet et al., 1996	
Australopithecus afarensis Johanson et al., 1978	*Australopithecus africanus tanzaniensis* und *Australopithecus africanus aethiopicus* Tobias, 1980
Australopithecus africanus Dart, 1925	*Australopithecus transvaalensis* Broom, 1936 *Plesianthropus transvaalensis* Broom, 1937 *Australopithecus prometheus* Dart, 1948
Paranthropus robustus Broom, 1938	*Australopithecus robustus* Oakley, 1954 *Paranthropus robustus robustus* Robinson, 1954 *Australopithecus robustus robustus* Campbell, 1964
Paranthropus crassidens Robinson, 1960	*Australopithecus crassidens* Oakley, 1954
Paranthropus boisei Robinson, 1960	*Zinjanthropus boisei* Leakey, 1959 *Australopithecus Zinjanthropus boisei* Leakey et al., 1964 *Australopithecus boisei* Tobias, 1967
Paranthropus aethiopicus Kimbel et al., 1988	*Paraustralopithecus aethiopicus* Arambourg and Coppens, 1968

H. habilis remained a 'premature species' for a very long time, as Tobias, the outstanding nestor of paleoanthropology, stated, evidence for a much older hominin – perhaps a direct forerunner of *Homo* – came from Hadar, in the Afar region of Ethiopia. In the 1970s, Johanson, Coppens, and Taieb conducted the famous Afar Research Expedition and for the first time discovered a nearly half-complete skeleton of an australopithecine; Lucy, as fossil A.L. 288-1 was nicknamed, became the best-known hominin, classified by Johanson et al. (1978) as *Australopithecus afarensis*.

In addition to the specimens from Hadar, fossils from Middle Awash, Fejej, and Omo (Ethiopia), East Turkana, Lothagam, Tabarin, and Chemeron (Kenya), and Laetoli (Tanzania) are attributed to *A. afarensis*, but some classifications remain unclear. Most interesting is the fact that the Garusi-Maxilla that was found by Kohl-Larsen (see above) belongs to the hypodigm, i.e., the former taxon *Praeanthropus* (*Meganthropus*) *africanus* is no longer valid (contra Strait et al. 1997).

Johanson et al. (1978) described *A. afarensis* as an ancient gracile australopithecine, and Rak (1983) explained its facial morphology as plesiomorphic in comparison to the other australopithecine taxa, interpreting the different facial features of *africanus* and *robustus/boisei* as functional adaptations within an evolutionary sequence.

A very recent discovery from the western side of Lake Turkana, which was named KNM-WT 17 000 or 'black skull' and was first attributed to the species *Paraustralopithecus aethiopicus*, was later classified as *Paranthropus aethiopicus*. Many experts questioned this explanation insofar as there are apparently two distinct genera of so-called robust and hyper-robust specimens. Today we have good arguments and justifications from evolutionary morphological research on cranial and dental remains of both taxa that *Australopithecus* and *Paranthropus* occupied different adaptive zones. The phylogeny is still the subject of intense debate, as gradualistic and cladistic approaches have created highly controversial models and scenarios.

The phylogenetic interpretation became much more complicated when White et al. (1994) described a 4.4-myr-old fossil that was found in Aramis (Ethiopia) as *Australopithecus ramidus sp. nova*. The finds are attributed to 17 individuals and consist of teeth, jaws, and cranial and postcranial remains. The fossils surprised the experts because there is indirect evidence that this early hominin was possibly habitually bipedal but still has arboreal-dwelling characteristics. Although paleontologists had expected such an intermediate pattern by reason of mosaic evolution, the unexpected aspect was that this species was a forest dweller, which raises tremendous problems concerning the credibility of the formerly well accepted hypothesis which states that the essential steps of hominization took place in a savannah habitat. The indication that *ramidus* was a forest dweller gave an impulse to rethink our current theories about how hominins became bipedal.

Very soon after the first publication on *Australopithecus ramidus*, harsh criticism by numerous taxonomists induced a revision of the classification. White et al. (1994) came to the conclusion that the former attribution to the genus *Australopithecus* could no longer be supported. A reevaluation resulted in a revision of the former diagnosis and classification of the fossils as *Ardipithecus genus novum* (White et al. 1995, Rothe et al. 1997). How to interpret the phylogenetic position of this oldest hominin taxon is still under debate, especially the question as to which species is the best representative of the so-called common ancestor. Although *Ar. ramidus* remains dated to about 4.4 myr, the additional fossil material from Ethiopia that was attributed to the same taxon was about a million year older (5.8–5.2 myr). The morphological differences between these specimens led Yohannes Haile-Selassie to argue for a subspecies differentiation; the taxon was assigned as *Ardipithecus ramidus kadabba*, while the younger specimens became *Ardipithecus ramidus ramidus* (Haile-Selassi 2001).

Until recently we knew very little about the phylogenetic tree during the late Miocene, the time span between 10 and 5 myr ago. But in this blank area there has been a tremendous increase in discoveries. Late in 2000 a team from the Collège de France and the Community Museums of Nairobi, led by Brigitte Senut and Martin Pickford, discovered fossil material from ca. 6.0-myr-old deposits in Tugen Hills. The material, which was nicknamed Millennium Man, has been assigned to a new genus and species, *Orrorin tugenensis*. The status and phylogeny of this earliest bipedal species are contested. Fossils even older than *Orrorin* have been found in the southern Sahara by Brunet and coworkers. The age of the tremendously interesting skull nicknamed Toumaï is 7.0–6.0 myr. The interpretation of the mosaic pattern of the neuro- and viscerocranium is difficult. For this reason *Sahelanthropus tchadensis* actually yields more problems than solutions (Brunet et al. 2002, Wolpoff et al. 2002), but this fossil provides evidence that human evolution was a pan-African phenomenon and not restricted to East and South Africa (Tobias 2003).

Some Plio–Pleistocene fossils have been classified to formerly unknown species of the human family, such as *Australopithecus anamensis*, *Australopithecus bahrelghazali* and *Australopithecus garhi*, *Australopithecus sp. indet.*, and *Kenyanthropus platyops*.

During the last decade fossils attributed to *A. anamensis* were found in Kanapoi and Allia Bay (northern Kenya) by Mary Leakey and Alan Walker, although the first fossil from the Kanapoi site was unearthed by Patterson in 1965. The distal humerus was analyzed in 1981 by Senut who ascribed a *Homo* status to the fossil. The new material discovered so far displays plesiomorph characteristics along with more derived features typical of later *Australopithecus* species (Leakey et al. 1995, 1998, 2000, Ward et al. 1999). The excavators are convinced that the special mosaic of features of *A. anamensis* belong closely to the ancestry of this genus. Taphonomic and paleoecological results allow the reconstruction of riverine woodlands and gallery forests between 3.8 and 4.2 myr. The morphology of *A. anamensis* shows that this species was still adapted to an arboreal biotope, but the tibia, for example, differs from those of all African apes in having indications of terrestrial locomotion. The knee is placed directly over its ankle joint, an obvious adaptation for bipedal walking (Leakey et al. 2000). The reconstruction of the paleobiotope at Kanapoi and Allia Bay delivers rich insights concerning the ecological niche of these perhaps earliest *Australopithecus* species.

Evidence that the early hominin phylogeny may be much more complicated than we believed hitherto comes from south African fossils that are older than *A. africanus* and similar in age to *A. afarensis*. These extremely interesting discoveries were made by R. J. Clarke (1999), who continued long-term excavations in the dolomite caves of Sterkfontein. While doing archaeozoological research on fossil bones of bovids, carnivores, and primates in September 1994, he found hominin foot bones that revealed ape-like and human characteristics. These included a first metatarsal that showed very strong indication that StW 573-ape man was adapted for tree climbing, while still demonstrating adaptations for terrestrial locomotion.

Nearly three years later, Clarke was lucky enough to find among material stored in his laboratory other bones of the same individual, but the rest of the skeleton was likely still encased in the cave breccia of the Silberberg Grotto. During 1998 he then made the first-ever discovery of a well preserved skull and an associated skeleton of *Australopithecus*. The species status as well as the sex are still unclear (*Australopithecus sp. indet.*), but the skull remains, which are embedded in a hard breccia, leave no doubt that this is a mature adult with a massive zygomatic arch, unlike the pattern known from *A. africanus*. The Sterkfontein hominin has been dated to

3.22–3.58 myr and is older than Lucy (3.2 myr), falling into the time span when *A. afarensis* existed (3.0–3.9 myr).

Besides the new discoveries from Sterkfontein, there is much digging activity at new fossil sites like Gondolin, Drimolen, and Gladysvale (Rep. South Africa), and we may expect instructive paleoanthropological news from these caves in the near future (Schmid 2000).

Very exiting news has been published by Asfaw et al. (1999) on fossil material from Ethiopia. The discovery from the Hata beds of Ethiopia's Middle Awash allows recognition of a new species named *Australopithecus garhi*. The 2.5-myr-old hominin cranial and dental remains demonstrate the occurrence of different locomotion patterns in association with an apparently australopithecine skull and dentition. The discoverers speculate that *A. garhi* descended from *A. afarensis* and is thus a candidate for an ancestor of early *Homo*. Arguments for this hypothesis are not clear, because the picture is blurred by the fact that the connection between the cranial and postcranial remains was not proven conclusively. We cannot be sure that the limb bones, which show a human-like ratio between the humerus and femur, and the lower and upper arm bones, whose proportions are ape-like, belong to the same species as the cranial remains. The skull differs from those of previous australopithecine species in the combination of features, notably the extremely large teeth, especially the rear ones, and the primitive skull morphology.

Finally, it is necessary to mention the Hominid Corridor Research Project in Northern Malawi. Schrenk, Bromage, and coworkers recovered an earliest *Homo* [*Homo rudolfensis*] and a *Paranthropus* specimen [aff. *Paranthropus boisei*] from two contemporaneous sites called Uraha and Malema. Faunal dating suggests an age of 2.5–2.3 myr for both fossils (Bromage and Schrenk 1995, Bromage et al. 1995, Kullmer et al. 1999, Schrenk and Bromage 1999, Schrenk et al. 1993, 1997). The hominins, as well as the faunal remains, link this area to the east and south African sites and allow reconstruction of a Pan African scenario of hominin evolution (see Section 6.2.3).

6.3.2.2 Chronological and Geographical Distribution

The taphonomic and diagenetic analyses of Taung and other southern African sites such as Sterkfontein, Makapansgat, Kromdraai, Swartkrans, Gondolin, Drimolen, and Gladysvale are still creating enormous problems concerning interpretation of the faunal context, because the australopithecines were discovered in dolomite caves, which they obviously did not occupy. There are many indications that these early hominins were actively transported to these places, e.g., by porcupines, or reached them by falling through the dolomite chimneys from the feeding trees of leopards. Faunistic comparisons with the well dated east African fossil sites and improvements in dating techniques have increased our knowledge recently. The east African hominin fossils were found in a stratigraphic context that allowed relative and absolute dating according to volcanic layers. The lava and tuff beds and their exact calibration were a great step forward in establishing a precise chronology of hominin evolution. Based on the comparison of faunal complexes (e.g., Bovidae, Suidae, Papionini), the data from, e.g., Olduvai, Koobi Fora, or Omo make it possible to

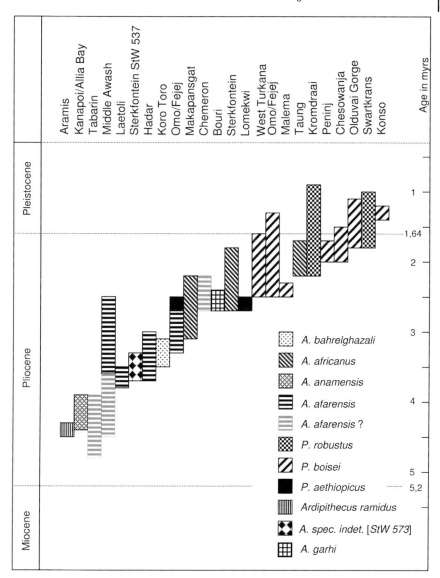

Figure 6.17 Chronology of sites where *Ardipithecus*, *Australopithecus*, or *Paranthropus* were discovered.

establish a more and more reliable chronology of the south African sites too. The chronology of those sites where *Ardipithecus*, *Australopithecus*, and *Paranthropus* were discovered (Figure 6.17) demonstrates the chronological diagram of early hominin fossils in comparison to the genus *Homo*. Within this sequence, *Ardipithecus ramidus* is the oldest one, followed by *A. anamensis*, which antedates *A. afarensis*. The still unclassified South African australopithecine that was described by Ron Clarke also overlaps in time with *A. afarensis* and *A. africanus*, but with a

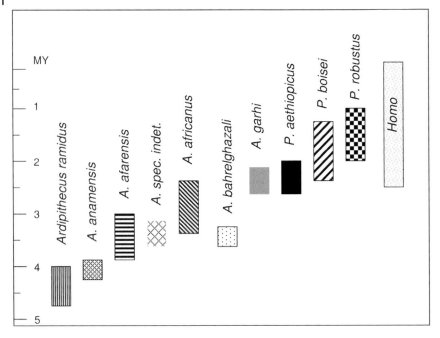

Figure 6.18 Chronological distribution of fossil hominin species.

clear tendency to younger periods. *A. bahrelghazali*, which was unearthed in Chad is contemporary with *A. afarensis* and *A. africanus*, and *A. garhi* overlaps only with *A. africanus* and *P. aethiopicus*. This hyper-robust species is distinctly older than *P. boisei* and *P. robustus*, which overlap in time but are not sympatric (Figure 6.18). The phylogenetic relationships of these species is discussed in Table 6.2 (see p. 145).

6.3.2.3 Morphological Patterns, Phylogenetic Relationship, and Paleoecology

What are the diagnostic features that characterize *Australopithecus* and *Paranthropus* and distinguish them as hominins from the panins/gorillins? From the vast catalogue given in Henke and Rothe (1994, S. 267) only some essential structures are mentioned here, indicating:

- Adaptations of habitual bipedality:
  - s-shaped vertebral column,
  - broader sacrum, enlarged hip joints, more human-like pelvic bones,
  - iliac crest,
  - gliding groove for the iliopsoas muscle at the pubis bone,
  - physiologically x-shaped legs, indicating a bipedal locomotion pattern.

- Very slight increase in cerebralization:
  - cranial capacity within the range of recent apes, but significantly evolved in comparison to the estimated body weights,

- enlarged neurocranial height and more rounded vault,
- more flexed cranial base,
- more anterior position of the occipital foramen,
- diminished nuchal plane.

- Obvious changes in dietary adaptation:
  - alveolar arc highly variable (parabolic to V-shaped; or V- to U-shaped),
  - diastema in the upper jaw missing,
  - frontal teeth small, postcanine teeth heavily enlarged (molarization of premolars),
  - homomorphic incisivi,
  - canines with low crowns, not jutting out of the occlusal plane,
  - P and M heavily enlarged, especially large occlusal areas in *Paranthropus*.

The morphological pattern of the australopithecines characterizes these early hominins as substantially adapted for bipedal walking, although their special locomotion pattern(s) would differ from the bipedality of recent humans. The australopithecine limb bones could resist the loads and stresses of upright stance and locomotion, although their pelvis girdle and femur were not fully shaped like those of humans.

Furthermore, there are obviously essential differences between the described species concerning their locomotion capabilities. For example, *A. anamensis* from the gallery forest at Allia Bay and Kanapoi and the newly described *A. sp. indet.* from Sterkfontein retained grasping functions of their extremities. These features indicate that they did not loose arboreal adaptations in total, perhaps using them in the context of their feeding or resting patterns (sleeping in the trees?) (M. Leakey et al. 1995, 2000).

The dental and cranial features of *Australopithecus* and *Paranthropus* are indicators of special feeding adaptations. The robust australopithecines were vegetarians with heavily specialized teeth (Henke and Rothe 1994, 1997, Walker 1981, Walker and Leakey 1978, 1988; Figure 6.19).

The morphology of *Ardipithecus ramidus*, the earliest member of the broadly accepted hominins, shows many similarities to that of chimpanzees, including thinner tooth enamel and smaller molars. Although the genera *Australopithecus* and *Paranthropus* differ essentially in morphology, thus making a differentiation plausible (which is why we divided them), other authors argue that such a decision is ill-advised because the robust australopithecines may not form a monophyletic group (Bilsborough 1992, Henke and Rothe 1998, Skelton and McHenry 1992, Skeleton et al. 1986, Strait et al. 1997, Ward et al. 1999). Below, a short morphological description is given only for the fully described species:

- *Australopithecus afarensis* (species time range: 3.9–3.0 myr; Figure 6.20):
  - ape-sized brain with an endocranial volume of 415 cc,
  - big snout; flat jaw joint; nasal sill,
  - highest point of the sagittal crest in the occipital region,
  - ape-like large central and small lateral incisors,

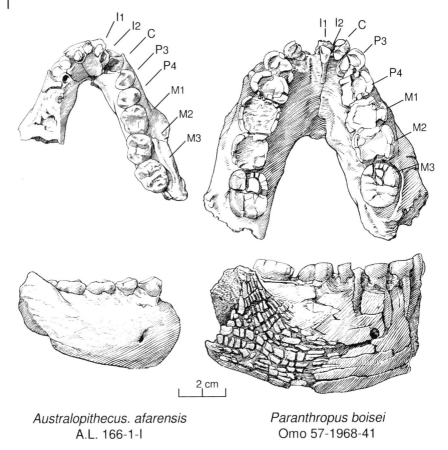

Figure 6.19 Morphological features of the dentition of *Australopithecus* and *Paranthropus*: two mandibles of each genus in occusal and dextrolateral view. Note the size differences, especially the enlarged postcanine chewing areas of *Paranthropus* (after Coppens 1978).

– reduced upper canines (human-like), but still large and with shear facets formed against the lower premolar (ape-like),
– premolars (ape-like); cusps variable, some resembling those of humans,
– parallel or convergent tooth rows (ape-like),
– large cheek teeth in relation to body weight,
– mostly human-like hip, thigh, knee, ankle, and foot,
– more apelike elongated and curved fingers and toes; relatively short thigh, backwardly facing pelvic bones,
– sexual dimorphism in body size greater than in modern humans, but less than in *Gorilla*,
– estimated weight: ♂ 45 kg, ♀ 29 kg,
– estimated body height: ♂ 151 cm, ♀ 105 cm.

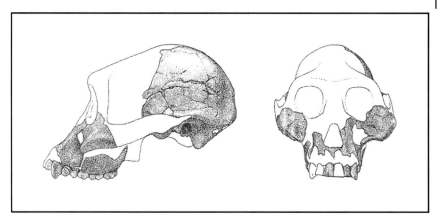

Figure 6.20 *Australopithecus afrarensis* (A.L. 444-2), lateral and frontal view
(adapted from Freeman and Herron 1998).

- *Australopithecus africanus* (species time range: 2.4–2.8 myr; Figure 6.21):
  - in relation to *A. afarensis*, more *Homo*-like cranio–dental features,
  - higher and more rounded vault than earlier species,
  - endocranial capacity 442 cc,
  - less prognathic face than *A. afarensis*,
  - larger cheek teeth than *A. afarensis*,
  - postcranium similar to *A. afarensis*; hand bones more *Homo*-like,
  - estimated weight: ♂ 41 kg, ♀ 30 kg,
  - body size resembles *A. afarensis* ♂ 138 cm, ♀ 115 cm.

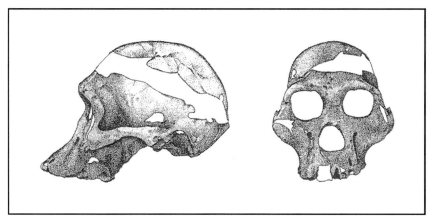

Figure 6.21 *Australopithecus africanus* (STs 5, Mrs. Ples), lateral and frontal view
(adapted from Freeman and Herron 1998).

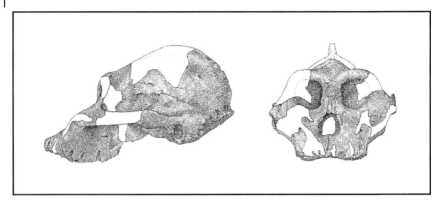

Figure 6.22 *Paranthropus aethiopicus* (KNM-WT 17 000), lateral and frontal view (adapted from Freeman and Herron 1998).

- *Paranthropus aethiopicus* (species time range: ~1.9–2.7 myr; Figure 6.22):
  - small cranial capacity (419 cc),
  - unflexed cranial base,
  - heart-shaped foramen magnum as in *P. boisei*,
  - flat jaw joint,
  - very strong prognathism; dished face with zygomatics raked forward,
  - smooth transition between the naso–alveolar clivus and the floor of the nose,
  - deep tympanic plate,
  - massive cheek teeth for heavy chewing activities,
  - postcranial features unknown.

- *Paranthropus robustus* (species time range: ~1.0–2.0 myr; Figure 6.23):
  - endocranial capacity (530 cc) relatively large in comparison with that of other Plio–Pleistocene non-*Homo* species,

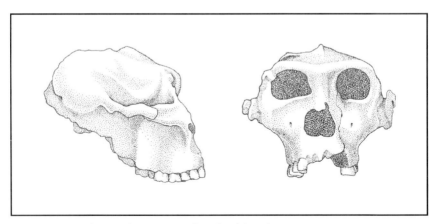

Figure 6.23 *Paranthropus robustus* (SK 48), lateral and frontal view (adapted from Freeman and Herron 1998).

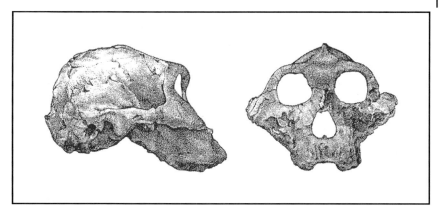

Figure 6.24 *Paranthropus boisei* (KNM-ER 406), lateral and frontal view (adapted from Freeman and Herron 1998)

- moderate prognathic, flat face with cheek bones raked forward, specialized for heavy chewing,
- robust jaws,
- cranial base strongly flexed,
- large postcanine teeth, molarization of premolars; cheek teeth larger than those of *Australopithecus*,
- relatively small-bodied with slight sexual dimorphism,
- estimated weight: ♂ 40 kg, ♀ 32 kg,
- body height ♂ 132 cm, ♀ 110 cm.

- *Paranthropus boisei* (species time range: 1.4–2.3 myr; Figure 6.24):
  - endocranial capacity 551 cc,
  - extremely specialized for heavy chewing; massive cheek teeth, jaws, and supporting architecture of the cranium to withstand chewing stress,
  - flat face,
  - strongly flexed cranial base,
  - deeper jaw joint,
  - estimated weight: ♂ 49 kg, ♀ 34 kg,
  - body size ♂ 137 cm, ♀ 124 cm.

The known fossil record can be interpreted in a variety of ways, but in spite of all the different approaches in the literature, one can recognize a number of time-successive events. These evolutionary trends within the mosaic evolution show that the first trend is the adoption of bipedalism maybe ca. 7–6 myr or at least 4 myr ago, followed by a change in the mastication apparatus, e.g., incisivation of the canines, molarization of the premolars, and enlargement of the molars (mega-dontism). Although dental enlargement and adaptation for heavy chewing cannot be observed in the *Homo* lineage, which exhibits steady brain enlargement, it is significant in the australopithecine lineage. How *A. afarensis*, whose postcranium (Figure 6.26) is best known of all australopithecines, moved, is still controversial.

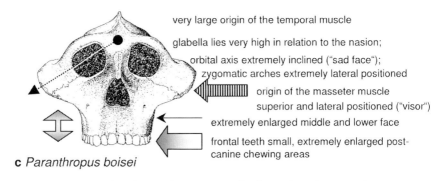

non-fused temporal lines indicate
a relatively small M. temporalis

low position of the glabella; horizontal orbital axis;
zygomatic arches small and gracile

small-sized origin of the masseter muscle

anterior pillar; nasoalveolar triangular frame
middle and lower face low
postcanine teeth with small chewing areas

a *Australopithecus* [*africanus*]

sagittal crest

glabella above the nasion

massive zygomatic arches
large-sized origin of the masseter muscle,
lateral position

anterior pillar, zygomaticomaxillary step and fossa

middle and lower face relatively large
frontal teeth diminished, premolars molarized,
chewing area of the molars extremely large

b *Paranthropus robustus*

very large origin of the temporal muscle

glabella lies very high in relation to the nasion;

orbital axis extremely inclined ("sad face");
zygomatic arches extremely lateral positioned

origin of the masseter muscle
superior and lateral positioned ("visor")

extremely enlarged middle and lower face

frontal teeth small, extremely enlarged postcanine chewing areas

c *Paranthropus boisei*

Figure 6.25 Features of the facial skeleton: (a) *Australopithecus*; (b) and (c) *Paranthropus*.

Although some authors see these early hominins as habitual bipeds, others suggest that they had a gait completely their own (Rak 1991). New biomechanical approaches seem to support the latter opinion, stating that *A. afarensis* was neither pre-dominantly arboreal nor fully bipedal (Aiello 1996, 1990, Aiello and Dean 1990).

A. africanus and the other gracile australopithecines such as *A. anamensis* and *A. garhi* may have preserved an adaptation pattern of arboreality in addition to the capability for terrestrial upright locomotion unlike that of apes and modern humans. Perhaps they spread out bipedally to forage on the ground and then slept in trees during the night. It may be expected that the *A. sp. indet.* from Sterkfontein will allow more precise information in the near future (Clarke 1999).

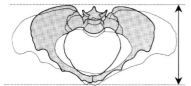

Superprojection of the pelvis of A.L. 288-1 and that of a recent female *H. sapiens*) with the same body weight (after RAK 1988).

Given the same sagittal diameter the iliac bones of "Lucy" are more laterally extended, and the transverse diameter of the pelvis is broader.

a

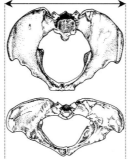

b Pelvis of *H. sapiens* (above) in comparison to the reconstructed pelvis A.L. 288-1; standardized to the same breadth (after Lovejoy 1988)

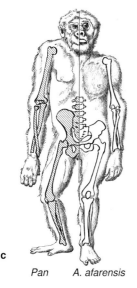

c

Pan   A. afarensis

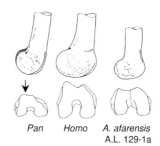

Pan   Homo   A. afarensis
                         A.L. 129-1a

Distal end of the femur in lateral and distal positions (from Tardieu 1983).

A. afarensis exhibits an elliptical and less posteriorly extended lateral condyle of the femur than *Homo*, but the fovea of the patella is *Homo*-like (➤).

Figure 6.26 (a) and (b) Morphological features of the pelvis of *A. afarensis* compared with that of *Homo*; (c) extremities of *A. afarensis* compared with those of *Pan* and *Homo* – *A. afarensis* exhibits no physiological X-leg position (after Zihlman 1982, from Henke and Rothe 1998).

Paranthropus probably moved in the same way as *Australopithecus*, although they had smaller heads and necks of the femur, the biomechanical relevance of which is not yet fully understood.

Concerning the dietary adaptations, there is good indication from the teeth and jaws that the gracile Australopithecines were mostly herbivorous. Their pattern and wear is consistent at least with some fruit-eating, whereas the less prognatic, flat broad faces of *P. robustus* and *P. boisei*, the massive mandibles, the increase in facial height, and the chewing areas of the premolars and molars, paralleled by a diminishing of the front teeth suggest a diet that involved much force between the teeth and a high amount of food processing. They could have been foli- and herbivorous or – more probable in a savannah – graminivorous. Research on the rate and formation of the hard outer enamel suggests faster growth and development than in modern humans and apes. The adaptive pattern of *P. aethiopicus*, which is represented only by a partial mandible from Omo and the nearly complete cranium KNM-WT 17 000, exhibits heavy chewing functions, but the combination of primitive traits with highly derived features is very difficult to interpret (Kimbel et al. 1988). No wonder that there are highly controversial hypotheses on the phylogenetic relationships. This is due to the fact that there could be significant convergent evolution, and we have no way to sort it out. For example, it could be that either heavy chewing resulted in the independent evolution of *P. aethiopicus*, *P. robustus*, and *P. boisei* or that other selection forces shaped *P. boisei*, *P. robustus*, and early *Homo* to resemble each other in encephalization, basicranial flexion, anterior dentition, and orthognathism (McHenry 1996). Whatever the true phylogeny may be – and there can be only one – the present record gives diverse views that are intensely discussed. And with every new species the Plio–Pleistocene hominin evolution becomes more complex. Only one point seems to be clear, that *Paranthropus*, which literally means 'parallel to man', was a well adapted primate genus that existed for a period of at least 1.5 million years alongside our own ancestral lineage and can be excluded from our direct ancestorship.

Several hypotheses regarding the phylogeny are shown in Figure 6.27. Hypothesis one shows *A. afarensis* as the common ancestor of two lineages, one leading to *Homo*, the other leading to *A. africanus*, from which *Paranthropus* splits. In this hypothesis, the south African gracile *Australopithecus* is the ancestor of all the robust forms. Hypothesis two views *A. africanus* as the common ancestor of the *Homo* and *Paranthropus* lineages. Hypothesis three also sees *A. afarensis* as the common ancestor, integrating *P. aethiopicus* as a possible ancestor of *P. robustus* and *P. boisei*. Hypothesis four may become important when we know more about the phylogenetic role of *A. anamensis* or the unclassified material from Sterkfontein. This scheme shows the 4.5-myr-old *A. anamensis* as ancestral to *A. afarensis* and *A. africanus*, but there are many question marks concerning further relationships (Tobias 1988).

Whatever the correct pedigree is, there is no doubt that, during a long period of the Plio–Pleistocene transition, the early Pleistocene, several hominin taxa lived contemporaneously and sympatrically. For this reason we can conclude that they must have occupied different ecological niches. The most probable hypothesis is that *Homo* was so tremendously successful because this species was the one who became tool users and tool makers and who pursued different subsistence strategies than the australopithecines. The reduced dentition and increased encephalization are seen as plausible indications that selection pressure on mental skills separated the australopithecines and *Homo*. The association with more complex subsistence strategies (e.g., scavenging, hunting, and gathering) on one hand and more complex social interaction on the other hand (home base, tradigenetic behavior, male parental investment) may have been the triggers.

Tobias (1989, p. 148) puts it this way: "Man-plus-culture makes the environment; environment-plus-culture makes man; therefore man makes himself."

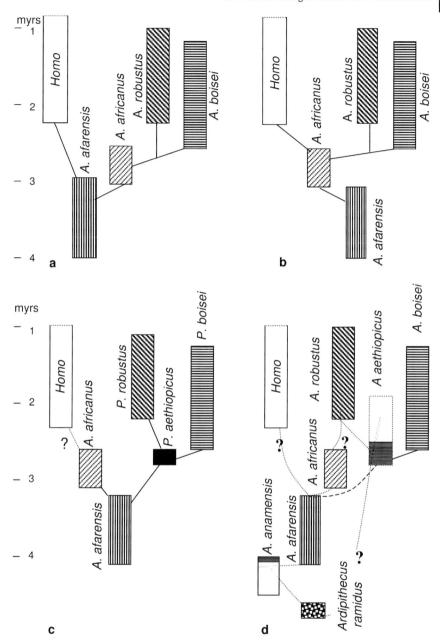

Figure 6.27 Pedigrees: (a) after Johanson and White (1979); (b) after Skelton et al. (1986); (c) after Grine (1993); (d) after Johanson and Edgar (1996).

6.3.3
The Earliest Evidence of the Genus *Homo*

There are few problems in paleoanthropology, says Howell (1978, p. 1985) that "have been more persistently troublesome than recognizing and defining the genus *Homo* because of its fragmentary hominid fossil record, the tendency to over-split hominid taxa and emphasize differences rather than similarities, and the lack of suitable methods to resolve the relative and absolute ages of fossil samples."

It is unrealistic to expect that the boundaries of the genus *Homo* are well defined (Wood 1992, 1996). Traditional expectations concerning *Homo* are full uprightness, reduction of the mastication apparatus as adaptation to changed food processing, encephalization, and evidence of cultural abilities. Morphological trends, like an enlarged brain and reduced dentition, do not meet an adequate definition of our own genus, which contains not only all living people, belonging to *Homo sapiens*, but also our closest fossil relatives as well.

Whether the fossil species attributed to *Homo* had already developed a language, a human-like society, or art can only be judged with great uncertainty, if at all. For example, allometrical effects may cover the real abilities, i.e., body sizes have to be taken into account to achieve an idea of the relative brain size. Current research on cultural behaviors and underlying cognitive and linguistic competences in early hominins is a highly interdisciplinary enterprise, which gives only very rough information concerning the first appearance of the genus *Homo*.

In spite of these problems, most anthropologists agree that early *Homo* emerged in Africa, where the oldest *Homo* specimens were described as *Homo habilis* in 1964 by L. Leakey, Napier and Tobias. Tobias (1989) claimed later that it was a "premature discovery", which means, after Stent (1972, p. 84) that "its implications cannot be connected by a series of simple logical steps to canonical, or generally accepted knowledge." The nomenclators of *Homo habilis* described the *species nova* as being more advanced than *A. africanus* and not as hominized as *H. erectus*. But even 15 years after the species was described for the first time, the majority of competent scholars in the field did not accept *H. habilis* as a valid taxon. The reason was that there was a lack of 'morphological space' between *A. africanus* and *H. erectus*, as Stringer (1986) surmised in his article on "The Credibility of *Homo habilis*", i.e., the new species was strongly criticized in the beginning, for many experts argued that the Olduvai specimens could be classified within existing taxa. Meanwhile, there seems to be worldwide acceptance of the fact that specimens with smaller endocranial capacities need not be excluded from membership in *Homo* and that the fossils are not a normal variant of either the putative australopithecine ancestor or the putative descent *H. erectus*. Within this material, attributed to 'early *Homo*' (Wood 1996), three species can be identified today, one of them resembles *Homo erectus* and is interpreted as either 'early African *Homo erectus*' or *Homo ergaster*, and the two other species are *Homo habilis* (*sensu stricto*) and *Homo rudolfensis*.

6.3.3.1 *Homo rudolfensis* and *Homo habilis* (*s. str.*)

Although Tobias (1991) made a strong case for only one species being represented among specimens attributed to *H. habilis*, other authors see evidence that there may be two species, one named *H. rudolfensis,* which was proposed in 1986 by the Russian anthropologist Valeri Alekseev for specimen KNM-ER 1470, and the other *H. habilis* [*s. str.*] (also Wood 1991, Stringer 1996, Tattersall 2000), which are sister taxa within a monophyletic *Homo* clade. However, this phylogenetic interpretation is only a little more parsimonious than a polyphyletic one, explaining the features typifying each species as parallel developments (also Bilsborough 1992). Anyway, whatever may be the correct taxonomic solution, there are strong arguments against a *H. habilis* [*sensu lato*] concept while the hypodigm remains unsure.

H. rudolfensis shows, apart from an increased average brain size of ca. 750 cc, features of the face and masticatory apparatus that parallel those of *Paranthropus,* e.g., marked orthognathy, broader midface than upper face, and large palate, but *H. habilis s. str.* shows a moderate average brain size of 610 cc and progressive features of cranium, face, and jaws (Figure 6.28).

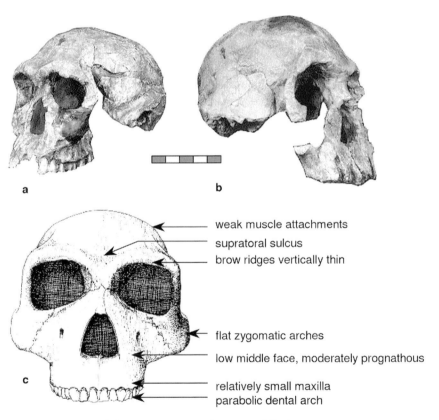

weak muscle attachments
supratoral sulcus
brow ridges vertically thin

flat zygomatic arches

low middle face, moderately prognathous

relatively small maxilla
parabolic dental arch

Figure 6.28 (a) *Homo habilis* (KNM-ER 1813); (b) *Homo rudolfensis* (KNM-ER 1470); (c) morphological features of *Homo habilis s.l.* (from Henke and Rothe 1998).

The postcranium of *H. rudolfensis* is evidently more derived, like that of later *Homo*, although the association with the skull fragments is not confirmed. In contrast, *H. habilis s. str.* shows a very plesiomorphic, australopithecine-like postcranium.

The described combination of australopithecine-like cranio–dental features with a derived postcranium in *H. rudolfensis* and of progressive cranium and dentition with primitive body proportions in *H. habilis s. str.* (a highly putative association) allows the conclusion that neither of the two species represents a reliable ancestor of later *Homo* (see *Homo ergaster*), because every interpretation has to take very unlikely evolutionary changes into account to explain these unusual mosaics.

6.3.3.2 Chronological and Geographical Distribution

Fossils representing *H. habilis* [*s.l.*] were first described from Olduvai, where they were uncovered from layers that have also yielded australopithecine skeletal material (Bed I and Bed II), but the largest contribution to the hypodigm comes from another site in east Africa, Koobi Fora, on the northeast shore of Lake Turkana (Figure 6.16).

Further remains of the species have been recovered from Members G and H of the Omo Shungura Formation. From the western shore of Lake Turkana, a cranial fragment from the Nachukui Formation has been described, and a fragmentary skull, Stw 53, was recovered from the South African cave of Sterkfontein in Member 5. The attribution of hominin material from Member 4 of Sterkfontein to *H. habilis s.l.* is uncertain, although material from Swartkrans Member I (Sk 847, Sk 27) was recently confirmed as belonging to this species.

In addition to the fossils from sub-Saharan sites, there are reports on *H. habilis s.l.* fossils from sites beyond Africa, the Near East and Asia, which have only little reliability and acceptance. The actual hypodigm concentrates especially on East Africa.

A newly unearthed mandible from Uraha (U501, Malawi), which was attributed to *H. rudolfensis* by Bromage et al. (1995) forms a link to the more northern sample (Schrenk and Bromage 1995, 1999).

The recognition of *H. habilis s.l.* and the dating of some specimens from Olduvai to about 2.0 myr BP was the first compelling evidence for the existence of Tertiary man in the sense of a species of the genus *Homo* (Tobias 1989). After improvements in dating methods and the discovery of new remarkable fossils, it is generally accepted that the genus *Homo* emerged before the end of the Pliocene. Stone tools, identified as Oldowan in character, have been traced back to about 2.6 to 2.7 myr BP; however, there is no proof that these implements testify to the presence of a particular hominin species. The definition of *H. habilis* is, essentially, an anatomical one, although ethological evidence may be added in support (Tobias 1989). The most recent occurrences of *H. habilis* are from Olduvai Bed II, dating to about 1.6 myr BP (Table 6.3).

The higher age of *H. rudolfensis* (2.5–1.8 myr) in comparison to *H. habilis s. str.* (2.1–1.6 myr) cannot be taken as evidence that this species is the better candidate for the direct *Homo* lineage, because the phylogenetic analysis has to be based on relevant diagnostic features.

Table 6.3a Earliest dispersal sites of the genus *Homo*
(see also Henke and Rothe 1994, Larick and Ciochon 1997).

Sites	Age (myrs)	Hominid fossils and stone-tool technology	Taxa
Hadar (Ethiopia)			
Kada Gona	2.7–2.5	core–flake (Omo)	
Kada Hadar	2.3–2.2	mandible core–flake (Omo/Oldowan)	*Homo sp. indet.*
Omo Shungura (Ethiopia)			
Member D	2.5–2.4	teeth	*Homo sp. indet.*
Member E	2.4–2.3	teeth core–flake (Omo)	*Homo sp. indet.*
Baringo (Kenya)			
Chemeron	2.4	temporal BC 1	*Homo sp. indet.*
Uraha (Malawi)			
Chiwondo beds	2.5–2.1	mandible UR 501	*H. rudolfensis*
Nachukui (West Turkana)			
Kalochoro	2.35–1.9	core–flake (Omo)	
Nariokotome, Lower Natoo	1.5	skeleton WT 15000	*H. ergaster*
Koobi Fora (East Turkana)			
Upper Burgi	1.90–1.88	mandible ER 1812 core–flake (Omo/Oldowan)	*H. ergaster*
		occiput ER 2598	*H. ergaster*
		pelvis ER 3228	*H. ergaster*
		skull ER 1470	*H. rudolfensis*
KBS	1.78	skull ER 3733	*H. ergaster*
Okote	1.65–1.50	skull ER 3883	*H. ergaster*
		mandible ER 992	*H. ergaster*
Swartkrans (South Africa)			
Member 1	1.7–1.8	skull SK 847	*H. ergaster*
Olduvai Gorge (Tanzania)			
Bed I	1.8	mandible OH 7 core–flake (Oldowan)	*H. habilis*
Lower bed II	1.6	skull OH 13 core flake (Developed Oldowan)	*H. habilis*
Olduvai Gorge (Tanzania)			
Upper bed II	1.25/1.4	skull OH 9 biface (Acheulian)	*H. erectus* [?]
Konso-Gardula (Ethiopia)			
KGA 4-KGA 12	1.4	mandible KGA 10-1 biface (Acheulian)	*H. ergaster*

Table 6.3b Early and later Eurasian dispersal sites
(see also Henke and Rothe 1994, Larick and Ciochon 1997).

Sites	Age (myrs)	Hominid fossils and stone-tool technology	Taxa
Yiron (Israel)	2,4 [?]	'tools'	
Erq el-Ahmar (Israel)	1.96–1.78	Upper Erq el-Ahmar, 2 core tools	
Ubeidiya (Israel) Ubeidiya form,	1.4	biface (Acheulian) lower layers	
upper layers		core–flake (Developed Oldowan)	
Riwat (Pakistan)	1.96–1.78	artifact horizon, 3 core tools	
Pabbi Hills (Pakistan)	2.0–0.90	fossil horizons, core–flake (Omo)	
Longgupo (China) Middle zone	1.96–1.78	mandible CV.939.1 2 core tools incisor CV.939.2	*Homo sp. indet.* [?] *Homo sp. indet.* [?]
Sangiran (Java, Indonesia) Pucangan formation	1.66	2 skulls S 27, S 31	*H. erectus*
Modjokerto (Java, Indonesia) Pucangan formation	1.81	skull Perning 1	*H. erectus*
Ngebung (Java, Indonesia) Kabuh Formation	0.75–0.25	core–flake (Sangiran Flake Industry)	
Dmanisi (Georgia) hominid locality	1.9 (?)	mandible, 3 crania on underlying basalt ca. 2.0 core–flake	*H. erectus* [?]
Atapuerca (Spain) Gran Dolina (TD) 4 Gran Dolina (TD) 6	1.6–0.75 0.99–0.78	core–flake frontal, maxilla, mandible core–flake	*H. antecessor*
Orce (Spain) Fuentenueva 3	1.07–0.99?	temporal core–flake	*Homo sp. indet.*
Isernia la Pineta (Italy) Sector I Sector II	0.99–0.78 0.8–0.5	core–flake	
Ceprano (Italy)	< 1.0	cranium	*H. erectus*

Table 6.3b (continued)

Sites	Age (myrs)	Hominid fossils and stone-tool technology	Taxa
Bodo (Ethiopia) Upper Bodo sand unit	0.6	skull biface (Acheulian)	*H. heidelbergensis*
Elandsfontein (South Africa) Saldanha Bay	0.7–0.4	skull	*H. heidelbergensis*
Mauer (Germany)	0.5	mandible biface (Acheulian)	*H. heidelbergensis*
Boxgrove (England) Quarry 1	0.5	tibia biface (Acheulian)	*H. heidelbergensis*

* Classification based on actual phylogenetic models.

6.3.3.3 Morphological Patterns, Phylogenetic Relationships, and Paleoecology

Tobias' (1989) review of the morphology of *H. habilis s.l.* lists the following critical morphological features of the first description, which have been strengthened and supplemented by subsequent studies:

- absolute and estimated relative brains size (average 640 cc) with spectacular advance over australopithecines; exaggerated encephalization,
- brow ridges vertically thin,
- relatively open-angled external sagittal curvature to occipital,
- thin-walled braincase,
- light pneumatization of cranial bones,
- face moderately prognathous, but less marked than in *A. africanus*,
- retreating chin, with a slight or absent mental trigone,
- foramen magnum slightly in front of the basis cranii,
- large canines in comparison with australopithecines and *H. erectus*,
- canines large compared with premolars,
- petrous pyramid of the temporal bone lying in nearly transverse and coronal plane,
- cheek-teeth with reduced crown diameters and crown area in comparison to those of australopithecines,
- molar crowns small buccolingually and elongated mesiodistally,
- third molars tending to be smaller than second molars,
- especially P3, P4, M1 showing buccolingual narrowing of the crowns,
- lateral aspect of the frontal lobe exhibiting a pattern of sulci, typical of *Homo sapiens*,
- well developed bulges in Broca's area and in the inferior parietal lobule (part of Wernicke's area),
- complex middle meningeal vascular pattern.

The postcranium exhibits a very controversial morphological pattern: on the one hand there are distinct similarities with *H. sapiens* (e.g., clavicle, broad terminal phalanges, capitate metacarpophalangeal articulations, stout and adduced big toe, well marked foot arches) and on the other hand distinguished differences (e.g., scaphoid, trapezium, trochlea surface of the talus, robust metatarsal III). The partial skeleton OH 62, a *H. habilis s. str.*, which was found by Johanson et al. (1987), has especially caused much discussion (Hartwig-Scherer and Martin 1991), because the estimated length and robustness of the humerus and forearm bones of OH 62 suggest that its proportions are remarkably ape-like, and the predicted weight/ stature relationships are also more ape-like (Aiello 1990, 1996), but the phylogenetic status of 'Lucy's child' remains uncertain.

Although Tobias (1988) gave an extremely detailed description of *H. habilis s.l.*, listing 334 cranial, mandibular, and dental features, the question of what autapomorphic features define *H. habilis* is still controversial (Bilsborough 1986, Chamberlain and Wood 1987, Henke and Rothe 1998, Stringer 1986, 1994, Walker and Leakey 1978, Wood 1985, 1987, 1991, 1992). The main reason for the uncertainty in the interpretation of the diversity of early *Homo* is that the fossil hominin remains that are formally or informally allocated to *H. habilis* or declared to have affinities with this species vary from one author to another; in other words, there are multiple taxon solutions. The claimed heterogeneity of the *H. habilis s.l.* material from Olduvai (e.g., OH 7, 13, 16, 24), Koobi Fora (e.g., KNM-ER 1470, 1590, 1805, 1813, 3732), Omo (L 894-1, Omo 75-14, Omo 222-2744), and Chemeron (KNM-BC1), as well as from Sterkfontein (e.g., Stw 53) caused different approaches to find a better-supported classification, but the split into two species is taxonomically ambiguous (Wood 1996, Henke and Rothe 1994, 1998).

Cladistics have made little contribution to the search for distinctive features, or autapomorphies, of *H. habilis*. For example, Delson et al. (1977, p. 273) stated "we have been able to identify no autapomorphic features of *Homo habilis*". Chamberlain and Wood (1987) concluded that when *H. habilis s. str.* and *H. rudolfensis* are separately included in a cladistic analysis, they are linked as sister taxa within a clade defined by the feature states of elongated anterior basicranium, higher cranial vault, mesiodistally elongated M_1 and M_2, and narrow mandibular fossa. The most complex cladistic analysis of early hominin relationship was conducted by Strait et al. (1997). Eight different approaches agreed in indicating that the robust australopithecines form a clade, that *A. afarensis* is the sister taxon of all hominins, and that the genus *Australopithecus*, conventionally defined, is paraphyletic. Concerning *H. habilis*, the relationships of *A. africanus* and *H. habilis* were unstable in the sense that their positions varied in trees that were marginally less parsimonious than the favored one. One solution is shown in Figure 6.29.

The paleoecological scenario that explains the observed phylogenetic pattern states that it is possible that bipedalism (and hence, the earliest hominins) evolved in response to changing ecological conditions in Africa during the late Miocene and early Pliocene. Vrba's (1988) faunal reconstructions indicate that hominin diversity between 2.5 and 1.5 myr BP was possibly associated with environmental desiccation. After 2.5 myr, hominin diversity is represented primarily by two distinct lineages,

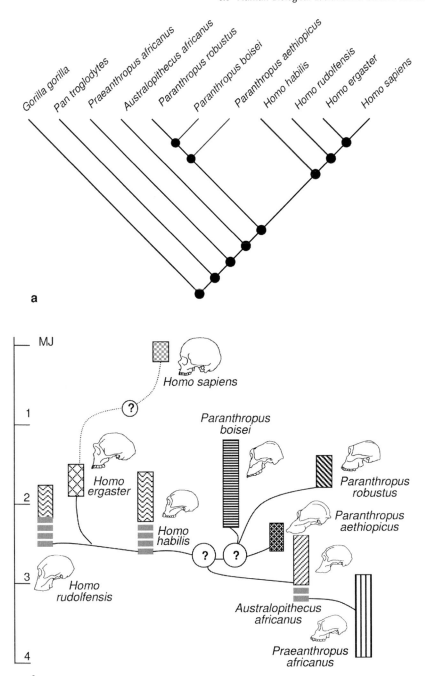

Figure 6.29 (a) Cladogram after Strait et al. (1997) demonstrating the most parsimonious solution of the relationships of the early hominins; (b) phylogenetic model after Strait et al. (1997).

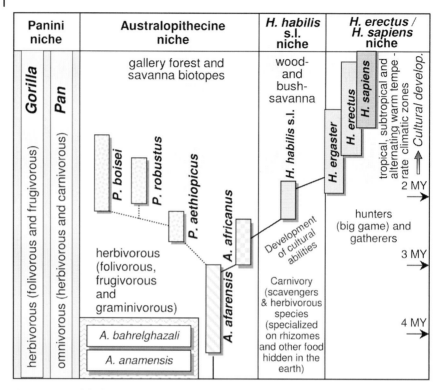

Figure 6.30 Model of niche separation (from Henke and Rothe 1998).

Paranthropus and *Homo,* which reacted to such desiccation by following different evolutionary trajectories (i.e., hypermastication vs. hypercephalization) (Strait et al. 1997, p. 56). The morphological changes demonstrate that the masticatory apparatus increased moderately in the early stages of human evolution. Subsequently it increased markedly in *Paranthropus* and decreased in *Homo,* a dichotomy that almost certainly represents a divergence in trophic adaptations, as can be seen from the adaptive niche model (Figure 6.30).

A cladistic analysis conducted by Wood and Collard (1999) showed that neither *H. habilis* nor *H. rudolfensis* can be assumed with any degree of reliability to be more closely related to *H. sapiens* than they are to species allocated to other genera.

6.3.4
Lower and Middle Pleistocene *Homo*

6.3.4.1 *Homo erectus* (incl. *Homo ergaster*)
At the end of the Basal Pleistocene (ca. 1.5 myr), *H. habilis* and *H. rudolfensis* disappear from the fossil record, followed somewhat later by *P. robutus* and *P. boisei.* The exact reason for their extinction is not known, but there are indications from high faunal turnovers that climatic fluctuations may have caused dramatic ecological

shifts. At the beginning of the Pleistocene, the first epoch of the Quaternary, which was characterized by a series of glacial and interglacial periods, a new hominin emerged, *Homo erectus*. There is a long-standing hypothesis that populations similar to this species were directly ancestral to the earliest members of the living species *H. sapiens*, whereas hypotheses concerning the link to hominin forerunners changed with the increase in the fossil record and are still under ongoing discussion (Franzen 1994a, Henke and Rothe 1995).

The first fossil finds of *H. erectus* were made in 1891 by Eugene Dubois in Central Java and were described as *Pithecanthropus*. The Dutch army doctor was convinced at that time that he had discovered the presumed 'missing link', the transitional form between apes and humans. The conviction that the new human form had been an erect bipedal creature resulted in the species name *erectus*. Because the Java man was the first non-European fossil in paleoanthropology, the discovery led to questioning of the European-centered world view, that had so far been supported by the famous Neanderthal fossils. From now on, Asia was expected to be the 'cradle of mankind' and became the center of the search for the earliest human fossils, until it became evident that hominins of a *H. erectus* grade existed in Europe and Africa too.

It became the favored phylogenetic hypothesis that *H. erectus* originated in Africa from an earlier species of the same genus, *H. habilis s.l.*, forming an intermediate position in the human family tree between the ancient forerunner and *H. sapiens*. This species was said to be the conqueror of the Old World, the first hominin to emigrate out of Africa and successively occupy Asia and Europe. But its evolutionarily intermediate position in the human family tree has been questioned more and more recently (e.g., Franzen 1994a, Henke and Rothe 1995, Rightmire 1990, Tattersall 1997, Tattersall and Schwartz 2000). Definitions of *H. erectus* still rest on the Far Eastern fossils from Zhoukoudian and Java (Jacob 1976, Howells 1980, Weidenreich 1943). The species *H. erectus* came to replace a variety of contemporaneous geographically distinguished genera, including '*Pithecanthropus*', '*Sinanthropus*', '*Meganthropus*', and '*Atlanthropus*'. A taxonomic revision by Campbell (1965) deleted older genera and species, lumping these Lower and Middle Pleistocene hominin taxa into a single species and separating them only on the subspecies level (*H. erectus erectus*; *H. e. modjokertensis*, *H. e. pekinensis*, *H. e. capensis*, *H. e. leakeyi*, *H. e. mauritanicus*, *H. e. heidelbergensis*, *H. e. ngandongensis*, *H. e. yuanmouensis*, *H. e. bilzingslebenensis*, *H. e. tautavelensis*, and others). But this taxonomic approach was obviously not the complete solution to all problems, because the more fossils were found and the more precise our datings became, the more complex the whole situation has become.

Franzen (1994b, p. 9) asked among others the following burning questions: "What really is *Homo erectus*? Is it a good species? Is it a phantom, a chimera behind which two or even more 'true' species may be hidden? ... How can *Homo erectus* be defined, particularly if it is not the result of a speciation event but just a transitional phase of phylogenetic development on the way to modern man? ... How can it be separated from 'archaic' *Homo sapiens*? ... And then again, is *Homo erectus* the result of a cladogenetic event or is it the result of continuous transition? Is it possible to distinguish between an Asian, and African and/or European branch of *Homo erectus*? Should the African and European branches, if they really exist, be called species of their own?"

The answers are as diverse as the questions, as is not unexpected in paleoanthropology, especially in regard to weaknesses in principal taxonomic approaches. Although some anthropologists regard *H. erectus* as a grade within a transitional phylogenetic model (e.g., Frayer 1992, Frayer et al. 1993, Thorne and Wolpoff 1981, 1992, Wolpoff 1984, 1989, 1992, 1996, 1996–97, 1999, Wolpoff and Caspari 1997a, b, Wolpoff et al. 1994), other authorities hold the view that *H. erectus* originated from a hominin branch of African origin – possibly *H. ergaster* – in Asia and remained restricted to the Far East (e.g., Tattersall 1995a, b, 1996, 1997, 2000, Tattersall and Schwartz 2000). Some experts claim that the Asian and African wings of *H. erectus* exhibit no autapomorphic features and that the early Europeans originated form an African species other than *H. erectus*, named *Homo heidelbergensis* (Rightmire 1990, 1992, 1998).

The recent discussion on the question of whether *H. erectus* was an ancestor of our own species or only an evolutionary side branch received new stimuli from fossils from Dmanisi (Georgia). From this site at the gates of Europe, Antje Justus, a coworker of Gerhard Bosinski (RGZM, Mainz, Germany) and Georgian staff unearthed in 1991 a fossil mandible and in 1999 two skulls whose unexpected high age (max. 1.9 myr; most reliable date 1.75 myr BP) and morphology indicate a much earlier Eurasian dispersal of *Homo* than was believed before (Bräuer et al. 1995, Henke et al. 1995, 1999a, Henke and Rothe 1999b, Gabunia et al. 2001). Further cranial and postcranial material led to the conclusion that hominins like those from Dmanisi may have played an important role in the peopling of Eastern Asia (Gabunia et al. 2001)

Because early *Homo* fossil material from Africa that has been classified as *H. ergaster* (Groves and Mazak 1975; formerly attributed as 'African' *H. erectus*) has a maximum age of only ca. 1.9 myr BP, and the hominin fossils from Dmanisi (Schmincke and van den Bogaard 1995, Gabunia et al. 2001), Sangiran and Modjokerto (Java, Swisher III 1994), and Longgupo, China (Huang et al. 1995, possibly pongid) may be around the same age or a little younger, there is a severe problem of explaining the very early Eurasian dispersal of finding out which species was the pioneering emigrant. The Dmanisi skulls, with their small cranial capacities and plesiomorphic features (e.g., postorbital constriction), will be key fossils in the current discussion.

6.3.4.2 Geographical and Chronological Distribution

For a better understanding of the complex phylogenetic pattern and deciphering of the 'muddle in the middle', as Rightmire (1998) characterized the problems concerning Late Lower and Middle Pleistocene human evolution, we have to analyze the regional and chronological hypodigm and discuss possible African origin and early Asian dispersal (Henke and Rothe 1994, 1999a, 1999b; Larick and Ciochon 1996; Wolpoff 1996–97). Only when species – in the sense of an evolutionary species (Wiley 1978) – have been adequately defined morphologically can appropriate comparisons be made and the distribution of character states across species be used to generate phylogenetic hypotheses. But until now, we do not have any consensus of the *H. erectus* hypodigm, which means that we have no agreement on the question of which fossils belong to the taxon that has been defined as *Homo erectus* (Franzen, 1994a, b, Henke and Rothe 1995, Howell 1996, Howells 1980, 1992). For this reason, those *Homo* fossils are listed in Table 6.3 which indicate the earliest appearance of *H. erectus* (or *H. ergaster*) or their possible forerunners *H. rudolfensis* and *H. habilis*, as well as presumable successors like *H. antecessor* or *H. heidelbergensis*, and *H. sapiens*. There is further information on the first wave of elementary tool technology in which association with early *Homo* is suspected but of course not proven. Information concerning the earliest traces of the genus *Homo* and the fossil sequence can be seen in Figure 6.31.

Because those hominins that were found in Java since 1891 and at Zhoukoudian, near Beijing., in the 1930s, now known as *H. erectus*, were clearly more archaic than fossils from Europe (e.g., Mauer) and Northwest Africa (e.g., Ternifine), it was initially stated that *Homo* emerged in East Asia and dispersed westward. Since around 1960, when specimens from different localities in the eastern Rift Valley and South Africa were assigned to 'early *Homo*' for the first time, the picture has changed. Especially have the hominin fossils around Lake Turkana, classified as *H. erectus* (now *H. ergaster*), proved to be the oldest ones and the more primitive forms too. Consequently, the now-preferred dispersal hypothesis sees the 'African *H. erectus*' as that species which immigrated to Asia and subsequently to Europe. Concerning the time span there was much debate, but no reliable hypothesis guessed a higher age than 1 myr BP. Advances in dating methods and new finds from Georgia, China, and Indonesia indicate that early *Homo* may have arrived in East Asia by ca. 2 myr ago (Figures 6.32 and 6.33).

Alan Turner stated recently that the effects of tectonic and climatic changes on the Levantine route during the Plio–Pleistocene suggests that a late Pliocene dispersal should be given serious consideration, because the *Homo* dispersal can be seen as part of the pattern of dispersion by members of the terrestrial mammalian fauna (Hemmer 1999, Torre et al. 1992).

From a paleoecological view, those hypotheses should not be neglected that propose that *H. erectus* may have reached Europe from Far East Asia and not directly from Africa via the Levant. To prove such dispersal scenarios, faunistic information should be taken much more into account, because of the coevolution of hominin predators with carnivores (Felidae, Canidae). Further information can be gained from the dispersal pattern of the mammals that they scavenged or hunted (Henke et al. 1999, Torre et al. 1992, Turner 1999).

The regional fossil records of the Lower and Middle Pleistocene hominins from Africa, Asia, and Europe demonstrate broadly similar morphological trends. There is – in the opinion of the gradualists – no convincing evidence to support a Middle Pleistocene speciation event leading to a distinct *H. sapiens* – quite the contrary, the proponents of the so-called multiregional theory of hominin evolution (see below) point out that there is morphological continuity between *H. erectus* and *H. sapiens*. For example, the Ngandong skulls from Java, whose age may be no

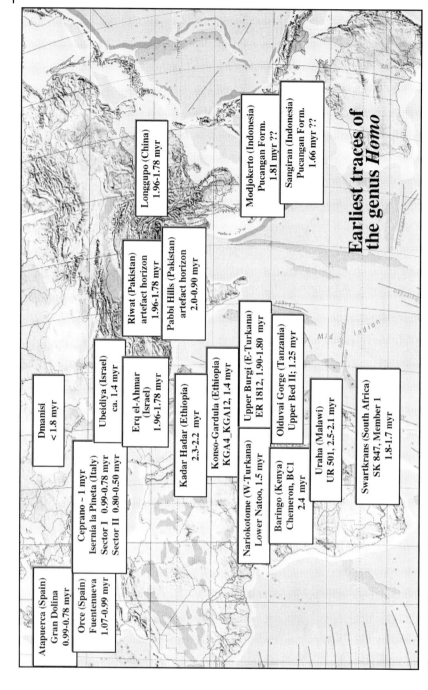

Figure 6.31 Earliest traces of the genus *Homo*.

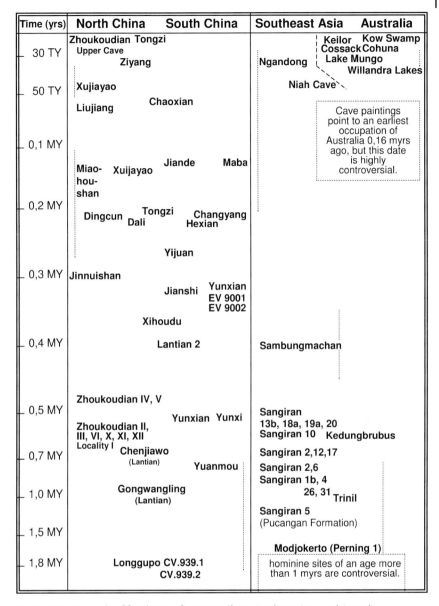

Time (yrs)	North China	South China	Southeast Asia	Australia
30 TY	Zhoukoudian Tongzi Upper Cave Ziyang		Ngandong	Keilor Kow Swamp Cossack Cohuna Lake Mungo Willandra Lakes
50 TY	Xujiayao	Chaoxian	Niah Cave	
	Liujiang			Cave paintings point to an earliest occupation of Australia 0,16 myrs ago, but this date is highly controversial.
0,1 MY		Jiande Maba		
	Miao-hou-shan Xuijayao			
0,2 MY	Dingcun Dali Tongzi Changyang Hexian			
	Yijuan			
0,3 MY	Jinnuishan	Jianshi Yunxian EV 9001 EV 9002		
	Xihoudu			
0,4 MY	Lantian 2		Sambungmachan	
0,5 MY	Zhoukoudian IV, V	Yunxian Yunxi	Sangiran 13b, 18a, 19a, 20 Sangiran 10 Kedungbrubus	
	Zhoukoudian II, III, VI, X, XI, XII Locality I Chenjiawo (Lantian)			
0,7 MY		Yuanmou	Sangiran 2,12,17 Sangiran 2,6 Sangiran 1b, 4	
1,0 MY	Gongwangling (Lantian)		26, 31 Trinil Sangiran 5 (Pucangan Formation)	
1,5 MY			Modjokerto (Perning 1)	
1,8 MY	Longgupo CV.939.1 CV.939.2		hominine sites of an age more than 1 myrs are controversial.	

Figure 6.32 Time scale of fossil sites of *Homo* in China, Southeast Asia, and Australia.

more than 34 000 years, have been described by some authors as *H. erectus* and by others as archaic *H. sapiens*. Due to the repeated occurrence of fossil specimens exhibiting a morphologically intermediate pattern between *H. erectus* and *H. sapiens*, which is obviously incompatible with a punctuational interpretation of human evolution, there is cause for much debate on stability and change in *H. erectus*.

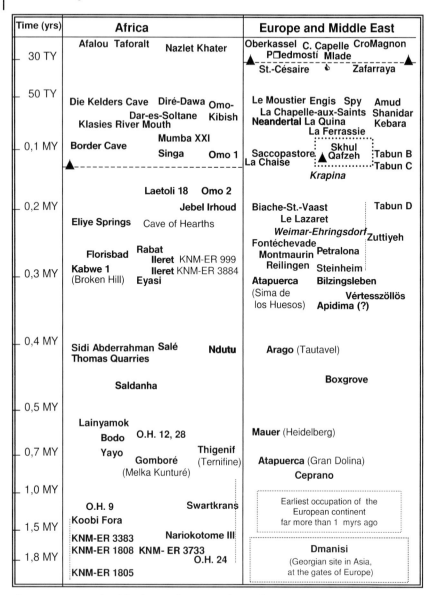

Time (yrs)	Africa	Europe and Middle East

Figure 6.33 Time scale of fossil sites of *Homo* in Africa and Europe.

6.3.4.3 Morphological Patterns, Phylogenetic Relationships, and Paleoecology

Those anthropologists who argue that a speciation event took place during the lower and middle Pleistocene within the genus *Homo* describe many morphological characteristics of *H. erectus* (incl. *H. ergaster*) (Figure 6.34), but those that are autapomorphic are rare and controversial (Andrews 1984, Bilsborough and Wood 1986, Bräuer and Mbua 1992, Henke and Rothe 1994, Howell 1986, Howells 1980,

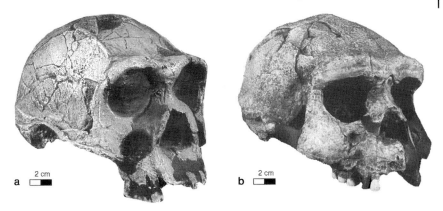

Figure 6.34 (a) *Homo ergaster* (KNM-ER 3733); (b) *Homo erectus* (Sangiran 17) (from Henke and Rothe 1998).

1993, Hublin 1986, Kennedy 1991, Rightmire 1990, 1992, 1998, Wolpoff 1996–97; Tattersall and Schwartz 2000). The splitting of the *Homo* hypodigm on the species level is, even if one does not take the extreme 'lumper' position of Wolpoff et al. (1994), explaining the variability by polymorphism and polytypism, which is methodologically highly suspect. As some 'splitters' confess, the decisions come much more from the stomach than the brain.

The following list describes some of the traits that set *H. erectus* apart from *H. habilis*/*H. rudolfensis* and *H. sapiens* as well (also Henke and Rothe 1994, 1998, 1999; Figure 6.35):

- cranial capacity ranging from 800 to 1225 cc,
- thick brow ridge (torus supraorbitalis), especially in the later forms,
- special neurocranial proportions: wide cranial base, vault walls relatively vertical in their lower portions; long, flat, low braincase,
- more arched than bell shaped contour of the braincase seen from the rear,
- occipital large and sharply angled,
- well marked nuchal plane bounded by a distinct nuchal ridge (occipital torus),
- temporal lines distinct and slightly raised, especially anteriorly,
- sagittal keeling and parasagittal depression in Asian skulls only,
- occipital ankles.

The separation of the African sample from the Asian is highly questioned. Wood (1984) for example describes the following autapomorphies of the Asian *H. erectus*:

- occipital torus with sulcus above,
- angular torus and mastoid crest,
- sulcus on frontal behind torus,
- proportions and shape of occipital bone,
- relatively large occipital arc.

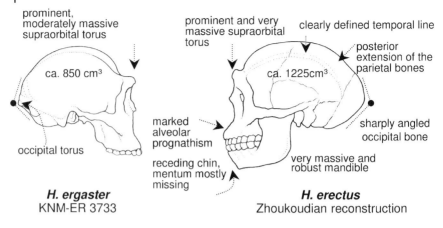

prominent,
moderately massive
supraorbital torus

ca. 850 cm³

occipital torus

H. ergaster
KNM-ER 3733

prominent and very
massive supraorbital
torus

clearly defined temporal line

posterior
extension of the
parietal bones

ca. 1225cm³

marked
alveolar
prognathism

sharply angled
occipital bone

receding chin,
mentum mostly
missing

very massive and
robust mandible

H. erectus
Zhoukoudian reconstruction

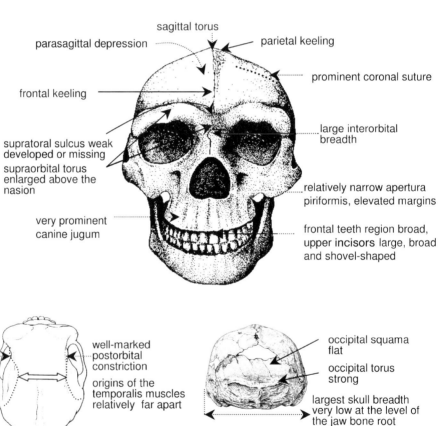

sagittal torus

parasagittal depression

parietal keeling

prominent coronal suture

frontal keeling

large interorbital
breadth

supratoral sulcus weak
developed or missing

supraorbital torus
enlarged above the
nasion

relatively narrow apertura
piriformis, elevated margins

very prominent
canine jugum

frontal teeth region broad,
upper **incisors** large, broad
and shovel-shaped

well-marked
postorbital
constriction

origins of the
temporalis muscles
relatively far apart

occipital squama
flat

occipital torus
strong

largest skull breadth
very low at the level of
the jaw bone root

Figure 6.35 Morphological features of the skull of *Homo ergaster* and *Homo erectus*,
demonstrated by the specimen KNM-ER 3733 and the reconstruction of the female
'*Sinanthropus*' from Weidenreich (1943), norma occipitalis of ZH XII (from Henke
and Rothe 1998).

He argues that *H. erectus* is an exclusively Asian taxon which possesses features not present in African specimens and not present in *H. sapiens* either. For this reason he sees better arguments for an African than for an Asian origin of *H. sapiens*.

Strong opposition comes from multiregionalists, because the supposed autapomorphic characters that underline the uniqueness of the Asian sample are not truly independent characters, because they are correlated within functional units of adaptation. No anthropologist denies that there are differences between the Asian and African *Homo* samples under discussion, but the point is whether these differences are sufficient to warrant taxonomic recognition at the species level (Bilsborough 1992).

That *H. erectus* was fully upright and bipedal is still expressed by the species name, but until the virtually complete skeleton KNM-WT 15 000 (Walker and Leakey 1993) was discovered, we knew very little about the *H. erectus* postcranium. The skeleton of the 12-year-old boy from Nariokotome, west Turkana, which has been dated to 1.6 myr BP, enables us to reconstruct stature, limb proportions, locomotion, maturation, and gestation. In adulthood the young boy, who measured 1.68 m, would have been 1.8 m tall and weighed ca. 47 kg. He was tall and thin, resembling present-day equatorial Africans. Rogers et al. (1996) discussed the behavioral implications of the archaeological and paleogeographical record and early *H. erectus* anatomy. Their summary is given in Figure 6.36.

The physiological changes (e.g., secondary altriciality, lower period of maturation, increase in need for food, increase in long-distance locomotor efficiency, and greater resistance to heat stress), combined with the implied behavioral changes (e.g., greater parental investment, larger home ranges) and the archaeological evidence for a changed behavioral ecology (e.g., lithic standardization by the reduction of single platform cores, use of large flakes for cores) suggests that the early *Homo* of the lower Pleistocene was less constrained than earlier hominins by the natural distribution of resources. This makes them an ideal candidate for emigrating pioneers.

For this reason and from the total morphological pattern, Wolpoff et al. (1994, p. 341) see "no distinct boundary between *H. erectus* and *H. sapiens* in time and space." They regard the lineage as a single evolutionary species, but other authorities describe different further speciations, stating that the emergence of *Homo* has not been a single linear transformation of one species into another (chronospecies), but rather a "meandering, multifaceted evolution" (Tattersall 2000).

A more moderate speciation model than that of Tattersall (2000, Figure 6.37) has been proposed by Rigthmire (1998, Figure 6.38), who describes a speciation process in Africa leading to *H. heidelbergensis*. This species is said to have given on the one hand rise to *H. neanderthalensis*, who split off in Europe, and on the other hand to *H. sapiens*, who emerged in Africa and successively occupied all other continents.

As has been expressed in the traditional views by subspecies names such as *H. erectus heidelbergensis*, *H. erectus bilzingslebenensis*, and *H. erectus tautavelensis*, it was widely believed that there was a European branch of *H. erectus*, but there was always much debate as to the validity of an European *H. erectus* taxon (Henke and Rothe 1996). Meanwhile, many paleoanthropologists believe that *H. erectus* never did reach Europe, but there are different causes for this change of view.

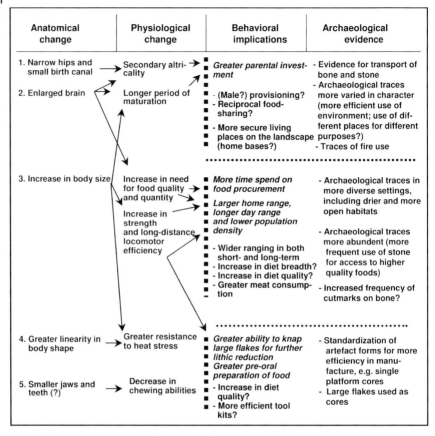

Anatomical change	Physiological change	Behavioral implications	Archaeological evidence
1. Narrow hips and small birth canal	Secondary altricality	■ *Greater parental investment*	- Evidence for transport of bone and stone
2. Enlarged brain	Longer period of maturation	■ - (Male?) provisioning? ■ - Reciprocal food- ■ sharing? ■ ■ - More secure living ■ places on the landscape ■ (home bases?)	- Archaeological traces more varied in character (more efficient use of environment; use of different places for different purposes?) - Traces of fire use
3. Increase in body size	Increase in need for food quality and quantity Increase in strength and long-distance locomotor efficiency	■ *More time spend on* ■ *food procurement* ■ ■ *Larger home range,* ■ *longer day range* ■ *and lower population* ■ *density* ■ ■ - Wider ranging in both ■ short- and long-term ■ - Increase in diet breadth? ■ - Increase in diet quality? ■ - Greater meat consump- ■ tion	- Archaeological traces in more diverse settings, including drier and more open habitats - Archaeological traces more abundant (more frequent use of stone for access to higher quality foods) - Increased frequency of cutmarks on bone?
4. Greater linearity in body shape 5. Smaller jaws and teeth (?)	Greater resistance to heat stress Decrease in chewing abilities	■ *Greater ability to knap* ■ *large flakes for further* ■ *lithic reduction* ■ *Greater pre-oral* ■ *preparation of food* ■ - Increase in diet ■ quality? ■ - More efficient tool ■ kits?	- Standardization of artefact forms for more efficiency in manufacture, e.g. single platform cores - Large flakes used as cores

Figure 6.36 Adaptation pattern of *H. erectus* (after Rogers et al. 1996; Henke and Rothe 1999b).

Although Wolpoff (1996–97) does not believe that the taxon *H. erectus* is valid in any event, others regard those European specimens that were formerly discussed as belonging to *H. erectus* as a species of their own, which originated in Africa around 0.6 myr ago. For taxonomic reasons it was named *H. heidelbergensis* and is – as proponents of this model claim – best described by a specimen from Broken Hill (Kabwe, Sambia) (Figure 6.39). Astonishingly enough, this skull was formerly attributed to 'archaic' *H. sapiens* and is an essential key fossil in the out-of-Africa hypothesis of Bräuer (1984).

Newly unearthed fossils from Bouri, Middle Awash (Ethiopia) afford unique insights into unresolved spatial and temporal relationships of *H. erectus*. The hominin calvaria and postcranial remains from the Dakanihylo Member of the Bouri Formation are ~1.0 myr old and are associated with abundant early Acheulean stone tools and a vertebrate fauna that indicates a predominantly savannah environment. Asfaw et al. (2002, p. 317) are convinced that the morphological attributes of the fossils "centre [them] firmly within *H. erectus*", and they see strong indications "that African *H. erectus* was the ancestor of *Homo sapiens*."

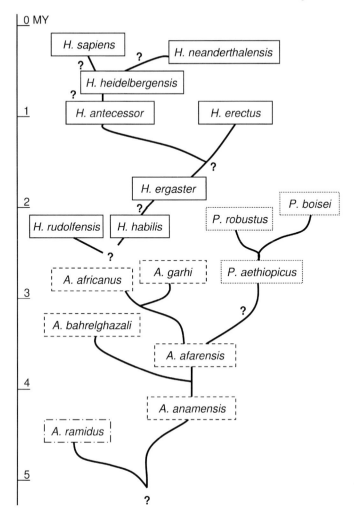

Figure 6.37 Speculative family tree showing the variety of hominin species that have populated the planet – some known only by a fragment of a skull or jaw. As the tree suggests, the emergence of *H. sapiens* has – as Tattersall states – not been a single linear transformation of one species into another but rather a meandering multifaceted evolution (after Tattersall 1999, redrawn).

The new fossils from Bouri shed light on newly described ca. 0.8 myr old fossils from the Gran Dolina of Atapuerca (Spain), which have been classified as *H. antecessor* (Arsuaga et al. 1999, Carbonell et al. 1999). The hypothesis that this taxon may have originated in Africa and given rise to *H. neanderthalensis*, which flourished between ca. 200 000 and 30 000 years ago while *H. sapiens* evolved as adelphotaxon of *H. heidelbergensis* in Africa, successively dispersing worldwide, is now even less reliable than before.

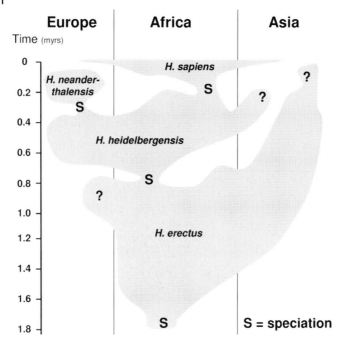

Figure 6.38 Speciation model proposed by Rightmire (1998, from Henke and Rothe 1999).

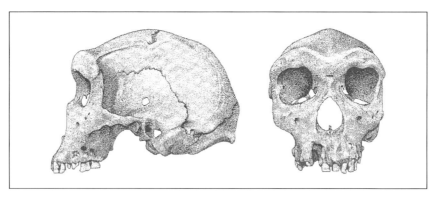

Figure 6.39 Specimen Kabwe 1 (formerly Broken Hill 1), which is classified by some authorities as *Homo heidelbergensis*.

6.3.5
Upper Pleistocene Human Evolution

In the Upper Pleistocene there is increased fossil evidence coupled with an expanded geographical range, paralleled by a more complex archaeological record. The increase in paleoanthropological and archaeological information within the Upper Pleistocene has a two-sided effect: on the one hand we have decidedly more evidence from the past, but on the other hand the complexity of the patterns is more and more difficult to unravel by our methodological skills. For this reason there is need to reflect on a compound of phylogenetic events to explain the phyletic changes and to interpret the regional diversity.

6.3.5.1 How Many Species?

As described above, there are highly differing scenarios of the origin of 'anatomically' modern human beings. (Tobias 1995, argues, as do Wolpoff and Caspari 1997b, that the term 'anatomically modern humans' should be discarded; we should speak of 'modern' or 'recent' humans, using a temporal not a morphological descriptor.) The irreconcilable standpoints have been hardened by totally different species concepts. The reproach of Tattersall (cited by Flanagan 1996, taken from Wolpoff 1999, p. 376) is that the multiregionalists are "linking everything from *H. erectus* to *Homo sapiens* into 'one big happy family' ... Paleontologists do not give other animals such a special treatment."

Wolpoff, the protagonist of the multiregionalists, comments by quoting the evolutionary geneticist A. Templeton: "We make far too much of our anatomical difference ... Biologists who study, say, fruit flies know that each population can look quite distinct ... and yet they are tempted to hastily split them into separate species. Why must we look at ourselves any differently?" (Wolpoff 1999, p. 376).

The crucial case of whether there was more than one species in the Upper Pleistocene comes from the long-standing unsolved debate on the Neanderthal problem: the fate of the Neanderthals is the longest lasting controversy in paleoanthropology and, in spite of contrary reports, especially comments on molecular biological results, there is currently no solution to the problem.

6.3.5.2 Geographical and Chronological Distribution

As can bee seen from Table 6.3 and Figures 6.40 to 6.42, there are a great amount of *Homo* fossils in Africa, Asia, and Europe from the late Middle and Upper Pleistocene whose morphology differs – in comparison to earlier specimens – only within a moderate range. There is a well described diachronic trend to less massive faces and larger neurocrania, approaching step by step the pattern of [anatomically] modern man in Africa. Because this gradual process is less convincingly verified in Asia and extremely controversial in Europe, so-called out-of-Africa hypotheses have been proposed (Stringer 1982, 1992, Bräuer 1984, 1992, Cann and Wilson 1992), which compete with the multiregional evolution model (Thorne and Wolpoff 1992, Wolpoff 1996–97, 1999, Wolpoff et al. 1984).

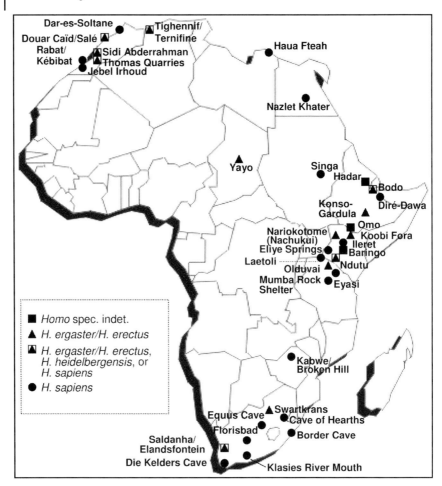

Figure 6.40 Fossil sites of hominins of the genus *Homo*
(classified as *Homo spec. indet., H. ergaster, H. erectus,* or *H. sapiens*)
(from Henke and Rothe 1998).

■ *H. spec. indet.*
◆ *H. erectus*
✪ *H. sapiens* (early form)
⊗ *H. sapiens sapiens*

Figure 6.41 Fossil sites of Pleistocene hominins in the Far East (from Henke and Rothe 1998).

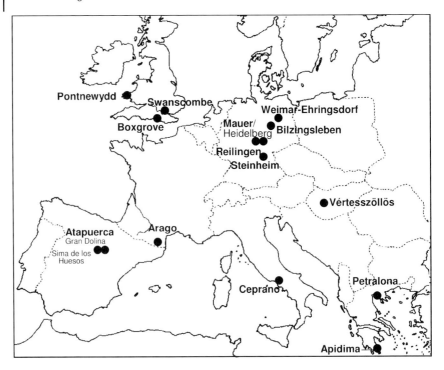

Figure 6.42 Fossil sites of early European hominins classified as *H. antecessor*, *H. erectus*, *H. heidelbergensis*, *H. neanderthalensis*, or *H. sapiens* (from Henke and Rothe 1998).

6.3.5.3 Multiregional Model versus Recent African Origin

Where did modern humans originate? This question divides paleoanthropologist. Two markedly contrasting theories on modern human origins have been intensely debated for around two decades (Bräuer and Smith 1992, Henke and Rothe 1994, 1999, Smith and Spencer 1984, Smith et al. 1989, Wolpoff 1996–97):

- The *multiregional evolution model* (MRE) traces all modern populations back to more than a million years ago when humans walked out of Africa for the first time. Since this phase there have been no speciations, but an interconnected web of ancient lineages existed in which the genetic contributions to all living peoples varied regionally and temporally (Thorne and Wolpoff 1981, 1992, Wolpoff 1989, 1992, 1996–97). The MRE has its historic base in the polycentric evolution hypothesis of Franz Weidenreich, proposing "that the conditions associated with the initial migrations of humans from Africa ... created the central and peripheral contrasts that affected the early establishment of regional features at the peripheries of the human range" (Wolpoff 1992, p. 26; Figure 6.43).

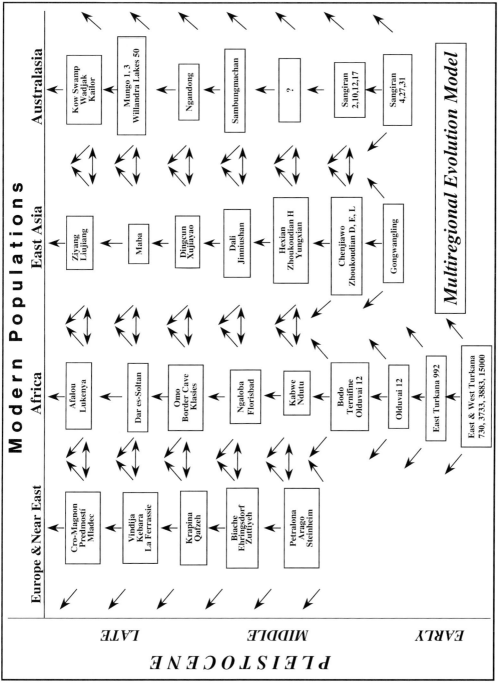

Figure 6.43 Multiregional evolution, illustrated as 'pattern' by David Frayer (the diagram does not show 'details', for further comments see Wolpoff 1999).

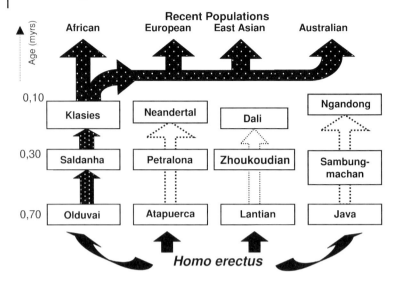

Figure 6.44 The out-of-Africa model argues that modern regional differences arose recently in populations that originated in Africa; existing populations in Europe, Asia, and Australia were replaced by these incoming populations (after Stringer 1990, from Henke and Rothe 1999a).

- The *recent African origin* (RAO) model suggests that modern humans descend from a very recent modern population that lived in Africa only around 200 000 years ago and replaced early humans elsewhere. This hypothesis has been called 'Garden of Eden model' by its opponents because it pleads for a single regional human origin. The main elements of this theory are derived from the chronological and morphological comparison of the African and non-African fossil specimens. The presence of regional continuity only in Africa and the first appearance of modern human features in this continent were taken as strong arguments for different replacement models (Protsch 1975; Afro–European *sapiens* hypothesis of Bräuer 1984). Although Bräuer's hypothesis supported the idea of a replacement *with* hybridization, other authorities spoke for a total replacement, for paleoanthropological (Stringer 1986, Stringer and Andrews 1988) or paleogenetic reasons (mtDNA research; Stoneking and Cann 1989, Wilson and Cann 1992). Especially have the genetic studies, which revealed that an African woman from ca. 200 000–150 000 years ago – called Eve or 'Lucky Mother' – was our common ancestor, gained much credit in public discussion and in the popular summary 'we are all Africans' (Figure 6.44).

In spite of contrasting assertions by the proponents of the different models, it can be concluded that none of them is unequivocally supported by the available data, although the out-of-Africa model gained tremendous support by the aDNA results from Neanderthal fossils. The voluminous literature on the topic of modern human origins allows us only to point to some of the pros and cons of the different models.

If we understand paleoanthropology as a hypothesis-testing natural science, we have to formulate predictions and to verify or falsify them. Slightly contrasting predictions of the out-of-Africa model have been formulated, e.g., by Stringer (1988), Bräuer (1984, 1992), Wolpoff (1984, 1992), or Smith et al. (1989, 1992). Some essential predictions that should be met in the RAO are:

- Transitional fossils that link archaic and modern humans should be found only in Africa, since the only transition to modern humans occurred there.
- Regional clade features should be of low antiquity except in Africa; presumably they should not predate the appearance of earliest modern humans anywhere in Eurasia.
- Modern humans should have a greater antiquity in Africa than anywhere else in the world (Smith 1992, p. 147).

There are unbridgeable controversies between the proponents of the RAO and the MRE models concerning the answers to the above predictions, but in spite of still-unsolved discrepancies there is consensus, e.g., in the underlying assumption of an ultimately African ancestor for all populations and the widespread movements of genes throughout the range of Pleistocene human populations, but within an extremely divergent chronological pattern ranging from the early-middle Pleistocene to the late Pleistocene.

The RAO (Stringer and Andrews 1988) model most strongly implies that modern humans did not assimilate or hybridize with the archaic humans. For this reason there should have been a total replacement of the Eurasian Neanderthals or archaic East Asian populations represented, e.g., by specimens from Dali, Maba, or Ngandong. In addition to those proponents who strictly toed the line of the RAO, arguing that all Eurasian archaic humans became extinct without making any contribution to modern human gene pools, there are other anthropologists, like Bräuer (1984, 1992), who plead for a 'hybridization and replacement model'. If there has been gene flow between regional archaic populations (e.g., Neanderthals) and modern *H. sapiens*, what is the acceptable position of this compromise model in terms of population genetics and taxonomy? The essential polarity between the RAO and the MRE is formed by a monocentric view on the one side and a polycentric view on the other side. The acceptance of hybridization cancels the principal position of the RAO. But does this simultaneously mean that a multiregional transition is accepted? The RAO with hybridization sees the movement of modern populations as a gradual process but does not view modern humans as the result of a biological speciation event. If the expanding modern populations assimilated some archaic genes into their gene pools, it recognizes that at least some degree of local continuity between archaic and modern humans existed. What makes this hypothesis a RAO model then? In Bräuer's view, the most dominant factor responsible for the emergence of modern humans in Asia and Europe remains the migration of African-derived populations into these parts of the Old World. The foremost factor is, for this reason, not regional continuity as in the MRE; instead, the RAO with hybridization regards examples of diachronic change other than in Africa as parallelisms or misinterpretations of the fossils. The defenders of the MRE (Wolpoff 1990, 1992, 1997–98, Thorne and Wolpoff 1992) obviously do not accept this explanation, but see plausible explanations within their model, which states:

- that no single definition of modern humans will apply in all regions,
- that modern human anatomical form does not necessarily have to appear earliest in Africa,
- that the earliest modern people in Eurasia will lack distinctly African regional features, since they are not the result of migrating Africans,
- that nonadaptive or regional clade features will be identifiable earliest in the most peripheral regions of the human range, before the appearance of modern humans, and latest in the central region (also Smith et al. 1989, p. 39).

The debate has received a new dynamic by the results of evolutionary human genetic research, which is outlined below. Before discussing the mtDNA analyses of recent samples (Cann and Stoneking 1989, Wilson and Cann 1992) and explaining the aDNA results from the classical Neanderthal specimens from Felderhofer Grotte (Krings et al. 1997), Mezmaiskaya Cave (Ovchinnikov et al. 2000), and other fossils, a view of the regional morphological patterns and the implications for the acceptance of a single origin model or a multiregional model is given.

6.3.6
Morphological Patterns and Paleoecology

Before the Afro–European *H. sapiens* hypothesis was advanced (Bräuer 1984), there were single statements on the possibility that [anatomically] modern humans were present earlier in Africa than in other continents (Stringer 1974, Protsch 1975). Upper Pleistocene hominins that served as key fossils came, e.g., from Border Cave, Klasies' River Mouth (Rep. South Africa), or Omo (Ethiopia). The cranium Omo I is very modern looking. It has a rounded vault, lacks a supraorbital torus, and exhibits an expanded parietal region, a canine fossa, and a prominent chin (Figure 6.45). Close morphological relationships have been described for Omo I with the so-called Cro-Magnon man. The affinities of these Upper Paleolithic people from Europe to the broadly [anatomically] modern populations from the Near East (Skhul, Qafzeh; Figure 6.46), which have been described as Proto-Cro-Magnons but are still linked with a Levallois–Mousterian culture, advanced the idea of a recent successive immigration of modern *Homo sapiens* from Southern and Eastern Africa via the Near East to Eastern Asia and Europe.

Although the chronological pattern of the paleontological and archaeological record has become more and more accurate as a result of the application of relative and absolute dating techniques, it became evident that the Neanderthals and broadly [anatomically] modern humans apparently coexisted in the Middle East for as long as 60 000 years (Figure 6.47). The strong affinities between the African modern *H. sapiens* and the Skhul–Qafzeh population on the one side, and the increasing acceptance of the idea that the nearly 'anatomically' modern populations from the

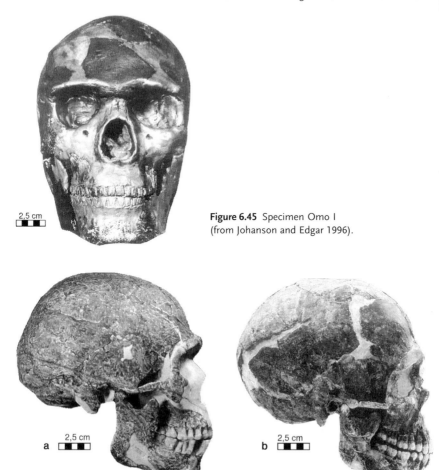

Figure 6.45 Specimen Omo I (from Johanson and Edgar 1996).

2,5 cm

a 2,5 cm

b 2,5 cm

Figure 6.46 'Proto-Cro-Magnons'. (a) Skhul v, (b) Qatzeh 9 (from Henke and Rothe 1998).

Near East are cranially and postcranially the most probable forerunners of Upper Paleolithic humans in Europe on the other, gave strong support to the single African origin model.

Arguments for a replacement were seen from the change of the Neanderthals to modern *H. sapiens* in Europe, which has been described as abrupt and obviously accompanied by a clear-cut cultural change (Henke 1990, 1992). New data demonstrate that the chronological overlap lasted much longer than formerly thought. Further, the Chatelperronian, which is regarded as the earliest Upper Paleolithic culture, was associated with Neanderthals, as proven in St. Césaire (France). In spite of these discrepancies within the fossil and cultural record, there are strong indications that the European modern *H. sapiens* differed from the Neanderthals in many ways. The Upper Paleolithics obviously had a great selective advantage

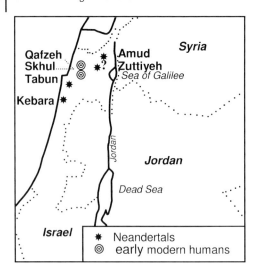

Figure 6.47 Fossil sites of Neanderthals and early modern humans in the Middle East (from Henke and Rothe 1998).

from their biology, which allowed them to become the sole pattern of human behavior after a relatively short period of evolutionary time (Henke 1989, Trinkaus and Shipman 1993, Bar-Yosef and Vandermeersch 1993).

Morphological and cultural evidence for a transition in Eastern and Southeast Asia is not very convincing, as the proponents of the replacement model claim. Although the fossil record is sparse, the advocates of the MRE see, e.g., in Indonesia, a continued regionality of the combination of Australasian skeletal features. The fact that the combination of features that distinguishes modern Australoids from other living human populations is – as Wolpoff (1996–1997, p. 566) mentions – "precisely that which distinguished their regional predecessors from their con-temporaries in East Asia, Africa and Europe."

Transitional sequences are also seen in the fossil record from China (Maba, Dali, Zhoukoudian). The confusing classification of the sparse Far Eastern fossil material as *H. erectus*, or alternatively within the very problematical term 'archaic' *H. sapiens* by the single origin proponents, demonstrates the difficulties to defining clear-cut species differences (Figure 6.41).

Taxonomical uncertainties, which are also evident in the alternative attribution of fossils from Africa to *H. erectus, H. rhodesiensis, H. heidelbergensis,* or archaic *H. sapiens* (e.g., Broken Hill (Kabwe, Sambia), Saldanha (Rep. of South Africa) or Omo 2 (Ethiopia) are strong indication for gradualism, because they can be explained within the model without problem.

Actually, there is little hope that the archaeological record will help to unravel the pattern of modern human origins (Klein 1989, 1992, Lewin 1998; Figure 6.48), though there are more and more indications for an African origin of modern cultures (McBrearty and Brooks 2000). The fact that modern *H. sapiens* preceded Neander-thals on Mount Carmel and followed a similar pattern of life for 60 000 years can be taken as an indication that biology alone cannot explain the cultural revolution that ensued (Bar-Yosef and Vandermeersch 1993), but comparison of the paleoecological

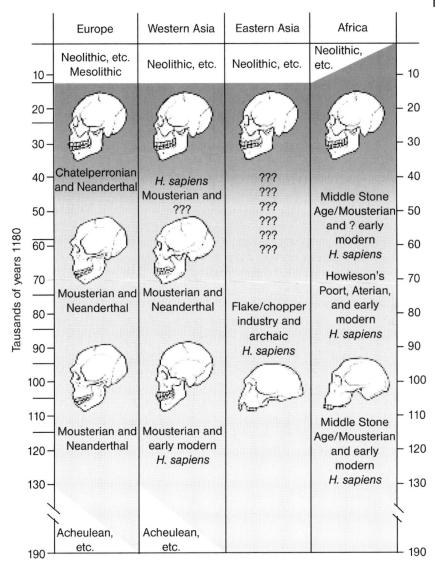

	Europe	Western Asia	Eastern Asia	Africa
10	Neolithic, etc. Mesolithic	Neolithic, etc.	Neolithic, etc.	Neolithic, etc.
40	Chatelperronian and Neanderthal	H. sapiens Mousterian and ???	??? ??? ???	Middle Stone Age/Mousterian and ? early modern H. sapiens
50			??? ??? ???	
70	Mousterian and Neanderthal	Mousterian and Neanderthal	Flake/chopper industry and archaic H. sapiens	Howieson's Poort, Aterian, and early modern H. sapiens
110	Mousterian and Neanderthal	Mousterian and early modern H. sapiens		Middle Stone Age/Mousterian and early modern H. sapiens
190	Acheulean, etc.	Acheulean, etc.		

Tausands of years 1180 (vertical axis)

Figure 6.48 Approximate chronological arrangement of the major cultural units and fossil human types since the penultimate glaciation (adapted from Klein 1992).

records of modern *Homo sapiens* (Upper Paleolithic humans) and Neanderthals gives good indications that there must have been essential differences in their behavior and skills, especially concerning how modern humans planned ahead. A prehistory of mind, which analyses the evolution of human cognitive abilities, has been reconstructed from the archaeological evidence, but a conclusive model that explains the very late and extremely impressive development of human capabilities in the Upper Paleolithic is still under discussion (Mithen 1996, Noble and Davidson 1996).

6.3.7
The Role of the Neanderthals

Since the discovery of fossils of a premodern human in the Neander Valley (near Düsseldorf, Germany) in 1856, the Neanderthals have provoked controversy. No early human relative has triggered more scientific debate or more inspired our fantasy than Neanderthal Man. They existed 200 000 years (or even much longer) ago, surviving under severe Ice Age glaciations until they dropped out of sight around 30 000 years ago, or as evidenced by new finds from Zafarraya (Spain) even down to 27 000 years (Hublin 1998).

Who were the Neanderthals? How did they live? What was their fate? (Henke et al. 1996, Henke and Rothe 1998, Tattersall 1995, Trinkaus and Shipman 1993, Stringer and Gamble 1993, Trinkaus 1989, Vandermeersch 1990, Wolpoff and Caspari 1997a).

There are many divergent answers to these essential questions: here, contradictory pairs, such as continuity versus discontinuity, gradualism versus replacement, regional evolution versus migration, characterize the diverging interpretations of the various authors regarding what happened toward the transition from the Middle to the Upper Paleolithic.

As can be seen from the huge number of popular publications as well, it has never been easy to be objective about Neanderthals, the most fascinating fossil human. As their craniofacial and postcranial morphology is undoubtedly different from that of modern humans, paleoanthropologists have asked about the adaptive and taxonomical relevance of their special features (Churchill 1998, Henke 1988, 1990, Henke and Rothe 1994, 1998, 1999, Henke et al. 1996, Trinkaus and Howells 1979). The first extraction of mtDNA from a Neanderthal-type specimen (Krings et al. 1997) initiated a totally new discussion focusing on the question of whether the Neanderthals can be definitely excluded from our direct ancestorship (Henke and Rothe 1999a, Lahr and Foley 1999, Relethford 1998, 2001).

6.3.7.1 **Geographical and Chronological Distribution**
Human fossils attributed to Neanderthals have been found from the Atlantic seaboard eastward across Europe and western Asia as far as Teshik Tash (Uzbekistan), and southward through the Crimea into the Middle East to Amud and Tabun (Israel) and Shandiar (Iraq) (Figure 6.49). Those specimens that were formerly discussed as North African or Far Asian Neanderthal remains have been excluded from the hypodigm. The inclusion of fossils that are older than 250 000 years – sometimes called Ante-Neanderthals (actually described by those paleoanthropologists who see different waves of immigration as *H. antecessor* and/or *H. heidelbergensis*) – is controversial, especially depending on the different taxonomical approaches. Within the Upper Pleistocene sample, a sequence of Pre-Neanderthals, early Neanderthals, and classical Neanderthals has been described. Furthermore, the European Neanderthals have been separated from the Middle (or Near) East Neanderthals. Within these, different subsamples of the western European Neanderthals best exhibit the pattern that has been described as classical Neanderthal.

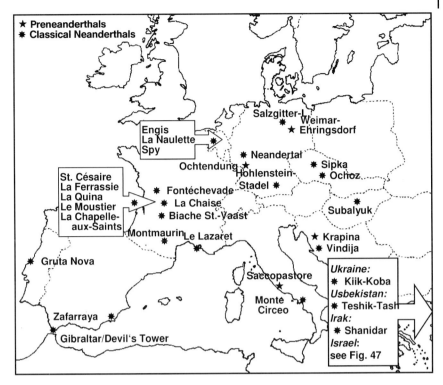

★ Preneanderthals
✻ Classical Neanderthals

Salzgitter-L✻
Weimar-
★ Ehringsdorf

Engis
La Naulette ✻
Spy

✻ Neandertal
Ochtendung ★
Hohlenstein-
Stadel ✻

✻ Sipka
✻ Ochoz

St. Césaire
La Ferrassie
La Quina
Le Moustier
La Chapelle-
aux-Saints

✻ Fontéchevade
✻ La Chaise
✻ Biache St.-Vaast

Subalyuk ✻

Montmaurin Le Lazaret
✻

Saccopastore

★ Krapina
✻ Vindija

✻ Gruta Nova

Ukraine:
✻ Kiik-Koba
Usbekistan:
✻ Teshik-Tash
Irak:
✻ Shanidar
Israel:
see Fig. 47

Monte
Circeo ✻

Zafarraya ✻

Gibraltar/Devil's Tower

Figure 6.49 Fossil sites of important pre-Neanderthals, as well as late or classical Neanderthals (after Henke and Rothe 1998).

6.3.7.2 Characteristic 'Neanderthal' Morphology

Craniofacial and postcranial features of the Neanderthal specimens that serve to identify them as a distinctive variant of ancient humans are described, e.g., by Vandermeersch 1985, Aiello and Dean 1980, Stringer and Gamble 1993, Henke and Rothe 1999a, Zollikofer et al. 1998 (Table 6.4).

Besides the above-mentioned features (Figure 6.50), very special morphological traits have been analyzed, e.g., by Hublin (1998; ear ossicles) or Schwartz and Tattersall (1999; internal nasal structure), which have been interpreted as indicating a long-lasting isolation of the Neanderthal population and a proof that Neanderthals were a species of their own: *Homo neanderthalensis*.

Adaptive significance of the 'Neanderthal' morphology

In morphological comparisons, the skeleton of the Neanderthals versus that of modern humans is surprisingly robustly built. The distinct muscle markings can be seen as an indication of a stronger development of the total muscle apparatus. Enlargements of muscle attachment areas are seen as an optimization of muscle forces. There is good evidence for adaptations toward an increased load-bearing capability in the lower extremities, which can be deduced from the characteristic robustness of the extremely strong bones such as the femur, as well as the tibia and

Table 6.4 Craniofacial and postcranial features of the Neanderthals (see Henke and Rothe 1994, 1998).

• High cranial capacity	• Subspherical vault from rear view
• Dolichocephalic, low skulls	• Occipital bunning ('chignon')
• Large face, both long and wide	• *Torus occipitalis* is always subdivided in the
• Large, double-arched supraorbital ridges	sagittal region into two lips limiting a
• High rounded orbits	supra-iniac fossa with a transverse
• High, wide and voluminous nose relative	elongation
to face	• External auditory meatus lying in the level
• Horizontally oriented distal nasal bones	of the zygomatic process
• The sub-orbital region comprises an	• Large and wide rib cage
extended maxilla and a flat malar, oblique	• Hand with strong grip and wide fingertips
backwards and outwards	• Long clavicle
• Inflated cheek bones are reduced and	• Large shoulder and elbow joints
retreating; no *fossa canina*	• Large hip joint, rotated outwards
• Expanded maxillary sinuses	• Long pubic bone
• Week chin	• Limb proportions stout
• Retromolar space	• Rounded, curved and thick-walled femur
• Large frontal teeth; shovel-shaped	shaft
incisors	• Short, flattened and thick-walled patella
• Taurodontism (large-sized pulp cavities)	• Wide and strong toe bones

even the patella. All three bones are equipped with proportionately more robustness and with rather massive articular ends and facets. Their wall thickness is strengthened, and the shape of the bones actually increases their resistance to fractures. Increased skeletal strength and leverage may be expected to positively co-occur in populations with an evolutionary history of high activity (Churchill 1998).

The accumulated evidence and results of the large number of studies performed with Neanderthal skeletons and bones point to many adaptations as being essential biological adaptations for a stronger physique and more intensive activities in a much wilder and more-unforgiving environment than we are used to.

In a study of traumatic injury patterns in Neanderthals, Trinkaus (1995) found evidence that they resembled the pattern of lesions found among modern rodeo performers, with a high frequency of head and neck injuries. He concluded that Neanderthals were more involved in close struggles with big game, whereas modern humans from the Upper Paleolithic used a less dangerous long-distance modern style of hunting (Frayer 1984, Henke 1990, Henke and Rothe 1999a).

Judging by the average height of modern Europeans during earlier times, the Neanderthal stature does not seem to justify the conclusion that the latter were particularly short, although their limb proportions were more cold-adapted, in being relatively short in their distal parts (Allen's rule). Body weight estimations of the Neanderthals indicate that they had a 30% larger body mass than recent *H. sapiens*, which can be attributed to either the more Northern latitude habitats, their greater muscularity, or both. Their height and weight can be used to estimate their surface areas, which can be interpreted as decidedly cold-adapted when compared to the relative surface areas of modern humans (Bergmann's rule; Coon 1939, 1962, 1982, Churchill 1998, Helmuth 1998, Henke 1995, Trinkaus and Shipman 1993; Figure 6.51).

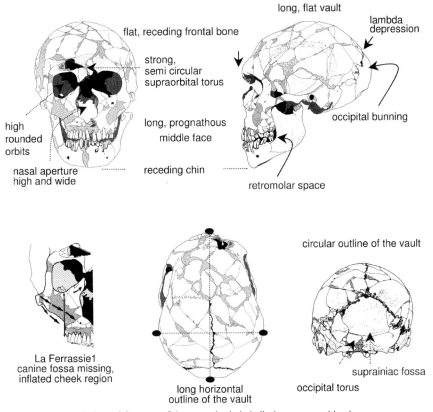

Figure 6.50 Morphological features of the Neanderthal skull, demonstrated by the specimen La Ferrassie 1 (after Heim 1976; Henke and Rothe 1998).

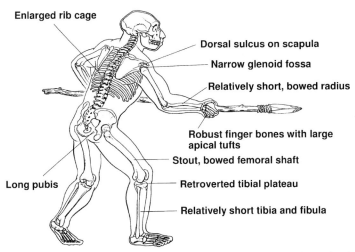

Figure 6.51 Neanderthal skeleton exhibiting characteristic features (from Churchill 1998).

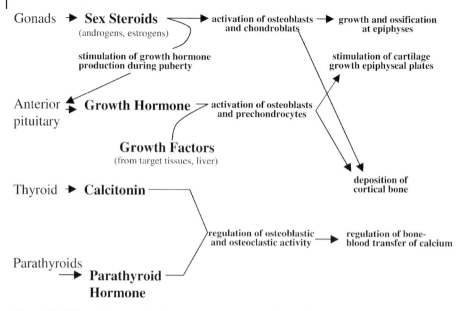

Gonads ➤ **Sex Steroids** ➤ activation of osteoblasts ➤ growth and ossification
(androgens, estrogens) and chondroblats at epiphyses

stimulation of growth hormone
production during puberty

stimulation of cartilage
growth epiphyseal plates

Anterior ➤ **Growth Hormone** ➤ activation of osteoblasts
pituitary and prechondrocytes

Growth Factors
(from target tissues, liver)

Thyroid ➤ **Calcitonin**

deposition of
cortical bone

regulation of osteoblastic ➤ regulation of bone-
and osteoclastic activity blood transfer of calcium

Parathyroids
➤ **Parathyroid**
Hormone

Figure 6.52 Effects of three endocrine axes on osteogenic and chondogenic
cells and their resulting effects on skeletal morphology. Only two of these
axes (sex steroids and growth hormone) have a major effect on the gross
morphology of skeletal elements (after Churchill 1998).

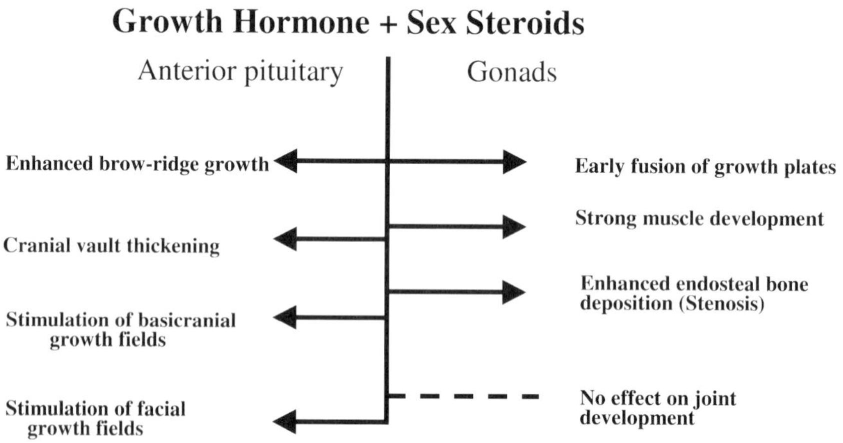

Growth Hormone + Sex Steroids

Anterior pituitary   Gonads

Enhanced brow-ridge growth ⬅➡ Early fusion of growth plates

➡ Strong muscle development

Cranial vault thickening ⬅

Enhanced endosteal bone
deposition (Stenosis) ➡

Stimulation of basicranial ⬅
growth fields

Stimulation of facial ⬅ - - - - No effect on joint
growth fields development

Figure 6.53 Proposed effects of an epigenetic shift affecting the combined
growth-hormone and sex-steroid axis on Neanderthal skeletal morphology.
In this model, increased production of osteogenically active hormones
produces a myriad of seemingly independent Neanderthal characteristics
in the skeleton, ranging from unique craniofacial morphology to great
postcranial robustness (after Churchill 1998).

When we look at the skulls of the Neanderthals, the pointed wedge-shaped face and, in particular, the long high nose were considered to be cold-adapted characteristics (Coon 1982). They could have insulated the cold-sensitive brain from the nasal cavity and provided a larger space for prewarming inhaled cold air. Disregarding that one cannot find any proof for this assumption, other physiological explanations for the nasal morphology, especially the capacious nasal cavity, have been proposed (Franciscus and Trinkaus 1995, Tattersall and Schwartz 1999).

The typical midfacial morphology of Neanderthals has been discussed, e.g., by Rak (1987) and Demes (1987), who argue that this suite of characters reflects the fact that the face is adapted to resist the bending and torsion in the sagittal plane that results form loading the anterior dentition. The special wear pattern of the incisors and the retromolar space also fit the 'teeth as tools' hypothesis. It may be that even the occipital bunning, the protrusion of the occipital bone of the neurocranium, can be explained as a biomechanical compensation pattern for the special masticatory and paramasticatory chewing activities of the Neanderthals.

Churchill (1998) recently gathered all available data on cold adaptation and heterochrony, a change in the form of descendants relative to their ancestors that is brought about by an alteration in the timing of developmental events. His sophisticated integrated approach, giving due respect to structuralist as well as adaptationist perspectives, tries to explain the morphology of the Neanderthals. Churchill's (1998, p. 49) inspiring hypothesis is that "it is possible that many Neandertal features were the correlated results of a change in a few regulatory genes." It is obvious that the heterochronic model is a very speculative hypothesis and largely untested, yet there is compelling paleontological, comparative, and experimental evidence from other animals that climate affects developmental rates and patterns. Endocrinological results show that sex steroids, growth hormones, calcitonin, and parathyroid hormones affect the activation and regulation of estrogenic and chondogenic cells (Figures 6.52 and 6.53). Churchill proposes that the sex steroids and growth hormones, which have a major effect on the gross morphology of skeletal elements, may have induced changes via an epigenetic shift, resulting in increased production of osteologically active hormones. These would have produced integrated changes of seemingly independent Neanderthal characteristics. If the change in a few regulatory genes may have caused the Neanderthal features, this possibility has deep implications for both cladistic and phenetic approaches to late Pleistocene phyletic evolution (also Schwartz 1999).

6.3.7.3 Paleoecological Remarks on the Neanderthals

Traditionally, the Neanderthals were portrayed in a dehumanized fashion, but a reexamination of the arguments and data indicates that they were not the fools or simpletons they were characterized as. There are reliable sources that indicate that they had skills for some specialized and structured use of space. Their settlement preferences demonstrate that they inhabited, e.g., rock shelters pointing south, and thereby providing protection from cold northerly winds, and were able to construct shelters. Furthermore, they used a sophisticated technology involving standard tools, basic bone tools, as well as evolved stone blades and other complex tools. In spite of these findings, archaeological evidence suggests that the elaborate-ness and efficiency of Neanderthal technology were apparently much poorer than those of anatomically modern man from the Upper Paleolithic. Trinkaus and Shipman (1996, p. 417) suggest, from all the archaeological and behavioral findings, that Neanderthals had "no choice but to accomplish the tasks of daily life through brute strength, incredible stamina, and dogged perseverance."

Neanderthal hunter–gatherers had a harsh lifestyle, as can be deduced from a high frequency of trauma. One particular paleopathological case from Shanidar is always cited as the most convincing evidence for a severe injury suffered in life, which was healed with the help of intensive social care. Those who were too weak to procure food by themselves received support from their companions. Further-more, scarce and fluctuating resources made altruistic support, a high degree of food sharing, and cooperative alliances imperative within Neanderthal society. The average life expectancy was extremely low; most adults died in their twenties and thirties. They were intentionally buried in plain graves (e.g., La Ferrassie, Le Moustier, Teshik-Tash, Shanidar, Kebara). The mortuary practices are taken as irrefutable evidence for the existence of rituals. Analysis of pollen and anthers from Shanidar Cave leave little doubt that flowers were used in some Neanderthal burials, constituting "an apparent example of symbolic and religious behavior" (Hayden 1993, p. 120).

The various indications of nursing, foresight capabilities, and mental templates, as well as evidence for strong social ties within a complex society, point toward an elaboration of different roles in Neanderthal society: "different individuals did different things" (Trinkaus and Shipman 1993, p. 418). Within the Neanderthal society, individuality reached a level that seems to have been previously unknown. A question often asked in this context is whether or not the Neanderthals possessed the capability of language, that is, a symbolic–verbal means of communication. The answer is hard to estimate: Crelin (1987) raised the important issue of how much language is required to make 'language' (also Lieberman et al. 1992). From an anthropological standpoint, it appears to be most unlikely that they were without language. But the fact that they buried their dead and that they may have cared for their sick and disabled comrades is only indirect evidence. Anatomical evidence comes from a fossil hyoid, a small bone lying above the larynx to which the tongue muscles attach, which was found in Kebara. The fossil is identical to the hyoid bone of modern humans, and the large noses, sinuses, and mouth of the Neander-thal face allow no dispute about their capability for vocal communication. But did Neanderthals have an articulate language, a complex system of syntax, grammar, and object naming? Did they have mental and symbolic capabilities? The essential differences between the Middle and Upper Paleolithic in ritual, symbolism, and art are often seen as a unbridgeable gap between Neanderthals and anatomically modern humans.

Overall, the paleoecology of the Neanderthals is indicative of new facts of behavior in the social and organizational field within human evolution; they cannot be interpreted as crude prototypes of modern *H. sapiens*, as Trinkaus and Shipman (1993, p. 419) notice: "they were themselves; they were Neanderthals – one of the most distinctive, successful, and intriguing groups of humans that ever enriched our family history." Although Tattersall (1995, p. 203) regards the Neanderthals as 'losers' in the evolutionary game, he notes that they were "highly successful for a long time, longer certainly than we have yet been, and they occupied a unique place in nature. ... It is profoundly misleading to see them simply as an inferior version of ourselves."

6.3.7.4 Neanderthal aDNA Sequences and the Origin of Modern Humans

The first isolation and sequencing of Neanderthal mtDNA by Krings et al. (1997) was published to lavish media fanfare on 11 July 1997 and was interpreted as powerful support for the out-of-Africa model. The cover of the scientific journal *Cell* announced in big letters: "Neanderthals Were Not Our Ancestors". Are the aDNA results really the ultimate solution to the long-standing Neanderthal problem? Besides those who claimed that debate about the phylogenetic role of the Neanderthals had once and for all come to a conclusion, there were very skeptical comments. In the beginning it was questioned whether the DNA found had remained intact and whether it may have been affected by degradation and contamination, a criticism that had already been refuted by the sophisticated design of the study: multiple controls indicated that the sequence was endogenous to the fossil. Others, e.g., Wolpoff, said that the researchers had jumped to conclusions too soon. Interesting enough, the paleogeneticists themselves were much more cautious than the newspapers with their flashy headlines. Their comments on their brilliant, innovative piece of work were very down-to-earth: DNA was extracted from the Neanderthal-type specimen, and by sequencing clones of short overlapping PCR products they were able to determine a hitherto unknown mitochondrial DNA sequence. Sequence comparisons with recent human mtDNA showed that the Neanderthal sequence falls outside the variation of modern humans, and phylogenetic analyses based on the mtDNA data supported the separate position of the Neanderthals. Compared with mtDNA from living humans, the Neanderthal specimen is different at 27 positions, considerably more than the average of 8 differences among recent humans. There is only very slight overlap, because the number of mtDNA differences between living humans and the Neanderthal specimen ranges from 22 to 36 substitutions, whereas the number of mtDNA differences among living humans ranges from 1 to 24 substitutions (Figure 6.54).

Furthermore, molecular clock calculations suggested that the common ancestor of Neanderthal and modern human mtDNAs was roughly four times longer ago (550 000 to 690 000 years) than the common ancestor of human mtDNAs.

The final conclusion of these results is "that Neandertals went extinct without contributing mtDNA to modern humans" (Krings et al. 1997, p. 19). However, they made the point that Neanderthals did not contribute any mtDNA to living humans, the possibility that Neanderthals contributed nuclear genes could not be ruled out.

Relethford (1998, p. 18) gives some additional arguments why there is need for more research, to put the 27-base pair difference into perspective. He asks whether the Neanderthal specimen is different "because he belonged to a different species,

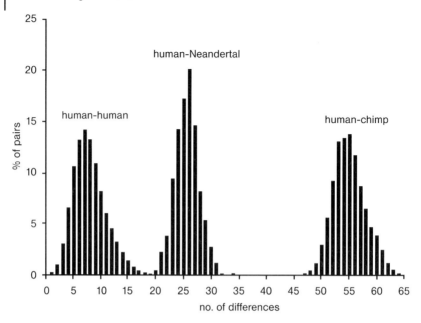

Figure 6.54 Distributions of pairwise sequence differences between humans, Neanderthals, and chimpanzees; x-axis = number of sequence differences; y-axis = percent of pairwise comparisons (from Krings et al. 1997).

because he lived many ten thousand years ago, because of demographic shifts over time, or because of recent natural selection?"

The paleogeneticists and population geneticists are working on the problem. Svante Pääbo and coworkers were able to extract a second mtDNA which confirmed the first analysis from 1997. More information that points into the same direction has been published by Ovchinnikov et al. (2000), who analyzed Neanderthal DNA from Mezmaiskaya (Northern Caucasus), and further results are based on the Vindija (Croatia) fossils. Although Caramelli et al. (2003) found clear evidence for a genetic discontinuity between Neanderthals and 24 000-year-old anatomically modern Europeans, based on fossil remains from the Paglicci Cave (Southern Italy), we need information about additional fossil specimens (e.g., Skhul, Qafzeh), further loci, and the possible roles of selection and neutrality (e.g., Hawks and Wolpoff 2001). There are many controversies relating to population genetics interpretations, especially concerning the factors that could affect the relationship between population size and census population size (e.g., Lahr and Foley 1998). As Relethford (1998, p. 19) notices, the breathtaking pace of discovery will soon change our present-day views. "Currently, neither phylogenetic model of modern human origins is unequivocally supported to the exclusion of the other" (Relethford 1998, p. 1; also Relethford 2001). In spite of much current debate, most of the molecular genetics "results suggest a pattern of genetic continuity in the modern human's genealogy from the Upper Paleolithic period to the present, but a clear discontinuity with respect to Neandertals of similar ages" (Caramelli et al. 2003, p. 6596).

6.3.8
Molecules and the Origin of Recent Humankind

The primary concept of an African origin of modern humankind was initially based on a cluster of crania that showed some morphological traits that were considered to be the mark of modern humanity (Bräuer 1984). Smooth-browed specimens from South African sites (e.g., Kanjera, Border Cave, Bushman Rock Shelter, Klasies River Mouth Cave), attributed to anatomically modern *Homo sapiens*, played a significant role in the 1970s, when improved dating techniques suggested that several of them were over 100 00 years old and thus apparently more ancient than the oldest anatomically modern *H. sapiens* from outside Africa. Protsch (1975, p. 297 f.) concluded "Based on ... the absolute dates [of African Late Pleistocene hominins,] I propose the worldwide evolution of all earliest anatomically modern fossil hominids from *Homo sapiens capensis* of Africa."

When further evidence for a modern-appearing people of very high Late Pleistocene age were obtained in the Middle East (Mugharet es-Skhul, Qafzeh; Bar-Yosef and Vandermeersch 1993), the proponents of the 'replacement model' looked upon these new data as proof of their out-of-Africa models (see above).

In spite of many conflicting interpretations of the fossils (e.g., Wolpoff 1992, 1996–1997, Tobias 1995), the morphognostical and morphometrical affinities to other Late Pleistocene representatives of *H. sapiens* were obvious. But one unsolved question remained: are the morphological similarities really indicative of direct genealogical relationships of African and non-African populations or just parallelisms? The debate based exclusively on the fossil record promised no solution of the controversy on 'continuity or replacement' (Bräuer and Smith 1992). Finally, new approaches in molecular genetics were heralded as being able to finally solve the puzzle of the origin of recent humankind. The source of new perspectives concerning the evolutionary history of our species and the genetic relatedness of human populations was the mtDNA of recent populations. Cann et al. (1987) proposed that all mtDNAs found in modern human populations are descended from a single common ancestor who lived in Africa 200 000 years ago. Those paleoanthropologists who espoused the out-of-Africa models enthusiastically welcomed the molecular data that supported their theories (Stringer 1988, Stringer and Andrews 1988).

Cann et al. (1987) studied DNA of mitochondria, i.e., cellular organelles responsible for generating energy, in specimens taken from various living human populations. mtDNA is inherited strictly through the maternal line without segregation or recombination. In their basic study, Wilson and colleagues analyzed restriction enzyme maps of individuals from all continents and produced a pattern of 182 different types (Figure 6.55). The inferred evolutionary relationship among these types implies an African origin of recent mankind. The conclusion from the mtDNA data was that a single woman, who lived in Africa as recently as the end of the Middle Pleistocene between 140 000 and 290 000 years BP (since then narrowed to ca. 200 000 years BP, Wilson and Cann 1992), was our common ancestor. Additionally, African populations display the greatest variation in mtDNA, which

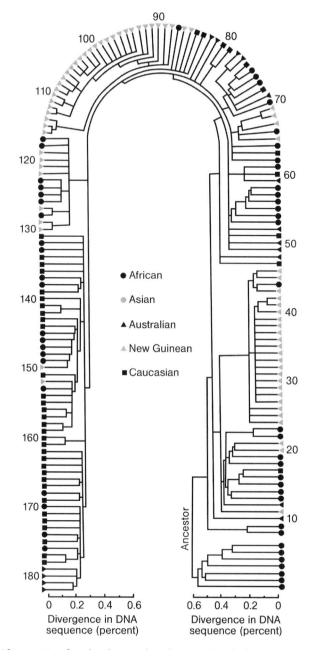

Figure 6.55 African origin of modern humans based on mitochondrial DNA analysis. Analysis of restriction enzyme maps of individuals from all geographic regions produced a pattern of 182 different types (outer edges). The inferred evolutionary relationships among these types implies an African origin. In addition, African populations display the greatest variation in mtDNA, which also supports an African origin (from Lewis 1998).

also supports the out-of-Africa model. The inheritance of mtDNA from only a single mother led them to name the founding mother of humanity 'Eve' or 'African Lucky Mother'. The Eve hypothesis met with some vehement criticism, because, right from the beginning, there was confusion (Wolpoff 1992). The biblical image of the African Eve was totally misleading, because the 'mitochondrial Eve' was not the literal mother of us all. She was merely the female from whom all of our mtDNA derives; many of the nuclear genes would have originated from other females – and from males as well (Relethford 1998, 2001). It simply refers to the fact that a single mtDNA type may have been present in a population of several hundred or even thousand females, most of whom left no mtDNA descendents due simply to chance.

In this context, the bottleneck principle plays an important role. Population geneticists speak of a bottleneck when the number of individuals in a population is drastically reduced and may afterward increase again. Bottlenecks are periods of intense selection (e.g., by a severe habitat change) or of very small population size during which only certain genes survive that eventually characterize the whole population. Compared to the original population, the new population possesses only a fraction of the former genetic variation that it in turn can pass on to future generations. Given a scenario of small populations sizes, small sibships, and high mortality rates in Pleistocene human populations, as well as a sex ratio of $1 : 1$ (males never transmit their mtDNA type), there is a high probability that mitochondrial diversity would be progressively reduced to a single type as a result of random factors (Bilsborough 1992, Relethford 1998, 2001, Templeton 1992, 1993).

Much criticism has been formulated, focused on (1) the sample, which was composed of African Americans, which functioned as a substitute for African populations, (2) the use of restriction analysis instead of more detailed comparisons between complete sequences, and (3) the lack of an outgroup, such as *Pan*, for a more accurate rooting of the human family tree. In response to the diverse criticism, the research was refined. The branching model remained principally the same as before, giving new support for an recent African origin of mankind – but not for long. The phylogenetic method of constructing the human tree was claimed to be inappropriate. The use of the computer program PAUP (Phylogenetic Analysis Using Parsimony), which ranks different trees according to the principle of parsimony, was heavily discredited as incorrect. Templeton's (1993) recalculations demonstrated that other parsimonious solutions are possible, which give no convincing support to the RAO and a recent branching of African and non-African populations.

As mtDNA research went on, and it became increasingly evident that, when working with DNA sequences only, it is extremely difficult to design tests that distinguish the African replacement model from the multiregional evolution model (Tishkoff et al. 1996, Wolpoff 1996). The trouble is that both models describe a species originating in Africa and spreading throughout Europe and Asia, then differentiating into regional subpopulations that nonetheless remain interconnected by gene flow. Both models include a common ancestry for all present-day humans that traces back to Africa. The essential difference is the time pattern of the dispersal.

The molecular evidence for a recent origin of *H. sapiens* was poorly grounded statistically, although the conclusions may be correct after all (Relethford 1998, 2001, Zischler et al. 1995).

One crucial point in the discussion is the molecular clock, the means of determining dates of evolutionary divergences using genetic similarities between extant species, based on the assumption that molecular evolution proceeds at a constant rate. Since rates of molecular evolution vary among lineages, one should be extremely cautious about extrapolating a calibration made for a particular taxonomic group to other species (Hillis et al. 1996).

Until now, there is no unequivocal answer from the genetic data, quite the opposite; so the situation concerning modern human origins remains problematic. Any attempt to force the genetic data from regional subpopulations into a phylogenetic branching model is invalid, because they are not separate evolutionary entities, as Relethford (1998) noted. Even if regional differences began only around 100 000 years ago as a "result of dispersal from Africa and subsequent branching, the continued action of gene flow makes the reconstruction of phylogenetic trees and the dating of population splits difficult at best" (Relethford 1998, p. 5). In spite of many discrepancies concerning the interpretation of the molecular data, it is becoming clearer that, due to the larger effective population size in Africa throughout this time and especially after an expansion, continued gene flow led to greater total African ancestry in all regions, as Relethford (2001) suggests. For this reason he proposes the 'mostly out of Africa' model, a multiregional model in which Africa contributes the most to accumulated ancestry in all regions (Figure 6.56), as the most parsimonious interpretation.

The male analog to the maternal mtDNA is the Y chromosome. The Y chromosome is paternally inherited, and, except for a small section, does not recombine (Hammer and Zegura 1996). Several attempts to estimate the coalescence date have been made (e.g., Whitfield et al. 1995, Underhill et al. 1997). Although some results are consistent with a recent African origin within the range of the mtDNA estimates, some data point to a much more recent date (37 000 to 49 000 years BP).

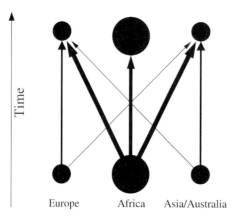

Figure 6.56 The 'mostly out of Africa' model (after Figure 9.3 in Relethford 2001).

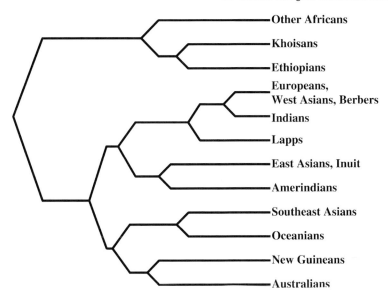

Other Africans
Khoisans
Ethiopians
Europeans, West Asians, Berbers
Indians
Lapps
East Asians, Inuit
Amerindians
Southeast Asians
Oceanians
New Guineans
Australians

Figure 6.57 Evidence from nuclear DNA supports an African origin. The data separate African and non-African populations and indicate an African origin (from Cavalli-Sforza 1991).

Besides the mtDNA and Y chromosomal data, which are often cited as convincing support for a recent African origin, there is further molecular evidence from the major histocompatibility complex (MHC) and from β-globulin and other polymorphisms, which have yielded controversial results.

Cavalli-Sforza's (1991) 'human diversity project' yielded many results that indicate an African origin (Figure 6.57). This holds also for new approaches: Cavalli-Sforza et al. (1994) analyzed microsatellites in nuclear DNA using two parts of the *CD4* gene on chromosome 12 from a global sample of 1600 individuals in 42 populations. The results demonstrated that the greatest diversity existed among sub-Saharan populations and that all populations outside of Africa were much more homogeneous.

A recent analysis of *Alu* insertion polymorphisms, a noncoding locus near the short tandem repeat (STR) locus on chromosome 12, estimated a date of approximately 1.4 myr ago for the human coalescence (Sherry et al. 1997).

In spite of tremendous discrepancies in interpretation of the molecular data (Tobias 1995, Relethford, 2001), the genetic evidence for modern human origins has most often been portrayed as support for a recent African origin model involving near-complete replacement (Freeman and Herron 1998, Lewin 1998). Because the evidence is not as conclusive as one might think, there is much more work needed on other loci and especially from the population structure perspective to achieve unequivocal results. Irrespective of the final outcome, there is no doubt that paleoanthropology has gained immensely from the paleogenetic approach.

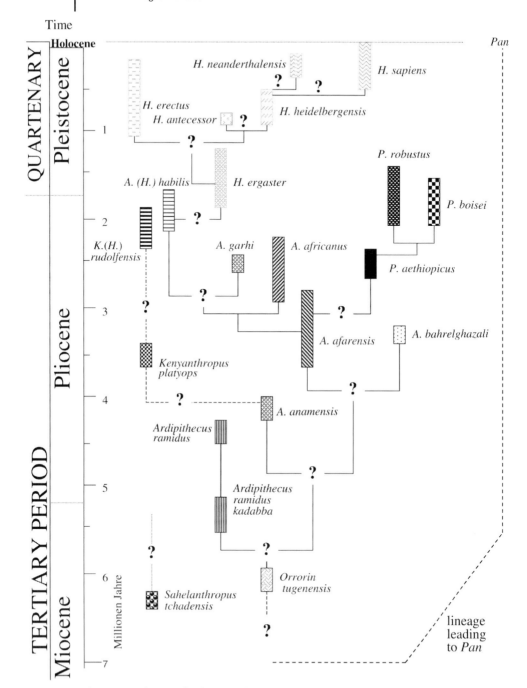

Figure 6.58 A hominin family tree as the current best guess of the so-called splitters (from Henke and Rothe 2003).

6.4
Concluding Remarks

This review of human biological evolution demonstrates that no conclusive agreement has been reached in the long-lasting controversies on human origins. On the one hand, there are obviously many biases and misreadings of the different viewpoints, which have increased the vigor of the debate. But on the other hand, there are essential differences in principal assumptions and preconditions (Tattersall 1996; Wolpoff and Caspari 1997a, b, Relethford 2001, Henke and Rothe 2003).

The crucial point for controversies are polarized positions concerning realistic criteria for species recognition in the human fossil record. There have been major advances in the acquisition and analysis of evidence for human evolution. But despite a tremendous increase in fossil evidence and a dramatic revolution in technical approaches, much work still needs to be done. This holds especially true for the conceptual framework for handling the resulting information from all kinds of paleoanthropological research. Although the scenario of human evolution is much more detailed and complex (Figure 6.58) than it was three decades ago; paradoxically, the more precise the picture became, the more questions arose. This is absolutely typical of science and not retrogressive at all.

In a general review it is not particularly useful to advocate a fixed position, although there are of course preferences for one or another finding which may become obvious to the close reader. Paleoanthropologists seek the doorways to the past and – as has been shown – have been very successful recently in developing new sets of keys. McHenry (1996, p. 86) puts it this way: "One needs to make the best of our tiny sample of life in the past. To be open to new discoveries and ideas, and to enjoy the pleasure of learning and changing."

6.5
References

AIELLO, L. C. (**1990**) Patterns of stature and weight in human evolution. *Am. J. Phys. Anthrop.* **81**, 186–187.

AIELLO, L. C. (**1996**) Terrestriality, bipedalism and the origin of language. In: RUNCIMAN, W. G., MAYNARD SMITH, J., DUNBAR, R. I. M. (Eds.), *Evolution of Social Patterns in Primates and Man*. London: The British Academy, 269–289.

AIELLO, L. C., DEAN, C. (**1990**) *An Introduction to Human Evolutionary Anatomy*. San Diego: Academic.

ANDREWS, P. (**1984**) On the characters that define *Homo erectus*. *Cour. Forsch. Senckenb.* **69**, 167–178.

ANDREWS, P. (**1986**) Molecular evidence for catarrhine evolution. In: WOOD, B. A., MARTIN, L. B., ANDREWS, P. J. (Eds.), *Major Topics in Primate and Human Evolution*. Cambridge: Cambridge University Press, 107–129.

ARAMBOURG, C., COPPENS, Y. (**1968**) Découverte d'un Australopithecien nouveau dans les gisements de l'Omo (Ethiopie). *S. Afr. J. Sci.* **64**, 58–59.

ARSUAGA, J.-L., MARTÍNEZ, I., LORENZO, C., GRACIA, A., MUNOZ, ALONSO, O., GALLEGO, J. (**1999**) The human cranial remains from Gran Dolina Lower Pleistocene site (Sierra de Atapuerca, Spain). *J. Hum. Evol.* **37**, 431–457.

ASFAW, B., WITHE, T., LOVEJOY, O., LATIMER, B., SIMPSON, S., SUWA, G. (**1999**) *Australopithecus garhi*: a new species of early hominid from Ethiopia. *Science* **284**, 629–635.

ASFAW, B., GILBERT, W. H., BEYENE, Y., KART, W. K., RENNE, P. R., WOLDEGABRIEL, G., VRBA, E. S., WHITE, T. (**2002**) Remains of *Homo erectus* from Bouri, Middle Awash, Ethiopia. *Nature* **416**, 317–320.

AX, P. (**1984**) *Das phylogenetische System*. Stuttgart, New York: Gustav Fischer Verlag.

AYALA, F. J. (**1997**) Vagaries of the molecular clock. *Proc. Natl. Acad. Sci. USA* **94** (15), 7776–7783.

BARRETT, P. H., GAUTEY, P. J., HERBERT, S., KOHN, D., SMITH, S. (Eds.) (**1987**) *Charles Darwin's Notebooks*. Cambridge: Cambridge University Press, 1836–1844.

BAR-YOSEF, O., VANDERMEERSCH, B. (**1993**) Koexistenz von Neandertaler und modernem *Homo sapiens*. *Spektrum der Wissenschaft* **6**, 32–39.

BEGUN, D. R., KORDOS, L. (**1997**) Phyletic affinities and functional convergence in *Dryopithecus* and other Miocene and living hominids. In: BEGUN, D. R., WARD, C. V., ROSE, M. D. (Eds.), *Function, Phylogeny and Fossils. Miocene Hominoid Evolution and Adaptations*. New York: Plenum, 291–316.

BEGUN, D. R., WARD, C. V., ROSE, M. D. (**1997**) Events in hominoid evolution. In: BEGUN D. R., WARD C. V., ROSE M. D. (Eds.), *Function, Phylogeny and Fossils. Miocene Hominoid Evolution and Adaptations*. New York: Plenum, 389–415.

BEGUN, D. R., WARD, C. V., ROSE, M. D. (Eds.) (**1997a**) *Function, Phylogeny and Fossils. Miocene Hominoid Evolution and Adaptations*. New York: Plenum.

BENEFIT, B. R. (**1999**) *Victoriapithecus*: the key to old world monkey and catarrhine origins. *Evol. Anthropol.* **7**, 154–174.

BILSBOROUGH, A. (**1986**) Diversity, evolution and adaptation in early hominids. In: AILY, G. N., CALLOW, P. (Eds.), *Stone Age Prehistory*. Cambridge: Cambridge University Press, 197–220.

BILSBOROUGH, A. (**1992**) Human Evolution. London: Blackie.

BILSBOROUGH, A., WOOD, B. A. (**1986**) The nature, origin, and fate of *Homo erectus*. In: WOOD, B. A., ANDREWS, P. (Eds.), *Major Topics in Primate and Human Evolution, Cambridge*. Cambridge University Press, 295–316.

BINFORD, L. R. (Ed.) (**1977**) *For Theory Building in Archaeology*. London: Academic Press.

BINFORD, L. R. (**1981**) *Bones: Ancient Man and Modern Myths*. New York: Academic Press.

BOCK, W. J., VON WAHLERT, G. (**1965**) Adaptation and the form-function complex. *Evolution* 19, 269–299.

BOESCH, C. (**1996**) The emergence of cultures among wild chimpanzees. In: RUNCIMAN, W. G., MAYNARD SMITH, J., DUNBAR, R. I. M. (Eds.), *Evolution of Social Patterns in Primates and Man*. London: The British Academy, 251–268.

BOESCH-ACHERMANN, H., BOESCH, C. (**1994**) Hominization in the rainforest: the chimpanzee's piece in the puzzle. *Evol. Anthropol.* **3**, 9–16.

Bräuer, G. (1984) The 'Afro-European *sapiens* hypothesis' and hominid evolution in East Asia during the late Middle and Upper Pleistocene. *Cour. Forsch. Senckenb.* **69**, 145–165.

Bräuer, G. (1992) Africa's place in the evolution of *Homo sapiens*. In: Bräuer, G., Smith, F. H. (Eds.), *Continuity or Replacement: Controversies in Homo sapiens Evolution*. Rotterdam: Balkema, 83–98.

Bräuer, G., Mbua, E. (1992) *Homo erectus* features used in cladistics and their variability in Asian and African hominids. *J. Hum. Evol.* **22**, 79–108.

Bräuer, G., Smith F. H. (Eds.) (1992) Continuity or Replacement. Controversies in *Homo sapiens* Evolution. Rotterdam: Balkema.

Bräuer, G., Henke, W., Schultz, M. (1995) Der hominide Unterkiefer von Dmanisi: Morphologie, Pathologie und Analysen zur Klassifikation. *Jahrb. des RGZM* **42**, 183–203.

Brenner, S., Hanihara, K. (Eds.) (1995) The Origin and Past of Modern Humans as Viewed from DNA. *Proceedings of the Workshop on the Origin and Past of Homo sapiens sapiens as Viewed from DNA: Theoretical Approach.* Singapore: World Scientific.

Bromage, T. G., Schrenk, F. (Eds.) (1995) Evolutionary history of the Malawi Rift. *J. Hum. Evol.* **28**, 1–120.

Bromage, T. G., Schrenk, F., Zonneveld, F. (1995) Paleoanthropology of the Malawi Rift: an early hominid mandible from the Chiwondo beds, northern Malawi. *J. Hum. Evol.* **28**, 71–108.

Broom, R. (1937) The Sterkfontein ape. *Nature* **139**, 326.

Broom, R. (1938) The Pleistocene anthropoid apes from South Africa. *Nature* **142**, 377–379.

Burger, J., Hummel, S., Herrmann, B., Henke, W. (1999) DNA preservation: a microsatellite-DANN study on ancient skeletal remains. *Electrophoresis* **20**, 1722–1728.

Burke, T., Dolf, G., Jeffreys, A. J. Wolff, R. (Eds.) (1991) *DNA Fingerprinting: Approaches and Applications*. Basel: Birkhäuser.

Brunet M., Guy F., Pilbeam, D., MacKaye, H. T., Likius A., Djimboumalbaye, A. et al. (2002) A new hominid from the upper Miocene of Chad, central Africa. *Nature*, **418**, 145–151.

Campbell, B. G. (1964) Quantitative taxonomy and human evolution. In: Washburn, S. L. (Ed.) *Classification and Human Evolution*. London: Metheun, 50–74.

Campbell, B. G. (1965) The nomenclature of the Hominidae. *Occas. Pap. Roy. Anthropol. Inst. London*, Vol. 22.

Campbell, B. G. (1998) *Human Evolution*, 4th ed. New York: Aldine de Gruyter.

Cann, R. L., Stoneking, M., Wilson, A. C. (1987) Mitochondrial DNA and human evolution. *Nature* **325**, 31–36.

Caramelli, D., Lalueza-Fox, C., Vernesi, C., Lari, M., Casoli, A., Mallegni, F., Chiarelli, B., Dupanloup, I., Bertranpetit, J., Barbujani, G., Bertorelle, G. (2003) Evidence for a genetic discontinuity between Neandertals and 24,000-year-old anatomically modern Europeans. *Proc. Natl. Acad. Sci. USA* **100**, 6593–6597.

CARBONELL, E., BERMÚDEZ, J.-M., ARSUAGA, J. L. (**1999**) Preface. *J. Hum. Evol.* **37**, 309–311.

CARTMILL, M. (**1992**) New views on primate origins. *Evol. Anthropol.* **1**, 105–111.

CAVALLI-SFORZA, L. L. (**1992**) Stammbäume von Völkern und Sprachen. *Spektrum der Wissenschaft* **1**, 190–198.

CAVALLI-SFORZA, L. L., MENOZZI, P., PIAZZA, A. (**1994**) The History and Geography of Human Genes. Princeton, NJ: Princeton University Press.

CHAMBERLAIN, A. T., WOOD, B. A. (**1987**) Early hominid phylogeny. *J. Hum. Evol.* **16**, 119–133.

CHURCHILL, S. M. (**1998**) Cold adaptation, heterochrony, and Neandertals. *Evol. Anthropol.* **7**, 46–61.

CIOCHON, R. L., CORRUCCINI, R. S. (Eds.) (**1983**) *New Interpretations of Ape and Human Ancestry*. New York: Plenum.

CLARK, G. A., WILLERMET, C. M. (Eds.) (**1997**) *Conceptual Issues in Modern Human Origins Research*. New York: Aldine de Gruyter.

CLARKE, R. J. (**1999**) First ever discovery of a well-preserved skull and associated skeleton of *Australopithecus. Beitr. z. Archäozool. Prähist. Anthrop. II*, 21–27.

COON, C. S. (**1939**) *The Races of Europe*. New York: Macmillan.

COON, C. S. (**1962**) *The Origin of Races*. London: Jonathan Cape.

COON, C. S. (**1982**) *Racial Adaptations. A Study of the Origins, Nature, and Significance of Racial Variations in Humans*. Chicago: Nelson-Hall.

CRELIN, E. S. (**1987**) *The Human Vocal Tract: Anatomy, Function, Development, and Evolution*. New York: Vantage.

CRONIN, J. E. (**1983**) Ape, humans, and molecular clocks: a reappraisal. In: CIOCHON, R. L., CORRUCCINI, R. S. (Eds.), *New Interpretations of Ape and Human Ancestry*. New York: Plenum, 115–137.

DART, R. (**1925**) *Australopithecus africanus*: the man-ape of South Africa. *Nature* **115**, 195–199.

DART, R. (**1948**) The Makapansgat proto-human *Australopithecus prometheus. Amer. J. Phys. Anthrop.* **6**, 259–284.

DARWIN, C. (**1859**) *On the Origin of Species by Means of Natural Selection*. London: Murray (dt. Übers.: SCHMIDT, H., 1982, 4. Aufl.).

DARWIN, C. (**1871**) *The Descent of Man and Selection in Relation to Sex*. London: Murray (dt. Übers.: NEUMANN, C. W., 1967).

DELSON, E., ELDREDGE, N., TATTERSALL, I. (**1977**) Reconstruction of hominid phylogeny: a testable framework based on cladistic analysis. *J. Hum. Evol.* **16**, 297–305.

DEMES, B. (**1987**) Another look at an old face: biomechanics of the Neanderthal facial skeleton reconsidered. *J. Hum. Evol.* **16**, 297–305.

DENNETT, C. D. (**1995**) *Darwin's Dangerous Idea. Evolution and the Meanings of Life*. New York: Simon and Schuster.

EFREMOV, I. A. (**1940**) Taphonomy: a new branch of paleontology. *Pan. Am. Geol.* **74**, 81–93.

ENARD, W., PRZEWORSKI, M., FISHER, S. E., LAI, C. S., WIEBE, V., KITANO, T., MONACO, A. P., PÄÄBO, S. (**2002**) Molecular evolution of *FOXP2,* a gene involved in speech and language. *Nature* **418**, 869–872.

ETTER, W. (**1994**) *Palökologie. Eine methodische Einführung.* Basel: Birkhäuser.

FLEAGLE, J. G. (**1988**) *Primate Adaptation and Evolution.* San Diego: Academic.

FLEAGLE, J. G., KAY R. F. (**1994a**) Anthropoid origins. past, present, and future. In: FLEAGLE, J. G., KAY R. F. (Eds.), *Anthropoid Origins.* New York: Plenum, 675–698.

FLEAGLE, J. G., KAY, R. F. (Eds.) (**1994b**) *Anthropoid Origins.* New York: Plenum.

FOLEY, R. A. (**1987**) *Another Unique Species. Patterns in Human Evolutionary Ecology.* Harlow: Longman.

FOLEY, R. A. (**1995**) *Humans before Humanity.* Cambridge, MA: Blackwell.

FRANCISCUS, R. G., TRINKAUS, E. (**1988**) Nasal morphology and the emergence of *Homo erectus. Am. J. Phys. Anthrop.* **75**, 517–527.

FRANZEN, J. L. (Ed.) (**1994a**) 100 Years of *Pithecanthropus.* The *Homo erectus* Problem. *Cour. Forsch.-Inst. Senckenberg* **171**. Frankfurt a. M.: Senckenberg.

FRANZEN, J. L. (**1994b**) The *Homo erectus* problem. In: FRANZEN, J. L. (Ed.), 100 Years of *Pithecanthropus.* The *Homo erectus* Problem. *Cour. Forsch.-Inst. Senckenberg* **171**, 9–10. Frankfurt a. M.: Senckenberg.

FRAYER, D. W. (**1984**) Biological and cultural change in the European Late Pleistocene and Early Holocene. In: SMITH, F. H., SPENCER, F. (Eds.), *The Origins of Modern Humans. A World Survey of the Fossil Evidence.* New York: Liss, 211–250.

FRAYER, D. W. (**1992**) Evolution at the European edge: Neanderthal and upper Paleolithic relationships. *Préhist. Europ.* **2**, 9–69.

FRAYER, D. W., WOLPOFF, M. H., SMITH F. H., THORNE, A. G., POPE, G. G. (**1993**) The fossil evidence for modern human origins. *Am. Anthrop.* **95**, 14–50.

FREEMAN, S., HERRON, J. C. (**1998**) *Evolutionary Analysis.* Englewood Cliffs, NJ: Prentice Hall.

GABUNIA, L., ANTÓN, S. C., LORDKIPANIDZE, D., VEKUA, A., JUSTUS, A., SWISHER, C. C. III (**2001**) Dmanisi and dispersal. *Evol. Anthropol.* **10**, 158–170.

GOODALL, J. (**1986**) *The Chimpanzees of Gombe.* Cambridge, MA: Harvard University Press.

GOODMAN, M. (**1962**) Immunochemistry of the primates and primate evolution. *Ann. NY Acad. Sci.* **102**, 219–234.

GOODMAN, M., BABA, M. L., DARGA, L. L. (**1983**) The bearing of molecular data on the cladogenesis and times of divergence of hominoid lineages. In: CIOCHON, R. L., CORRUCCINI, R. S. (Eds.), *New Interpretations of Ape and Human Ancestry.* New York: Plenum, 67–86.

GRINE, F. E. (**1993**) Australopithecine taxonomy and phylogeny: historical background and recent interpretation. In: CIOCHON, R. L., FLEAGLE, J. G. (Eds.), *The Human Evolution Source Book.* Englewood Cliffs, NJ: Prentice Hall, 198–210.

GROVES, C. P., MAZAK, V. (**1975**) An approach to the taxonomy of the Hominidae: gracile Villafranchium hominids of Africa. *Can. Min. Geol.* **20**, 225–247.

HAILE-SELASSIE, Y. (**2001**) Late Miocene hominids from Middle Awash, Ethiopia. *Nature* **412**, 178–181.

HAMMER, M. F., ZEGURA, S. L. (**1996**) The role of the Y chromosome in human evolutionary studies. *Evol. Anthropol.* **5**, 116–134.

HARTWIG, W. C. (Ed.) (**2002**) *The Primate Fossil Record.* Cambridge: Cambridge Univ. Press.

HARTWIG-SCHERER, S., MARTIN, R. D. (**1991**) Was 'Lucy' more human than her 'child'? Observations on early hominid postcranial skeletons. *J. Hum. Evol.* **21**, 439–449.

HAWKS, J., WOLPOFF, M. H. (**2001**) Paleoanthropology and the population genetics of ancient genes. *Amer. J. Phys. Anthropol.* **114**, 269–272.

HAYDEN, B. (**1993**) The cultural capacities of Neandertals: a review and re-evaluation. *J. Hum. Evol.* **24**, 113–146.

HEIM, J. L. (**1976**) Les hommes fossiles de la Ferrassie (Dordogne) et la problème de la definition des Néandertaliens classiques. *L'Anthropologie* **74**, no. 1/2.

HELMUTH, H. (**1998**) Body height, body mass and surface area of the Neandertals. *Zschr. Morph. Anthrop.* **82**, 1–12.

HELMUTH, H., HENKE, W. (**1999**) *The Path to Humanity*. Toronto: Canadian Scholars' Press.

HEMMER, H. (**1999**) Die Feliden aus dem Epivillafranchium von Untermaßfeld bei Meiningen (Thüringen). In: KAHLKE, R. D. (Ed.), *Das Pleistozän von Untermaßfeld bei Meiningen (Thüringen), Teil II*. Römisch-Germanisches Zentralmuseum, Mainz. Monographien.

HENKE, W. (**1988**) Die Menchen der letzten Eiszeit: Zur Frage der Differenzie-rung der endpleistozänen Hominiden Europas. *Anthrop. Anz.* **46**, 289–316.

HENKE, W. (**1990**) *Jungpaläolithiker und Mesolithiker Europas*, Habil. Thesis. University of Mainz, Germany, Fachbereich (Microfiche) Biologie.

HENKE, W. (**1992**) Die Proto-Cromagnoiden: Morphologische Affinitäten und phylogenetische Rolle. *Anthropologie* **30**, 1–36.

HENKE, W. (**1995**) Spätpleistozäne und frühholozäne Hominidenmorphologie und Klima. In: ULLRICH, H. (Ed.), Man and Environment in the Palaeolithic. *Etudes et Recherches Archéologiques de L'Université de Liège*, 111–136.

HENKE, W. (**1998**) Current aspects of dental research in palaeoanthropology. In: ALT, K. W., RÖSING, F. W., TESCHLER-NICOLA, M. (Eds.), *Dental Anthropology*. Vienna: Springer, 179–200.

HENKE, W., KIESER, N., SCHNAUBELT, W. (**1996**) *Die Neandertalerin. Botschafterin der Vorzeit*. Gelsenkrichen, Schwelm: Edition Archaea.

HENKE, W., ROTH, H., ALT, K. W. (**1999**) Dmanisi and the early Eurasian dispersal of the genus *Homo*. In: ULLRICH, H. (Ed.), *Lifestyles and Survival Strategies in Pliocene and Pleistocene Hominids*. Gelsenkirchen, Schwelm: Edition Archaea. 138–155.

HENKE, W., ROTH, H., SIMON, C. (**1995**) Qualitative and quantitative analysis of the Dmanisi mandible. In: RADLANSKI, R. J., RENZ, H. (Eds.), *Proceedings of the 10th International Symposium on Dental Morphology, Berlin 1995*. Berlin: Brünne, 6–10.

HENKE, W., ROTHE, H. (**1994**) Paläoanthropologie. Berlin: Springer.

HENKE, W., ROTHE, H. (**1995**) *Homo erectus*: valides Taxon der europäischen Hominiden? *Bull. Soc. Suisse d'Anthrop.* **1**, 15–26.

HENKE, W., ROTHE H. (**1997a**) Zahnphylogenese der nicht-menschlichen Primaten. In: ALT, K. W., TÜRP, J. C. (Eds.), *Die Evolution der Zähne. Phylogenie, Ontogenie, Variation*. Berlin: Quintessenz, 229–278.

HENKE, W., ROTHE, H. (**1997b**) Zahnphylogenese der Hominiden. In: ALT, K. W., TÜRP, J. C. (Eds.), *Die Evolution der Zähne. Phylogenie, Ontogenie, Variation.* Berlin: Quintessenz, 279–360.

HENKE, W., ROTHE, H. (**1998**) *Stammesgeschichte des Menschen. Eine Einführung.* Heidelberg: Springer.

HENKE, W., ROTHE, H. (**1999a**) Die phylogenetische Stellung des Neandertalers. *Biologie in unserer Zeit* **29**, 320–329.

HENKE, W., ROTHE, H. (**1999b**) Migrationen früher Hominini: Überlegungen zu Eurytopie, Exogenie und Expansion in Verbindung mit tiergeographischen Befunden. *Betr. z. Archäozool. Prähist. Anthrop.* **II**, 28–35.

HENNIG, W. (**1950**) *Grundzüge einer Theorie der Phylogenetischen Systematik.* Berlin: Deutscher Zentralverlag.

HENNIG, W. (**1966**) *Phylogenetic Systematics.* Urbana, IL: University of Illinois Press.

HERRMANN, B., HUMMEL, S. (Eds.) (**1994**) *Ancient DNA: Recovery and Analysis of Genetic Material from Paleontological, Archaeological, Museum, Medical, and Forensic Specimens.* New York: Springer.

HILL, A. (**1975**) *Taphonomy of contemporary and Late Cenozoic East African Vertebrates*, Ph. D. thesis. London: University of London.

HILLIS, D. M., MABLE, B. K., MORITZ, C. (**1996**) Applications of molecular systematics: the state of the field and a look to the future. In: HILLIS, D. M., MORITZ, C., MABLE, B. K. (Eds.), *Molecular Systematics.* Sunderland, MA: Sinauer.

HOLROYD, P. A., CIOCHON, R. L. (**1994**) The Asian origin of Anthropoidea revisited. In: FLEAGLE, J. G., KAY, R. F. (Eds.), *Anthropoid Origins.* New York: Plenum, 143–162.

HOWELL, F. C. (**1978**) Hominidae. In: MAGLIO, V. J., COOKE, H. B. S. (Eds.), *Evolution in African Mammals. Cambridge*, MA: Harvard University Press, 154–248.

HOWELL, F. C. (**1986**) Variabilité chez *Homo erectus* et problème de la présence de cette espèce en Europe. *L'Anthropologie* **90**, 447–481.

HOWELL, F. C. (**1996**) Thoughts on the study and interpretation of the human fossil record. In: MEIKLE, W. E., HOWELL, F. C, JABLONSKI, N. G. (Eds.), *Contemporary Issues in Human Evolution.* San Francisco: California Academy of Sciences, 1–45.

HOWELLS, W. W. (**1980**) *Homo erectus*: who, when and where: a survey. *Yrb. Phys. Anthrop.* **23**, 1–23.

HOWELLS, W. W. (**1993**) *Getting Here. The Story of Human Evolution.* Washington, DC: Compass.

HUANG W.-P., CIOCHON, R., YÚMIN, G., LARICK, R., QIREN, F., SCHWARCZ, H., YONGE, C., DE VOS, J., RINK, W. (**1995**) Early *Homo* and associated artefacts from Asia. *Nature* **378**, 275–278.

HUBLIN, J. J. (**1986**) Some comments on the diagnostic features of *Homo erectus*. *Anthropos (Brno)* **23**, 175–187.

HUBLIN, J. J. (**1998**) Die Sonderevolution der Neandertaler. *Spektrum der Wissenschaft* **7**, 56–63.

HUMMEL, S. (**2002**) *Ancient DNA Typing. Methods, Strategies and Applications.* Berlin: Springer.

HUXLEY, T. H. (**1863**) *Evidences as to Man's Place in Nature.* London: William and Norgate (dt. Übers.: CARUS, J. V.).

JACOB, T. (**1976**) Early population in the Indonesian region. In: KRIK, R. L., THORNE, A. G. (Eds.), The Origin of the Australians. *Austra. Inst. Aboriginial Studies, Canberra,* 81–93.

JOHANSON, D. C. (**1989**) A partial *Homo habilis* skeleton from Olduvai Gorge, Tanzania: a summary of preliminary results. In: GIACOBINI, G. (Ed.), Hominidae. *Proc. 2nd Int. Congr. Hum. Paleiont, Turin 1987.* Milan: Jaca Books, 155–166.

JOHANSON, D. C., EDGAR, B. (**1996**) *From Lucy to Language.* New York: Nevraumont.

JOHANSON, D. C., WHITE, T. D. (**1979**) A systematic assessment of early African hominids. *Science* **202**, 321–330.

JOHANSON, D. C, WHITE, T. D., COPPENS, Y. (**1978**) A new species of the genus *Australopithecus* (Primates; Hominidae) from the Pliocene of Eastern Africa. *Kirtlandia* **28**, 1–14.

JOHANSON, D. C., MASAO, F. T., ECK, G. G., WHITE, T. D., WALTER, R. C., KIMBEL, W. H., ASFAW, B., MANEGA, P., NDESSOKIA, P., SUWA, G. (**1987**) New partial skeleton of *Homo habilis* from Olduvai Gorge, Tanzania. *Nature* **327**, 205–209.

JONES, S., MARTIN, R. D., PILBEAM, D. (Eds.) (**1992**) *The Cambridge Encyclopedia of Human Evolution.* Cambridge: Cambridge University Press.

KENNEDY, G. E. (**1991**) On the autapomorphic traits of *Homo erectus*. *J. Hum. Evol.* **20**, 375–412.

KEYSER, A. W. (**2000**) The Drimolen skull: the most complete australopithecine cranium and mandible to date. *South African Journal of Science* **96**, 189–193.

KIMBEL, W. H., WHITE, T. D., JOHANSON, D. C. (**1988**) Implications of KNM-WT 17000 for the evolution of 'robust' *Australopithecus*. In: GRINE, F. E. (Ed.), *Evolutionary History of the 'Robust' Australopithecines.* New York: Aldine de Gruyter, 259–268.

KLEIN, R. G. (**1989**) *The Human Career: Human Biological and Cultural Origins.* Chicago, University of Chicago Press.

KLEIN, R. G. (**1992**) The archeology of modern human origins. *Evol. Anthropol.* **1**, 5–14.

KRINGS, M., STONE, A., SCHMITZ, R. W., KRAINITZKI, H., STONEKING, M., PÄÄBO, S. (**1997**) Neandertal DNA sequences and the origin of modern humans. *Cell* **90**, 19–30.

KULLMER, O., SANDROCK, O., ABEL, R., SCHRENK, F., BROMAGE, G., JUWAYEYI, M. (**1999**) The first *Paranthropus* from the Malawi Rift. *J. Hum. Evol.* **37**, 121–127.

LAHR, M. M., FOLEY, R. A. (**1998**) Towards a theory of modern human origins: geography, demography, and diversity in recent human evolution. *Yrbk. Phys. Anthrop.* **41**, 137–176.

LARICK, R., CIOCHON, R. L. (**1996**) The African emergence and early Asian dispersal of the genus *Homo*. *Amer. Sci.* **84**, 538–551.

LEAKEY, L. S. B. (**1959**) A new fossil skull from Olduvai. *Nature* **201**, 967–970.

LEAKEY, L. S. B, TOBIAS, P. V., NAPIER, J. R. (**1964**) A new species of the genus *Homo* from Olduvai Gorge. *Nature* **202**, 7–9.

LEAKEY, M. G., FEIBEL, C. S., McDOUGALL, I., WALKER, A. (**1995**) New four-million-year-old hominid species from Kanapoi and Allia Bay, Kenya. *Nature* **376**, 565–571.

LEAKEY, M. G., FEIBEL, C. S., McDOUGALL, I., WARD, C., WALKER, A. (**1998**) New specimens and confirmation of an early age for *Australopithecus anamensis*. *Nature* **393**, 62–66.

LEAKEY, M. G., WARD, C. V., WALKER, A. C. (**2000**) *Australopithecus anamensis*: a new hominid species from Kanapoi. In: SCHULTZ, M., CHRISTIANSEN, K., GREIL, H., HENKE, W. et al. (Eds.) *Schnittstelle Mensch – Umwelt in Vergangenheit, Gegenwart und Zukunft*. Proceedings 3. Kongress der Ges. für Anthropologie 1.–3. Oktober 1998 in Göttingen. Göttingen: Cuvillier, 5–8.

LEWIN, R. (**1993**) *Human Evolution: An Illustrated Introduction*, 3rd ed. Cambridge, MA: Blackwell.

LEWIN, R. (**1998**) *The Origin of Modern Humans*. New York: Scientific American Library.

LIEBERMAN, P., LAITMAN, J. T., REIDENBERG, J. S., GANNON, P. J. (**1992**) The anatomy, physiology, acoustics, and perception of speech: essential elements in analysis of the evolution of human speech. *J. Hum. Evol.* **23**, 447–467.

MARTIN, R. D. (**1986**) Primates: a definition. In: WOOD, B. A., MARTIN, R. D., ANDREWS, P. (Eds.), *Major Topics in Primate and Human Evolution*. Cambridge: Cambridge University Press, 1–31.

MARTIN, R. D. (**1990**) *Primate Origin and Evolution: A Phylogenetic Reconstruction*. London: Chapman and Hall.

MARTIN, R. D. (**1995**) Hirngröße und menschliche Evolution. *Spektrum der Wissenschaft* **9**, 48–55.

MAYR, E. (**1969**) *Principles of Systematic Zoology*. New York: McGraw-Hill.

MAYR, E. (**1975**) *Grundlagen der zoologischen Systematik*. Berlin: Parey.

McBREARTY, S., BROOKS, A. S. (**2000**) The revolution that wasn't: a new interpretation of the origin of modern human behavior. *Hum. Evol.* **39**, 453–563.

McHENRY, H. M. (**1996**) Homoplasy, clades, and hominid phylogeny. In: MEIKLE, W. E., HOWELL, F. C., JABLONSKI, N. G. (Eds.), *Contemporary Issues in Human Evolution*. San Francisco: California Academy of Sciences, 77–92.

MITHEN, S. (**1996**) *The Prehistory of the Mind: The Cognitive Origins of Art, Religion, and Science*. London, New York: Thames and Hudson.

MULLIS, K. B. (**1990**) Eine Nachfahrt und die Polymerase-Kettenreaktion. *Spektrum der Wissenschaft* **6**, 60–67.

NOBLE, W., DAVIDSON, I. (**1996**) *Human Evolution, Language and Mind: A Psychological and Archaeological Inquiry*. Cambridge: Cambridge University Press.

OAKLEY, K. P. (**1954**) Dating the *Australopithecus* of Africa. *Amer. J. Phys. Anthrop.* **12**, 9–23.

OVCHINNIKOV, I. V., GÖTHERSTRÖM, A., ROMANOVA, G. P., KHARITONOV, V. M., LIDÉN, K. GOODWIN, W. (**2000**) Molecular analysis of Neanderthal DNA from the northern Caucasus. *Nature* **404**, 490–493.

PIANKA, E. R. (**1983**) *Evolutionary Ecology*, 3rd ed. New York: Harper and Row.

PILBEAM, D. (**1972**) *The Ascent of Man: An Introduction to Human Evolution*. New York: Macmillan.

PITTENDRIGH, C. S. (**1958**) Adaptation, natural selection, and behavior. In: ROE A., SIMPSON, G. G. (Eds.), *Behavior and Evolution*. New Haven, CT: Yale University Press, 390–416.

PREMACK, D. (**1976**) On the study of intelligence in chimpanzees. *Curr. Anthropol.* **17**, 516–521.

PROTSCH, R. R. R. (**1975**) The absolute dating of Upper Pleistocene subsaharian fossil hominids and their place in human evolution. *J. Hum. Evol.* **4**, 297–322.

RAK, Y. (**1983**) *The Australopithecine Face*. New York: Academic.

RAK, Y. (**1986**) The Neanderthal: a new look at an old face. *J. Hum. Evol.* **15**, 151–164.

RAK, Y. (**1991**) Lucy's pelvic anatomy: its role in bipedal gait. *J. Hum. Evol.* **20**, 283–290.

RELETHFORD, J. H. (**1998**) Genetics of modern human origins and diversity. *Annu. Rev. Anthropol.* **27**, 1–23.

RELETHFORD, J. H. (**2001**) *Genetics and the Search for Modern Human Origins*. New York: Wiley-Liss.

RIGHTMIRE, G. P. (**1990**) *The Evolution of Homo erectus: Comparative Anatomical Studies of an Extinct Human Species*. Cambridge: Cambridge University Press.

RIGHTMIRE, G. P. (**1992**) *Homo erectus*: ancestor or evolutionary side branch? *Evol. Anthropol.* **2**, 43–49.

RIGHTMIRE, G. P. (**1998**) Evidence from facial morphology for similarity of Asian and African representatives of *Homo erectus*. *Am. J. Phys. Anthrop.* **106**, 61–85.

ROBINSON, J. T. (**1954**) The genera and species of the Australopithecinae. *Am. J. Phys. Anthrop.* **12**, 181–200.

ROBINSON, J. T. (**1960**) The affinities of the new Olduvai Australopithecine. *Nature* **186**, 456–458.

ROGERS, M. J., FEIBEL, C. S., HARRIS, J. W. K. (**1996**) Deciphering early hominid land use and behaviour: a multidisciplinary approach from the Lake Turkana basin. In: MAGORI, C. C., SAANANE, C. B., SCHRENK, F. (Eds.), Four Million Years of Hominid Evolution in Africa: Papers in Honour of Dr. Mary Douglas Leakey's Outstanding Contribution in Palaeoanthropology. *Kaupia* **6**, 9–19.

ROSS, C. F., KAY, R. F. (Eds.) (**2004**) *Anthropoids Origins. New Visions*. Cambridge: Cambridge Univ. Press.

ROSS, C. F., WILLIAMS, B., KAY, R. F. (**1998**) Phylogenetic analysis of anthropoid relationships. *J. Hum. Evol.* **35**, 221–306.

ROSS, P. E. (**1992**) Nucleic acids and proteins trapped in ancient mummies and still more ancient bones can serve as time capsules of history: molecular biologists are beginning to unlock their secrets. *Scientific American* **5**, 82–91.

ROTHE, H., WIESEMÜLLER, B., HENKE, W. (**1997**) Phylogenetischer Status des fossilen Neulings *Ardipithecus ramidus*: eine kritische Evaluation gegenwärtiger Konzepte. In: ÜBERSEEMUSEUM BREMEN, KÖNIG, V., HOHMANN, H. (Eds.), *Bausteine der Evolution*. Gelsenkirchen, Schwelm: Edition Archaea, 159–168.

SARICH, V. M. (**1983**) Appendix: retrospective on hominoid macromolecular systematics. In: CIOCHON, R. L., CORRUCINI, R. S. (Eds.), *New Interpretations of Ape and Human Ancestry*. New York: Plenum, 137–150.

SARICH, V. M., WILSON, A. C. (**1967**) Immunological time scale for hominoid evolution. *Science* **158**, 1200–1203.

SAVAGE-RUMBOUGH, S., LEWIN, R. (**1995**) *Kanzi, der sprechende Schimpanse. Was den tierischen vom menschlichen Verstand unterscheidet*. München: Droemersche Verlagsanstalt, Th. Knaur Nachf.

SAVAGE-RUMBOUGH, E. S., RUMBOUGH, D. M., BOYSEN, S. (**1978**) Symbolic communication between two chimpanzees (*Pan troglodytes*). *Science* **201**, 641–644.

SCHMID, P. (**2000**) Neueste Entdeckungen in den *Australopithecus*-Fundstellen Südafrikas. In: SCHULTZ, M., CHRISTIANSEN, K., GREIL, H., HENKE, W. et al. (Eds.) *Schnittstelle Mensch – Umwelt in Vergangenheit, Gegenwart und Zukunft*. Proceedings 3. Kongress der Ges. für Anthropologie 1.–3. Oktober 1998 in Göttingen. Göttingen: Cuvillier, 33–39.

SCHMIDT-KITTLER, N., VOGEL, K. (Eds.) (**1991**) *Constructional Morphology and Evolution*. Berlin: Springer.

SCHMINCKE, H.-U., VAN DEN BOGAARD, P. (**1996**) Die Datierung des Mašavera-Basaltlavastroms. *Jahrb. RGZM* **42**, 1995, 75–76.

SCHMITZ, J., ZISCHLER, H. (**2002**) A novel family of tRNA-derived SINEs in the colugo and two retrotransposable markers separating dermopterians from primates. *Mol. Phylogenet. Evol.* **28**, 341–349.

SCHMITZ, J., OEHME, M., SURYOBROTO, B. ZISCHLER, H. (**2003**) The colugo (*Cynocephalus variegatus*, Dermoptera): the primates' gliding sister. *Mol. Biol. Evol.* **19**, 2308–2312.

SCHRENK, F., BROMAGE, T. G. (**1999**) Climate change and survival strategies of early *Homo* and *Paranthropus* in the Malawi Rift. In: ULLRICH, H. (Ed.), *Hominid Evolution: Lifestyles and Survival Strategies*. Gelsenkirchen, Schwelm: Edition Archaea, 72–88.

SCHRENK F., BROMAGE T. G., BETZLER C. G., RING, U. (**1993**) Oldest *Homo* and Pliocene biogeography of the Malawi Rift. *Nature* **365**, 833–836.

SCHWARTZ, J. H. (**1999**) The Origin and Identification of Species. *Anthropologie (Brno)* **37**, 211–220.

SCHWARTZ, J. H., TATTERSALL, I., LAITMAN, J. T. (**1999**) New thoughts on Neanderthal behaviour: evidence from nasal morphology. In: ULLRICH, H. (Ed.), *Hominid Evolution: Lifestyles and Survival Strategies*. Gelsenkirchen/Schwelm: Edition Archaea, 166–186.

SENUT, B. (**1981**) Outlines of the distal humerus in hominoid primates: application to some Plio–Pleistocene hominids. In: CHIARELLI, B., CORRUCCINI, R. S. (Eds.), *Primate Evolutionary Biology*. Berlin: Springer, 81–92.

SHERRY, S. T., HARPENDING, H. C., BATZER, M. A., STONEKING, M. (**1997**) *Alu* evolution in human populations: using the coalescent to estimate effective population size. *Genetics* **147**, 1977–1982.

SHIPMAN, P. (**1981**) *Life History of a Fossil. An Introduction to Taphonomy and Palaeoecology*. Cambridge, MA: Harvard University Press.

SIMONS, E. L. (**1992**) The primate fossil record. In: JONES, S., MARTIN, R. D., PILBEAM, D. (Eds.), *The Cambridge Encyclopedia of Human Evolution*. Cambridge: Cambridge University Press, 197–208.

SIMONS, E. L., PILBEAM, D. R. (**1965**) Preliminary revision of the Dryopithecinae (Pongidae, Anthropoidea), Folia. *Primat.* **3**, 81.

SIMPSON, G. G. (**1961**) *Principles of Animal Taxonomy*. New York: Columbia University Press.

SKELTON, R. R., McHENRY, H. H. (**1992**) Evolutionary relationships among early hominids. *J. Hum. Evol.* **23**, 309–349.

SKELTON, R. R., McHENRY, H. M., DRAWHORN, G. M. (**1986**) Phylogenetic analysis of early hominids. *Curr. Anthrop.* **27**, 21–43.

SMITH, F. H. (**1992**) The role of continuity in modern human origins. In: BRÄUER, G., SMITH, F. H. (Eds.), *Continuity or Replacement. Controversies in Homo sapiens Evolution*. Rotterdam: Balkema, 145–156.

SMITH, F. H., SPENCER, F. (Eds.) (**1984**) *The Origins of Modern Humans: A World Survey of the Fossil Evidence*. New York: Liss.

SMITH F. H., FALSETTI, A. B., DONNELLY, S. M. (**1989**) Modern human origins. *Ybk. Phys. Anthrop.* **32**, 35–68.

SNEATH, P. A. H., SOKAL, R. R. (**1973**) *Numerical Taxonomy: The Principles and Practice of Numerical Classification*. San Francisco: Freeman.

STENT, G. S. (**1972**) Prematurity and uniqueness in scientific discovery. *Scientific American* **227**, 84–93.

STONEKING, M., CANN, R. L. (**1989**) African origin of human mitochondrial DNA. In: MELLARS P., STRINGER C. B. (Eds.), *The Human Revolution: Behavioural and Biological Perspectives on the Origins of Modern Humans*. Edinburgh: Edinburgh University Press, 17–30.

STRAIT, D. S., GRINE F. E., MONIZ, M. A. (**1997**) A reappraisal of early hominid phylogeny. *J. Hum. Evol.* **32**, 17–82.

STRINGER, C. (**1974**) Population relationships in later Pleistocene hominids: a multivariate study of available crania. *J. Archaeol. Sci.* **1**, 317–342.

STRINGER, C. B. (**1984**) The definition of *Homo erectus* and the existence of the species in Africa and Europe. *Cour. Forsch. Senckenb.* **69**, 131–144.

STRINGER, C. B. (**1986**) The credibility of *Homo habilis*. In: WOOD, B. A., MARTIN, L. B., ANDREWS, P. (Eds.), *Major Topics in Primate and Human Evolution*. Cambridge: Cambridge University Press, 266–294.

STRINGER, C. B. (**1988**) The dates of Eden. *Nature* 331, 565–566.

STRINGER, C. B. (**1990**) The emergence of modern humans. *Scientific American* **263**, 98–104.

STRINGER, C. B. (**1992**) Replacement, continuity and the origin of *Homo sapiens*. In: BRÄUER, G., SMITH F. H. (Eds.), *Continuity or Replacement. Controversies in Homo sapiens Evolution*. Rotterdam: Balkema, 9–24.

STRINGER, C. B. (**1996**) Current issues in modern human origins. In: MEIKLE, W. E., HOWELL, F. C., JABLONSKI, N. G. (Eds.), *Contemporary Issues in Human Evolution*. San Francisco: California Academy of Sciences, 116–134.

STRINGER, C. B., ANDREWS, P. (1988) Genetic and fossil evidence for the origin of modern humans. *Science* **239**, 1263–1268.

STRINGER, C. B., GAMBLE, C. (1993) *In Search of the Neanderthals*. London: Thames and Hudson.

SWISHER, C. C. III, CURTIS, G. H., JACOB, T., GETTY, A. G., SUPRIJO, A., WIDIASMORO (1994) Age of the earliest known hominids in Java, Indonesia. *Science* **263**, 1118–1121.

SZALAY, F. S., DELSON, E. (1979) *Evolutionary History of the Primates*. New York: Academic.

TATTERSALL, I. (1986) Species recognition in human paleontology. *J. Hum. Evol.* **15**, 165–175.

TATTERSALL, I. (1995a) *The Fossil Trail: How We Know What We Think We Know about Human Evolution*. New York: Oxford University Press.

TATTERSALL, I. (1995b) *The Last Neanderthal: The Rise, Success, and Mysterious Extinction of Our Closest Relatives*. New York: Nevraumont.

TATTERSALL, I. (1996) Paleoanthropology and preconception. In: MEIKLE, W. F., HOWELL, F. C., JABLONSKI, N. G. (Eds.), *Contemporary Issues in Human Evolution*. San Francisco: California Academy of Sciences, 47–54.

TATTERSALL, I. (1997) *Puzzle Menschwerdung. Auf der Spur der menschlichen Evolution*. Berlin: Spektrum.

TATTERSALL, I. (2000) Once We Were Not Alone. *Scientific American* **2**, 56–63.

TATTERSALL, I., SCHWARTZ, J. (2000) *Extinct Humans*. New York: Westview.

TATTERSALL, I., DELSON, E., van COUVERING, J. (Eds.) (1988) *Encyclopedia of Human Evolution and Prehistory*. New York: Garland.

TEMPLETON, A. R. (1992) Human origins and analysis of mitochondrial DNA sequences. *Science* **255**, 737.

TEMPLETON, A. R. (1993) The 'Eve' hypothesis: a genetic critique and reanalysis. *Am. Anthropol.* **95**, 51–72.

TEMPLETON, A. R. (1997) Out of Africa? What do genes tell us? *Curr. Opinions Genet. Dev.* **7**, 841–847.

THORNE, A. G., WOLPOFF, M. H. (1981) Regional continuity in Australasian Pleistocene hominid evolution. *Am. J. Phys. Anthrop.* **55**, 337–349.

THORNE, A. G., WOLPOFF, M. H. (1992) Multiregionaler Ursprung des modernen Menschen. *Spektrum der Wissenschaft* **6**, 80–87.

TISHKOFF, S. A., DIETZSCH, E., SPEED, W, PAKSTIS, A. J., KIDD, J. R., CEUNG, K., BONNÉ-TAMIR, B., SANTACHIARA-BENERECETTI, A. S., MORAL, P., KRINGS, M., PÄÄBO, S., WATSON, E., RISCH, N., JENKINS, T., KIDD., K. K. (1996) Global patterns of linkage disequilibrium at the *CD4* locus and modern human origins. *Science* **271**, 1380–1387.

TOBIAS, P. (1967) *Olduvai Gorge, Vol. 2. The Cranium and Maxillary Dentition of Australopithecus (Zinjanthropus) boisei*. Cambridge: Cambridge Univ. Press.

TOBIAS, P. V. (1980) "*Australopithecus afarensis*" and *A. africanus*: Critique and alternative hypothesis. *Paleontol. Afr.* **23**, 1–17.

Tobias, P. V. (**1988**) Numerous apparently synapomorphic features in *Australopithecus robustus*, *Australopithecus boisei* and *Homo habilis*: support for the Skelton–McHenry–Drawhorn hypothesis. In: Grine, F. E. (Ed.), *Evolutionary History of the 'Robust' Australopithecines*. New York: Aldine de Gruyter, 293–308.

Tobias, P. V. (**1989**) The status of *Homo habilis* in 1987 and some outstanding problems. In: Giacobini, G. (Ed.), *Hominidae: Proc. 2nd Int. Congr. Hum. Paleont., Turin, 1987*. Milan: Jaca Books, 141–149.

Tobias, P. V. (**1991**) *Olduvai Gorge, Vol. 4, parts V–IX. The Skulls, Endocasts and Teeth of Homo habilis*. Cambridge: Cambridge University Press.

Tobias, P. V. (**1995**) Africa-derived skulls and Africa-derived mitochondrial DNA: towards a reconciliation. In: Brenner, S., Hanihara, K. (Eds.), *The Origin and Past of Modern Humans as Viewed from DNA. Proceedings of the Workshop on the Origin and Past of Homo sapiens sapiens as Viewed from DNA: Theoretical Approach*. Singapore: World Scientific, 189–215.

Tobias, P. (**2003**) Twenty questions about human evolution. *Int. J. Anthropol.* **18**, 9–63.

Torre, D., Ficcarelli, G., Masini, F., Rook, L., Sala, B. (**1992**) Mammal dispersal events in the early Pleistocene of western Europe. *Cour. Forsch. Senckenb.* **153**, 51–58.

Trinkaus, E. (**1983**) *The Shanidar Neanderthals*. New York: Academic.

Trinkaus, E. (Ed.) (**1989**) *The Emergence of Modern Humans: Biocultural Adaptations in the Later Pleistocene*. Cambridge: Cambridge University Press.

Trinkaus, E. (**1995**) Patterns of trauma among Neanderthals. *J. Archaeol. Science* **22**, 841–852.

Trinkaus, E., Howells, W. W. (**1979**) The Neanderthals. *Scientific American* **241**, 122–133.

Trinkaus, E., Shipman, P. (**1993**) *Die Neanderthaler*. Spiegel der Menschheit. München: Bertelsmann.

Turner, E. (**1999**) The problem of interpreting hominid subsistence strategies at Lower Palaeolithic sites: Miesenheim I: a case study from Central Rhineland of Germany. In: Ullrich, H. (Ed.), *Lifestyles and Survival Strategies in Pliocene and Pleistocene Hominids*. Gelsenkirchen, Schwelm: Edition Arachae, 365–382.

Tulp, N. (**1641**) *Observationum medicarum*. Amsterdam: libri tres, 274–279.

Tyson, E. (**1699**) *Orang-outang, sive Homo sylvestris*. London: Bennet.

Underhill, P. A., Jin, L., Lin, A. A., Mehdi, S. Q., Jenkins, T. et al. (**1997**) Detection of numerous Y chromosome biallelic polymorphisms by denaturing high-performance liquid chromatography. *Genome Res.* **7**, 996–1005.

Vandermeersch, B. (**1990**) Les Néanderthaliens et les premièrs Hommes modernes. In: Huys, M. et al. (Ed.), *5 Millions d'Années. L'Aventure Humaine*. Palais de Beaux-Arts de Bruxelles, 68–86.

Vogel, C. (**1975**) Praedispositionen und Praeadaptationen der Primaten-Evolution im Hinblick auf die Hominisation. In: Kurth, G., Eibl-Eibesfeldt, I. (Eds.), *Hominisation und Verhalten*. Stuttgart: Gustav Fischer, 1–31.

Vogel, C. (**1999**) Anthropologische Spuren (Hrsg. von Volker Sommer), *Zur Natur des Menschen*. Stuttgart: Hirzel.

Vrba, E. S. (1988) Late Pliocene climatic events and hominid evolution. In: Grine, F. E. (Ed.), *Evolutionary History of the 'Robust' Australopithecines.* New York: Aldine de Gruyter, 405–426.

Walker, A. C. (1981) Diet and teeth: dietary hypotheses and human evolution. *Phil. Trans. Royal Soc., London, Ser. B* 292, 57–64.

Walker, A. C., Leakey, R. E. F. (1978) The hominids of East Turkana. *Scientific American* 239, 54–66.

Walker, A. C., Leakey, R. E. F. (1988) The evolution of *Australopithecus boisei.* In: Grine, F. E. (Ed.), *The Evolutionary History of the 'Robust' Australopithecines.* New York: Aldine de Gruyter, 247–258.

Walker, A. C., Leakey, R. E. F. (Eds.) (1993) *The Nariokotome Homo erectus skeleton.* Cambridge, MA: Harvard University Press.

Ward, C. V., Leakey, M. G. Brown, B., Brown, F., Harris, J., Walker, A. (1999) South Turkwell: a new Pliocene hominid site in Kenya. *J. Hum. Evol.* 36, 69–95.

Weidenreich, F. K. (1943) The skull of *Sinanthropus pekinensis*: a comparative study on a primitive hominid skull. Paleont. *Sinica, neue Serie D* 10, 1–485.

Wiesemüller, B., Rothe, H., Henke, W. (2003) *Phylogenetische Systematik. Eine Einführung.* Berlin: Springer.

White, T. D., Suwa, G., Asfaw, B. (1994) *Australopithecus ramidus,* a new species of early hominid from Aramis, Ethiopia. *Nature* 371, 306–312.

White, T. D., Suwa, G., Asfaw, B. (1995) Corrigendum. *Nature* 375, 88.

Whitfield, L. S., Sulston, J. E., Goodfellow, P. N. (1995) Sequence variation of the human Y chromosome. *Nature* 378, 379–380.

Wiley, E. O. (1978) The evolutionary species concept reconsidered. *Syst. Zool.* 27, 17–26.

Wilson, A. C., Cann, R. L. (1992) Afrikanischer Ursprung des modernen Menschen. *Spektrum der Wissenschaft* 6, 72–79.

Wolpoff, M. H. (1984) Evolution of *Homo erectus*: the question of stasis. *Palaeobiology* 10, 389–406.

Wolpoff, M. H. (1989) Multiregional evolution: the fossil alternative to Eden. In: Mellars, P., Stringer, C. (Eds.), *The Human Revolution: Behavioural and Biological Perspectives on the Origins of Modern Humans.* Edinburgh: University of Edinburgh Press, 62–108.

Wolpoff, M. H. (1992) Theories of modern human origins. In: Bräuer, G., Smith, F. H. (Eds.), *Continuity or Replacement. Controversies in Homo sapiens Evolution.* Rotterdam: Balkema, 25–63.

Wolpoff, M. H. (1996a) Neandertals are a race of *Homo sapiens. J. Hum. Evol.* 32, A25.

Wolpoff, M. H. (1996b) Interpretations of multiregional evolution. *Science* 274, 704–706.

Wolpoff, M. H. (1996) *Human Evolution.* New York: McGraw-Hill.

Wolpoff, M. H. (1999) *Paleoanthropology,* 2nd ed. Boston: McGraw-Hill.

Wolpoff, M. H., Caspari, R. (1997a) *Race and Human Evolution. A Fatal Attraction.* New York: Simon and Schuster.

WOLPOFF, M. H., CASPARI, R. (**1997b**) What does it mean to be modern? In: CLARK, G. A., WILLERMET, C. M. (Eds.), *Conceptual Issues in Modern Human Origins Research*. New York: Aldine de Gruyter, 28–44.

WOLPOFF, M. H., THORNE, A. G., JELINEK, J., YINYUN, Z. (**1994**) The case for sinking *Homo erectus*. 100 years of *Pithecanthropus* is enough! In: FRANZEN, J. L. (Ed.), 100 Years of Pithecanthropus: The *Homo erectus* Problem. *Cour. Forsch. Senckenb.* **171**, 341–361.

WOLPOFF, M., SENUT, B., PICKFORD, M., HAWKS, J. (**2002**) *Sahelanthropus* or 'Sahelpithecus'? *Nature* **419**, 581–582.

WOOD, B., COLLARD, M. (**1999**) The changing face of genus *Homo*. *Evol. Anthropol.* **8**, 195–207.

WOOD, B. A. (**1984**) The origin of *Homo erectus*. *Cour. Forsch.-Inst. Senckenberg* **69**, 99–111.

WOOD, B. A. (**1985**) Early *Homo* in Kenya and its systematic relationships. In: DELSON, E. (Ed.), *Ancestors: The Hard Evidence*. New York: Liss, 206–214.

WOOD, B. A. (**1987**) Who is the real '*Homo habilis*'? *Nature* **327**, 187–188.

WOOD, B. A. (**1991**) *Koobi Fora Research Project IV: Hominid Cranial Remains from Koobi Fora*. Oxford: Clarendon.

WOOD, B. A. (**1992**) Origin and evolution of the genus *Homo*. *Nature* **355**, 783–790.

WOOD, B. A. (**1996**) Human evolution. *BioEssays* **18**, 945–954.

ZISCHLER, H., GEISERT, H., VON HAESELER, A., PÄÄBO, S. (**1995**) A nuclear 'fossil' of the mitochondrial D-loop and the origin of modern humans. *Nature* **378**, 489.

ZOLLIKOFER, C. P. E., PONCE DE LÉON, M. S., MARTIN, R. D. (**1998**) Computer-assisted paleoanthropology. *Evol. Anthropol.* **6**, 41–54.

ZUCKERKANDL, E., PAULING, L. (**1962**) Molecular disease, evolution and genetic heterogeneity. In: KASH, M., PULLMAN, B. (Eds.), *Horizons in Biochemistry*. New York: Academic.

7

Evolution on a Restless Planet: Were Environmental Variability and Environmental Change Major Drivers of Human Evolution?

Peter J. Richerson, Robert L. Bettinger, and Robert Boyd

7.1
Introduction

Two kinds of factors set the tempo and direction of organic and cultural evolution, those external to biotic evolutionary process, such as changes in the earth's physical and chemical environments, and those internal to it, such as the time required for chance factors to lead lineages across adaptive valleys to a new niche space (Valentine 1985). The relative importance of these two sorts of processes is widely debated. Valentine (1973) argued that marine invertebrate diversity patterns responded to seafloor spreading as this process generated more or less niche space. He suggested that natural selection is a powerful force and that earth's biota are in near equilibrium with the niches available on the geological time scale. Walker and Valentine (1984) modeled the evolution of species assuming a logistic speciation rate limited by internal factors and a diversity-independent death rate caused by ongoing environmental change. Fitting this model to the observed evolution of shelled marine invertebrates suggests that the lag between extinctions and the evolution of new species leaves perhaps 30% of ecological niches unfilled. In this model, the biota lag environmental change by perhaps a few million years.

However, as Valentine (1985) notes, if adaptive landscapes have whole suites of niches protected by deep maladaptive valleys, the waiting time for some pioneering species to cross the divide may be very long, generating the rare events that set new body plans and generate major adaptive radiations. Eldredge and Gould (1972) and Gould (2002) championed the idea that internal processes such as genetic and developmental constraints, coupled with the complexity of the adaptive landscape, resulted in a highly historically contingent evolutionary process. On Gould's account, most of the history of life had to do not with a relatively close tracking of a changing environment but with the halting evolutionary exploration a deeply fissured niche space, mostly by rapid bursts of evolution as a fissure was crossed, followed by long periods of stasis. Note that if the adaptive landscape is deeply fissured for any reason, evolution may take on a progressive character (Stewart 1997). Imagine that the original simple forms of life began at the foot of a large mountain range of

Handbook of Evolution, Vol. 2: The Evolution of Living Systems (Including Hominids)
Edited by Franz M. Wuketits and Francisco J. Ayala
Copyright © 2005 Wiley-VCH Verlag GmbH & Co. KGaA, Weinheim
ISBN: 3-527-30838-5

adaptive topography. Potentially, the whole history of life has been a halting and episodic process of moving first onto to local optima of the near foothills and subsequently filtering across adaptive chasms to higher peaks deeper into the complex topography. Perhaps we have not yet come anywhere near to reaching the highest peaks in the topography on earth, even after perhaps 3.5 billion years of life on our planet.

More complex scenarios are possible. Vermeij (1987) argues that much evolution is driven by the top-down biotic process of predator–prey coevolution, but that the degree of escalation of predator attack strategies and prey defenses is limited by external factors, especially those that control productivity. Vermeij does not commit himself on the issue of how closely the predator–prey escalation process tracks external environmental change.

Discussions of the large-scale patterns of evolution typically assume that the overall environmental framework of the earth is static and that changes in features like the size of brains represent a series of progressive changes from simpler to more complex organisms. Billions of years have transpired since the origins of life on earth, and about 540 million years have transpired between the abundant fossil animals of the Cambrian and the evolution of humans. If the earth's environment has been essentially constant since the origin of life or even since the beginning of the Cambrian, the growth of organic complexity by natural selection and other evolutionary processes such as species selections would have to have been so limited by internal processes as to be exceedingly slow.

On the other hand, since the discovery of seafloor spreading 40 years ago, the role of external factors in macroevolutionary processes has become much clearer (Valentine 1973). Today we have a reasonably clear picture of past continental configurations and past biogeochemistry, especially the chemistry of the oceans and atmosphere (Holland 1984; Scotese 2003). Past environments were very different from those of today. For example, during the late Paleozoic and again during the late Mesozoic, high oxygen concentrations in the atmosphere supported gigantic flying forms, including dragonflies a meter long and pterosaurs weighing perhaps 100 kg. (Graham et al. 1995; Dudley 2000). Everyone accepts, we suppose, that external processes are important regulators of the rate and direction of evolution in the very long run, and everyone accepts that evolution is not an instantaneous process. At the most extreme, life on earth could not begin to evolve until the earth formed, and new species do not evolve in one generation. But the gap is very wide between those that argue that most of the history of life, at least since the late Precambrian, is mainly regulated by internal processes and those that think that, for the most part, the earth's biota are in near equilibrium with existing environmental conditions, aside from a few empty niches resulting from relatively short-term constraints operating on evolutionary processes.

Discussions of human evolution are especially prone to adopt a progressive explanation (e.g., Bronowski 1973; Lenski and Lenski 1982). The implicit hypothesis seems to be that humans are adaptively superior to all other animals, so superior as to constitute a new grade of the evolution of life, equivalent in magnitude to the evolution of multicellular animals (Maynard Smith and Szathmary 1995). If humans or similar creatures had evolved in the Cretaceous, say, the subsequent biotic history of the earth would have been transformed.

Recent discoveries in paleobiogeochemistry and paleoclimatology suggest that we should take a hard look at the internal and progressive account of human origins. Cenozoic, especially Plio-Pleistocene, climate changes probably exerted selective pressures favoring large brains and systems of cultural evolution. The climate deterioration of the last few million years resulted in high-amplitude variation in climates on the scale of the classic glacial advances and retreats (21 000–100 000) years. These variations are too slow to favor adaptation by behavioral flexibility and culture. During the last few glacial cycles, however, the ice ages were accompanied by high-amplitude fluctuations on time scales of a millennium or less. Theory suggests that these are the time scales that can favor the evolution of costly systems for adapting to variable environments by individual learning and cultural transmission. On millennial and submillennial time scales, range changes and organic evolution are too slow to approximate adaptive equilibrium in many habitats. Evolution thus favors big-brained creatures that can adapt rapidly by behavioral means. Human culture is particularly suited to high-frequency fluctuating environments, because humans can use cumulative cultural evolution to develop quite fancy adaptations to particular environments on time scales of a few decades to a few millennia. Even so, the 11 000-year-long Holocene period of climate stability, which led to the origins of plant cultivation and eventually to the exceedingly complex societies and technology of the present, shows that the rate of cultural evolution lags behind environmental change by ten or more millennia.

7.2
The Evolution of Cenozoic Environments

The last 65 million years of earth history have been quite dynamic. Zachos et al. (2001) and Barrett (2003) provide recent summaries. The Cenozoic began with the southern continents still close to Antarctica. Warm currents penetrated to high latitudes in the Southern Hemisphere, maintaining a temperate, if highly seasonal, environment in Antarctica. The Northern Hemisphere was also ice-free. As India, South America, and Australia drifted northward, a seaway circling the Antarctic continent opened and the circum-Antarctic current systems developed. These current systems insulated Antarctica from warming currents, and ice sheets began to develop on that continent as early as about 38 million years ago. The collision between India and Asia began to raise the Tibetan Plateau around 35 million years ago. Ice and high-elevation dry plateaus are very bright, reflecting sunlight back to space. Thus, the net heat income of the whole earth declined, cooling the entire globe. A number of other still poorly understood processes modulated the effects of continental drift and ocean circulation reorganization. The earth warmed abruptly at the end of the Oligocene about 26 million years ago, and cooling did not resume until the mid-Miocene about 15 million years ago.

From the mid-Miocene onwards, mean temperatures have dropped and oscillations of climate have increased in several steps. Opdyke (1995) provides a summary the data (Figure 7.1). During the latter half of the Miocene, the climate variation was dominated by a 23 000-year quasi-cycle driven by the wobble of the earth's axis of rotation (causing precession of the equinoxes). By changing the seasonal heat income of different regions of the earth, changes in this and other orbital parameters influence the earth's climate. For example, when high-latitude summers are cool in the Northern Hemisphere, large glaciers form on high-latitude land masses. The direct effects of the orbital parameters are rather modest, and they must have their main effects via complex, poorly understood climate feedbacks (Broecker 1995; Bradley 1999). Continental drift and other geological factors have recently changed the way the earth's climates respond to orbital fluctuations. In the late Pliocene, a little more than three million years ago, a two million year period of steady cooling of the earth's climate began. At the same time, the variation in climate increased sharply and came to be dominated by the 41 000-year quasi-cycle caused by variation in the tilt of the earth's axis of rotation. Then, after about one million years ago in the middle of the Pleistocene, the overall cooling trend stopped, but the climate variation increased dramatically, now in tune especially with the 100 000-year quasi-cycle of the eccentricity of the earth's orbit (the degree to which the earth's orbit around the sun is elliptical rather than circular). The low-resolution record from Antarctica and the Southern Ocean shows a further distinct change about 420 000 years ago (Becquey and Gersonde 2002; EPICA 2004). The causes of these changes

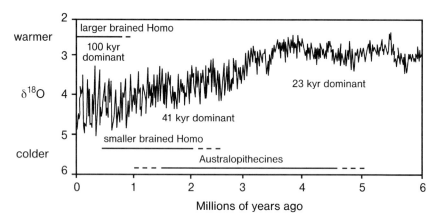

Figure 7.1 A composite marine core record of climate deterioration since the mid-Miocene. During cold periods the oceans are enriched in the heavy ^{18}O isotope, which serves as a proxy for paleotemperature. Periods during which different orbital quasi-cycles dominate the variation in paleotemperatures are indicated. The time lines for important groups of hominin taxa are indicated. Australopithecines include both gracile and robust forms from *Australopithecus ramidus* to *A. robustus*. The line for small-brained *Homo* includes *H. rudolfensis*, *H. erectus*, and *H. ergaster*. The line for large-brained *Homo* includes *H. heidelbergensis*, *H. neanderthalensis*, and *H. sapiens*. Redrawn from Opdyke (1995), Hominin time lines adapted from Klein (1999).

are at present poorly understood, but the upshot is the periodic advance and retreat of high-latitude and mountain glaciers on a massive scale, cycles of aridity in the tropics, changes in ocean current systems, and a host of related environmental changes. For example, drops in mean global temperature affect the earth's water budget. Cold periods mean less water vapor generated by evaporation from tropical oceans and less rainfall over the continents. Deserts and grasslands have expanded greatly at the expense of forests in the last few million years, and wetter and drier habitat types expand and contract dramatically over glacial–interglacial cycles. Whole suites of new mammalian species evolved to live in these drier environments, including our early bipedal ancestors (deMenocal 1995; Potts 1996a,b; deMenocal 2004).

7.3
Climate and the Evolution of Large Brains and Cultural Artifacts

The most important factor in the evolution of hominins and other large-brained creatures was likely the development of variation at quite short time scales. Animal learning and, as we shall see, human culture are adaptations to environmental changes on time scales of less than one to perhaps a few hundred generations. Animals adapt to changes on the order of thousands of generations by range changes and organic evolution rather than by mechanisms of phenotypic flexibility. The variation driven by the relatively long time scales of the orbitally tuned glacial quasi-cycles has undoubtedly been important in the evolution of many species (Vrba et al. 1995), but the advanced cognitive skills of humans and the rich culture that it supports must be driven by the need to adapt to much more rapid variation. That is, sophisticated nervous systems are mechanisms for adaptive phenotypic flexibility. Large brains are very costly organs and must pay their keep (Aiello and Wheeler 1995). Ordinary learning is an individual-level adaptation and hence is especially suited to environments that vary from generation to generation or to changes that occur within a generation. Human social learning (culture) is a complex and sophisticated system. It is suited to rapid adaptation on time scales ranging from less than a generation to tens or even hundreds of generations, depending upon how the cultural system is structured. High-resolution ice cores raised from the Greenland ice cap in the early 1990s were the first to document a stunning amount of variation on millennial and submillennial scales as illustrated in Figure 7.2 (Johnsen et al. 1992; Alley 2000). Such high-frequency, high-amplitude variation has characterized the last five glacials, whereas the much briefer interglacials have been much less variable, roughly like the current interglacial (McManus et al. 1999). Our knowledge of the onset of high-frequency high-amplitude part of the climate record awaits the collection of longer high-resolution cores from environments like anoxic ocean basins and deep, old lake sediments. No doubt such cores will be produced in the next few years.

We gain some insight into the evolution of high-frequency climate variation by using mammalian brain size evolution as an index of climate variation on the time

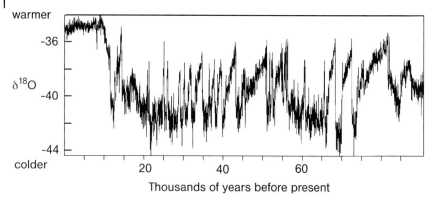

Figure 7.2 The Greenland ice paleotemperature proxy record. During periods of high ice volume ^{18}O is depleted in ice as it accumulates in the ocean. These data are filtered (averaged) using a 150-year low-pass filter, so that variations on the time scale of ≤ 150 years are not portrayed. The Holocene is the little-varying last 11 000 years. Redrawn from Ditlevsen et al. (1996).

scales relevant to phenotypic flexibility. Jerison's (1973) classic study of fossil endocasts of mammal brains indicated a gradual increase in average mammalian brain size throughout the Cenozoic, ending with a sharp increase from the Miocene to the Recent period. The time resolution of this record is very coarse, but it roughly follows the decline in global temperatures and the increase in orbital-scale variation in climate. Ditlevsen et al. (1996) argue that orbital-scale variation should generate a 'variance cascade'. By this argument, increases in variance at low frequencies cause increases in variation at higher frequencies, potentially at scales right down to day-to-day variations in weather. To the resolution of the brain size data, Ditlevsen et al.'s hypothesis seems borne out. Brain sizes increases in mammals roughly parallel the decline in mean global temperature, with much of the increase concentrated in the Pleistocene as high-frequency variation in climate increased dramatically.

The evolution of our lineage over the last few million years has been the subject of intensive study. DeMenocal (1995, 2004) summarized the evidence for the relationship between the evolution of climate and human ancestors. Early hominins, up to and including the apparently many species of *Australopithecus*, resemble modern humans in acquiring bipedal posture but were not encephalized above the range of current nonhuman apes. Australopithecines lived in increasingly arid environments with considerable fluctuation induced by the 21 000-year precessional quasi-cycle. The dominance of the 41 000-year tilt quasi-cycle began about 2.8 million years ago, not long before the first Oldowan stone tool tradition appears at around 2.6 million years ago. The earliest fossils attributed to our genus *Homo* appeared 2.5–2.3 million years ago. The Acheulean stone tool tradition, significantly more sophisticated than the Oldowan, appeared about 1.6–1.7 million years ago, not long after an intensification of the variation of climate in tune with the tilt cycle. *Homo erectus* and related species were once thought to have had larger brains than australopithecines, but recent evidence that their bodies were large suggests that their degree of encephalization was only modestly advanced over that of the australopithecines and earlier hominins, which in turn were not particularly large-brained compared to living apes. *Homo erectus*-grade humans were the first hominins to spread out of Africa, reaching the Caucasus by about 1.75 million years ago (Vekua et al. 2002). This is significant if we suppose that adaptations to temporally varying environments also lead to a propensity to adapt to a wide range of contemporaneous environments. The australopiths were certainly diverse in form, but the version of *Homo* that spread to Eurasia was very similar to the contemporaneous African form, often referred to as the species *Homo ergaster*, considered by some to be a variant of *Homo erectus*. The developmental rate of fossil hominins can be estimated from growth lines in their teeth. A slow developmental rate would allow more scope for learning and culture acquisition. *Homo erectus/ergaster* had a developmental rate only slightly slower than that of living apes and earlier fossil hominins. Later fossils, including *H. heidelbergensis* and *neanderthalensis*, show development rates intermediate between those of *H. erectus* and modern humans (Dean et al. 2001; Ramirez Rozzi and Bermudez de Castro 2004).

The onset of the dominance of the 100 000-year eccentricity quasi-cycle was followed by the evolution of distinctly large-brained hominins, usually referred to as the species *Homo heidelbergensis* or *rhodesiensis*, whose earliest fossils date to around 500 000 years ago in Africa. These populations made tools of a distinctly modern cast compared to the Acheulean industry, although only after about 350 000 years ago following the intensification of the 100 000 quasi-cycle. McBrearty and Brooks (2000) review the classification of the fossils. Human skeletons sufficiently modern to be classified as *H. sapiens*, but of a distinctly archaic subspecies, date back to at least 154 000 years ago in Africa (White et al. 2003). McBrearty and Brooks (2000) outline the progressive modernization of the human toolkit between about 250 000 and 50 000 years ago in Africa (Figure 7.3). By about 50 000 years ago, African tool assemblages contained most of the features that characterize modern hunter–gather assemblages. Some elements of this complex occurred quite early in the sequence, including blade-making technology and artifacts derived from blades such as spearpoints, as well as grindstones and associated pigment processing. Bone tools appeared in the middle of the sequence, and symbolic artifacts like beads and artwork were late. About the time the artifact assemblage became fully modern, modern humans spread out of Africa into Eurasia, replacing earlier populations there (Klein 1999).

If the emerging picture of the parallels between ongoing climate deterioration and the evolution of cognitive and cultural complexity is not misleading, it suggests that hominin evolution could well have been driven by climate and that the lag between the onset of a new climate regime and a biotic response in these lineages was quite short, on the order of $\leq 100\,000$ years. Given that mammals in general evolve rapidly and that in favorable situations many new species of vertebrates can evolve in 100 000 years (Verheyen et al. 2003), the idea that hominin evolution has closely tracked increases in high-frequency climate variation is plausible. The climate data are rapidly improving, as are those regarding fossil hominins and their artifacts.

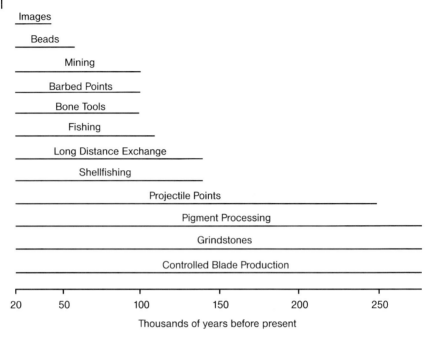

Figure 7.3 Time lines for the acquisition of key components of the artifact assemblage that came to characterize anatomically modern humans in Africa. The spread of modern humans out of Africa took place just after the final modernization. After McBrearty and Brooks (2000).

External hypotheses will become progressively more testable in the near future. The idea that more ancient increases in brain size reflect so-far-undetected increases in high-frequency climate variation reaching back to the beginning of the Cenozoic is obviously more speculative. However, if brain size and associated behavioral sophistication did respond fairly quickly to rapid increases in climate variation in the Plio-Pleistocene, the much slower changes over the preceding 63 million years certainly could be a proxy for climate change.

7.4
Learning and Social Learning as Responses to Variability Selection

Potts (1996a,b; 1998a,b) has made a general case that climate variation on various scales has been one of the most important selective agents on hominins over the last few million years. We think that theoretical models of individual and social learning show in detail how variability selection has favored the evolution of a culture-bearing species (Richerson and Boyd 2004, Chapter 4; see also Calvin 2002). Individual learning and the many other forms of individual-level phenotypic flexibility are the main means that most species use to adapt to the highest-frequency components of variable environments. Very slow environmental variation favors range changes and organic evolution. Theoretical models suggest that social learning is an effective method of coping with environmental variation with a significant generation-to-generation autocorrelation (Boyd and Richerson 1985, Chapter 4). Even on the within-generation time scale, horizontal transmission of socially learned variants can be highly adaptive. Humans are well known to bias their acquisition of variants using a number of decision-making strategies that enhance the effectiveness of social learning. For example, a conformist bias of the form "adopt the commonest behavior in your set of cultural models" is adaptive under a wide variety of conditions (Henrich and Boyd 1998). A basic feature of many models is that a cultural system of inheritance tracks variable environments more rapidly than genes because decision-making forces supplement natural selection. Individuals' own learning and biased selection of cultural variants, summed over many individuals, has the population-level consequence of increasing the frequency of adaptive cultural variants even if these decision-making effects are weak. In very slowly changing environments, any extra costs related to having a social learning system, for example the requirement for a large brain, prevent one from arising, since genes track slow change well enough. In very rapidly changing environments, each individual's environment becomes random with respect to every other's, and attempting to use the behavior of others as a clue to one's own adaptive behavior is useless. Over a broad range of intermediate environmental change between these extremes, social learning is favored.

The models suggest that social learning should be common in animals and that it should have become more common as climates deteriorated. Jerison's (1973) data on brain size evolution suggest that many mammalian lineages evolved larger brains in response to Plio-Pleistocene climate deterioration. Reader and Laland (2002) surveyed primates and correlated field estimates of brain size, innovation, social learning, and tool use. All correlations were strongly positive. Other data suggest that individual and social learning should be correlated, because almost all the known nonhuman systems of social learning depend mostly on individual learning. Nonhumans do not have much, if any, capacity to imitate. Rather, individuals are attracted to the locations or cues that other individuals use to find food, avoid predators, and the like. Then they learn themselves exactly what is good to eat or what should be avoided, mainly by the same individual learning strategies they would use in the absence of social cues. The effect is exemplified by the well studied case of Norway rat social learning (Galef 1996). Rat colonies act like information exchange centers, but all that is exchanged is the smell of foods that a rat has consumed. Rats are prone to sample foods that smell the same as those they have smelled on others in the nest. If the food item is reinforcing, they add the new item to their diet. If not, they do not.

Human social learning differs from that of other animals because we are fast, accurate imitators. Tomasello (1996) and Whiten (2000) evaluated the abilities of chimpanzees and children to imitate using the same tasks demonstrated by human models. Although chimpanzees show some ability to imitate, children as young as two years old do better. Humans are also much more prone to teach each other than are apes. The human ability to acquire complex cultural elements by imitation means that populations of humans can evolve complex adaptive traditions relatively quickly. Innovations made by one person can be added to those of others, so the cultural complexity builds up over time. Other animals, dependent mainly on individual learning, cannot acquire traditions more complex than each individual is capable of mastering on its own. Humans evolve traditional artifacts, skills, and social systems of far greater complexity than any one individual can learn unaided. Kayaks, the calculus, and constitutional government are examples. The runaway success of the human species is due to our ability to evolve a diverse array of technologies and social organizations suited to particular environments. Even without agriculture, humans spread to most of the earth's terrestrial surface, including environments far different from our tropical homeland.

7.5
Internal Constraints on the Evolution of Human Culture

Even if all we have argued about the relationship between climate and the evolution of brain size in mammals generally, and hominins in particular, is true, we have to account for why humans but not other species became highly cultural. Given that humans are a spectacular evolutionary success across such a wide range of environments, why have many other species not converged upon the same adaptation? We can think of four internal constraint hypotheses that might explain why humans are such an exceptional species.

7.5.1
Time

Modern humans evolved only in the last few hundred thousand years, and the last elements of our behavioral repertoire perhaps evolved as recently as 50 000–70 000 years ago. Lags on the order of tens to hundreds of thousands of years are reasonable for the evolution of complex adaptations. Perhaps many species will converge on the complex culture adaptation in the future. Arguably, some other lineages such as whales are some way down this path (Moore 1996; Rendell and Whitehead 2001).

7.5.2
Preadaptations

The hominin lineage had at least two long-standing features that seem to have predisposed humans to adopting culture. The upright posture left the hands free to respond to selection for tool-making. The relatively large brain of our ancestral apes likely gave them an initial advantage in the evolution of a brain large enough to manage imitation and complex cultural traditions. Most students of hominin evolution accept that large brains and free hands played a role in the evolution of humans. Other features have been nominated. For example, Maryanski and Turner

(1992) argue that apes have relatively flexible social systems that could easily respond to selection to become entrained by culturally evolved norms to create highly variable social systems. Social organization does stand beside technology as a major pillar of our system of adaptation by cultural traditions.

7.5.3
Functional Constraints

The capacity to learn complex cultural traditions by imitation and teaching may have difficulty increasing when it is rare (Boyd and Richerson 1996). Suppose that imitation and teaching require a costly investment in brain tissue. When imitation and teaching are rare, populations are unable to sustain complex traditions. Thus, when the capacity is rare, its adaptive advantage – the rapid, low-cost acquisition of complex, locally adapted technology and social organization – is absent. The first culture-capable individual would find no culture to imitate. Even worse, the evidence suggests that human societies have to be rather large to sustain the complex traditions of typical hunter–gatherers (Henrich 2004). On Tasmania, technology became dramatically simplified compared to that in Australia after the flooding of the Bass Strait made it an island and reduced its population to a few thousand. Perhaps humans evolved the intelligence necessary to acquire culture initially as the byproduct of another adaptation that could increase when rare. That human intelligence originally evolved to play sophisticated social games rather than for imitation is one such suggestion (Dunbar 1998).

7.5.4
Cultural Diversification

Major evolutionary innovations typically lead to flocks of species exploiting the innovation, often including convergent adaptations from diverse lineages (Price 2003). Humans are an exception in that a single species has achieved unprecedented dominance of the globe. However, in an ecological and cultural sense, humans *do* constitute a highly diversified adaptive radiation. Men and women tend to divide labor based on different skills traditionally assigned to one or the other. In simple societies, social and technological traditions vary enormously by place and time. In modern societies, the skills needed to practice different professions differ so extremely that few people can master more than one. Once our species became good at rapid cultural adaptation, we probably tended to fill many of the niches where adaptation by tradition was favored, confronting other species with effective competition for the 'cognitive niche' (Tooby and DeVore 1987). Until the late Pleistocene, hominins were a diverse lineage with typically several species extant at any one time. The probable extirpation of Neanderthals and other archaic hominin species by modern humans reduced us to one biological species not long after we became fully modern in the biological sense. Modern humans, however, show much more variation in both stylistic and functional features of their artifacts than earlier hominins. Interestingly, many of the most sophisticated nonhuman social learners are not primates, but birds (Moore 1996) and whales (Rendell and Whitehead 2001), groups of animals with relatively little niche overlap with humans.

The constraints and opportunities reviewed here are probably sufficient to explain why the hominin lineage rather than some other led to the evolution of modern humans. The future of human evolution is much harder to know. Are we a breakthrough species that amounts to the evolution of a new grade of organism (Maynard Smith and Szathmáry 1995)? Or are we an ice-house flower with a costly investment in a cultural system of inheritance that may have no utility in the environments of the future?

7.6
Cultural Evolution in the Late Pleistocene and Holocene

Cultural evolution over the past 50 000 or so years exhibits a roughly progressive dynamic. At the beginning of this period, all humans lived in small-scale hunting and gathering societies. Today, human populations are vastly larger. Both technology and social organization are much more sophisticated. This pattern seems to encourage progressive interpretations without leading to explicit hypotheses about mechanisms that would sustain progress. The details of cultural evolutionary trajectories in contrast lead to several explicit hypotheses invoking external controls of cultural evolution.

7.6.1
Holocene Climates and the Origins of Agriculture

The late Pleistocene cultures of modern humans progressed only slightly, to judge by the excellent archaeological record from Europe. Almost all the progress relative to our ancestral hunting and gathering adaptation is due to the subsistence revolution made possible by plant and animal domestication beginning about 11 000 years ago. Agriculture is a Holocene phenomenon. Holocene climates have been much more stable than those of the last glacial. They also have been wetter on average and have higher carbon dioxide concentrations. We have argued that last-glacial climates probably made agriculture impossible (Richerson et al. 2001). In contrast, Holocene climates have made agriculture compulsory in the long run. Agricultural societies are typically larger and richer than hunting and gathering communities. Thus, farmers are normally able to out-compete hunter gatherers for land and other resources.

7.6.2
Geographic Regulation of Rates of Holocene Progress

The advance of agriculture was highly variable, leading to societies highly dependent on agriculture in the southwestern Eurasia by about 9000 years ago and, in the same region, to the first states about 5000 years ago. In other areas – Western

North America, Australia, and Argentina are the most extreme examples – agriculture came late or not at all to areas highly suitable for agriculture. Each major region had a distinctive pattern of agricultural evolution and of evolution of associated technology and social organization. For example, the development of agriculture and state-level social organization in North and South America has many parallels to the trajectory in Western Asia, but the rate of development was rather slower. Diamond (1997) cogently argues that much of this variation is due to geographical factors. Evolution was fastest in Eurasia, the largest continent with the greatest scope for recombining innovations made by different peoples. Western Asia was especially well endowed with large-seeded grasses and animals preadapted to domestication. The New World was poor in both. Africa harbors more pathogens of humans and livestock than any other continent, and these act as brakes on the expansion of human populations.

7.6.3
Collapses of Civilizations and Other Holocene Events

Complex societies are not particularly stable systems. On the millennial time scale, collapses are quite common. In some instances, such the lowland Mayan city-states, the collapse was rapid, and depopulation was long-lasting and nearly complete. Climate change is one of the commonest explanations for such collapses, and ever-better data give archaeologists the chance to test such hypotheses more rigorously. For example, the Mayan collapse occurred during an extended drought (Curtis et al. 1996; Haug et al. 2003). The vagaries of climate affected the fortunes of less complexly organized societies as well; for example Arctic hunter–gatherers were particularly susceptible to changes in temperature (Jordan 1984; Dumond 1987). At lower latitudes drought is reckoned the cause of widespread cultural disruption among coastal and desert hunter–gatherers and agriculturalists in the western U.S. during what is termed the Medieval Climatic Anomaly (Lindsay 1986; Stine 1994; Jones et al. 1999; Kennett and Kennett 2000; Coltrain and Leavitt 2002) In as profound a way, but much earlier in time, the rapid onset of the extremely cold, dry Younger Dryas likely put an end to Natufian experiments with food production (Richerson et al. 2001).

Suffice it to say that the case for external controls on cultural evolution is quite strong. The record is not consistent with simple progress regulated entirely by internal constraints. Nonetheless, on the millennial time scale, even such a fast-evolving system as culture is liable to meet some internal constraints to its evolution. Soltis et al. (1995) estimated the rates of change of sociopolitical evolution under cultural group selection, based on the rates of social extinction of clan-scale units observed by anthropologists in the Highlands of New Guinea. Very roughly, such rates of group selection support the sweep of a newly favored cultural practice in about 1000 years. Very roughly, the rate of progress of sociopolitical organization in human cultures in the Holocene has had a millennial time scale. Certainly, no trajectory of environmental change parallel to the growth of technical and socio-political complexity in the Holocene is currently known.

7.7
Anthropogenic Effects on Climate Evolution

Ruddiman (2003) presents evidence that the 'Anthropocene' geological era (the time during which humans have been a dominant geochemical/geophysical force) began, not in the last few hundred years, but about 8000 years ago with agriculture. According to his reading of the data, the current Holocene interglacial should have been a brief spike of warmth, as the Antarctic ice data suggest that the last several interglacials were, rather than an 11 600-year-long period of near-constant warmth. The earth is perhaps 0.8 °C warmer than an entirely natural trajectory would predict. CO_2 concentrations began to rise 8000 years ago as early farmers began to clear forests to make their fields. About 5000 years ago rice farmers began to create wetlands on a scale the led to increasing production of CH_4. Ruddiman estimates that before the industrial era began, CO_2 concentrations were 40 ppm higher than natural levels and CH_4 levels 250 ppb higher. The rising concentrations of greenhouse gases due to agriculture were sufficient to counteract the expected cooling trend, leading to the 11 600-year-long span of the Holocene interglacial. This warming is sufficient at high latitudes (2 °C) to have prevented the spread of glaciers in the Canadian Arctic during the last millennium. These glaciers are the nucleus of the North American ice age ice sheet. Ruddiman also argues that the reforestation that accompanied the decline in agriculture in the wake of the Black Death and other plagues after 1300 CE had a sufficient impact on greenhouse gas levels, sufficient to explain the Little Ice Age.

Currently, humans are on a path to produce the strongest climate perturbation in the last 65 million years (Barrett 2003 gives a striking graphical summary). The rapid warming of the earth (and/or the actions that we will have to take to prevent it) will necessarily produce a huge adaptive response, mostly cultural on the part of humans and mostly genetic and in the form of range adjustments on the part of other species. The form this evolution will take is difficult to predict, but the broad outlines of the challenge are straightforward. Climate scientists warn us that at least some of the changes in response to anthropogenic forcing will be large and abrupt (National Research Council 2002). The good news is that human culture arose, if our arguments here are correct, to cope with rapid unpredictable high-amplitude climate variation. We are quite likely to adapt to the 'inevitable surprises' of anthropogenic climate change one way or another. The bad news is that in the past both complex and simple societies have often collapsed in the face of relatively minor climate challenges. Modern science and technology do give us unprecedented tools to understand climate change and devise responses. Modern social systems are also powerful systems for converting science and technology into effective widely supported public policy. Current attempts to reach a global consensus on managing climate change have been shaky at best, but the hour is still early. A reasonably safe assertion is that our immediate descendants are going to live in interesting times.

7.8
Conclusions

Clearly, both internal and external processes have played roles in human evolution. We have here emphasized the possibility that ongoing climate deterioration has been the main driver of hominin evolution. The succession of adaptive grades of hominins with increasingly large brains and more sophisticated culture appear to have evolved as glacial climates have increasingly gripped the earth and as the amplitude of glacial–interglacial cycles has increased. That many mammalian lineages have experienced parallel brain size increases reduces the need to search for causes specific to the hominins. The hypothesis that the direct cause of large brains generally and human culture in particular is increasing high-frequency environmental variation currently lacks strong empirical confirmation; it is rather in the spirit of using theory to search for plausible proxies for past climates. We expect that much richer data on this aspect of climate change will be forthcoming over the next decade. Paleoanthropological data come more slowly, but increases in our sophistication are growing steadily. The hypothesis will be tested more stringently by these data. The cultural evolutionary events that have dominated the past few tens of thousands of years, perhaps the last few hundred thousand, also show dramatic responses to external events.

Internal processes must have shaped the details of biotic and cultural evolution in hominins. The fact that our lineage is the only one to have become so highly cultural, despite similar selection pressures from climate variation acting on most species, suggests that the genetic, developmental, and functional constraints operating in our lineage were weaker than in any other. Or, to put it positively, our lineage was uniquely preadapted to evolve complex culture in ways that are at least in part fairly obvious. The abrupt Pleistocene–Holocene transition and the relatively stable climates that characterize the Holocene constitute a natural experiment to probe the constraints limiting rates of cultural evolution (Richerson and Boyd 2001). Evidently, 11 600 years has been too short a time for cultural evolution to reach an equilibrium adaptation to the Holocene environment. Granted, cultural evolution in the Holocene produced a massive adaptive radiation of cultures. But as impressively swift as it can be, internal constraints make cultural evolution a descent-with-modification process with plenty of similarities to, as well as differences from, organic evolution.

Whether hominin evolution can be considered progressive is a complex issue, given the many difficult issues raised by the concept of evolutionary progress (Nitecki 1988). Culture supports language, knowledge systems like science, social systems of great scale and sophistication, and technological adaptations of radically new sorts. The case for treating human culture as an evolutionary event on the scale of the evolution of multicellularity is strong. A skeptic might worry about the durability of human civilization. Suppose we define progress as evolution in a direction consistent with fulfilling humanistic goals by processes that are strong enough to override any external challenges and internal constraints that our evolution is likely to face in the next few millennia. Our sense is that progressive evolution by this definition is possible. Science and technology have proven highly progressive in a different sense; they have greatly improved our understanding of the natural and social worlds and have thereby given us many practical tools for managing these worlds. Institutional reformers in the modern era have devised social systems that, however imperfectly, put the tools of science and technology to the service of meeting human wants and needs. For the first time in the history of the earth, a species has the tools with which it can aspire to control its own evolution. Whether these processes are powerful enough to overcome such constraints as the proliferation of sectarian ideologies, the concentration of political power in jealous nation-states armed with weapons of mass destruction, and the technical requirements of managing the global environment remains to be seen. Human evolution is currently an adventure in progress struggling to be an event. We have some individual and collective control over our fate, but not nearly enough control to make our future lives boring!

7.9
References

AIELLO, L. C., WHEELER, P. (**1995**) The expensive-tissue hypothesis: the brain and the digestive system in human and primate evolution. *Current Anthropology* **36**, 199–221.

ALLEY, R. B. (**2000**) *The Two-mile Time Machine: Ice Cores, Abrupt Climate Change, and Our Future*. Princeton: Princeton University Press.

BARRETT, P. (**2003**) Palaeoclimatology: Cooling a continent. *Nature* **421**, 221–223.

BECQUEY, S., GERSONDE, R. (**2002**) Past hydrographic and climatic changes in the Subantarctic Zone of the South Atlantic: the Pleistocene record from ODP Site 1090. *Paleogeography, Paleoclimatology, Paleoecology* **182**, 221–239.

BOYD, R., RICHERSON, P. J. (**1985**) *Culture and the Evolutionary Process*. Chicago: University of Chicago Press.

BOYD, R., RICHERSON, P. J. (**1996**) Why culture is common but cultural evolution is rare. *Proceedings of the British Academy* **88**, 73–93.

BRADLEY, R. S. (**1999**) *Paleoclimatology: Reconstructing Climates of the Quaternary, Second Edition*. San Diego: Academic Press.

BROECKER, W. S. (**1995**) *The Glacial World According to Wally*. Palisades, NY: Eldigio Press.

BRONOWSKI, J. (**1973**) *The Ascent of Man*. Boston: Little, Brown.

CALVIN, W. H. (**2002**) *A Brain for All Seasons: Human Evolution and Abrupt Climate Change*. Chicago: University of Chicago Press.

COLTRAIN, J. B., LEAVITT, S. W. (**2002**) Climate and diet in Fremont prehistory: economic variability and abandonment of maize agriculture in the Great Salt Lake Basin. *American Antiquity* **67**, 453–485.

CURTIS, J. H., HODELL, D. A., BRENNER, M. (**1996**) Climate variability on the Yucatan Peninsula (Mexico) during the past 3500 years, and implications for Maya cultural evolution. *Quaternary Research* **46**, 37–47.

DEAN, C., LEAKEY, M. G., REID, D., SCHRENK, F., SCHWARTZ, G. T., STRINGER, C., WALKER, A. (2001) Growth processes in teeth distinguish modern humans from *Homo erectus* and earlier hominins. *Nature* **414**, 628–631.

DEMENOCAL, P. B. (1995) Plio-Pleistocene African climate. *Science* **270**, 53–59.

DEMENOCAL, P. B. (2004) African climate change and faunal evolution during the Pliocene-Pleistocene. *Earth and Planetary Science Letters* **220**, 3–24.

DIAMOND, J. (1997) *Guns, Germs, and Steel: The Fates of Human Societies*. New York: Norton.

DITLEVSEN, P. D., SVENSMARK, H., JOHNSEN, S. (1996) Contrasting atmospheric and climate dynamics of the last-glacial and Holocene periods. *Nature* **379**, 810–812.

DUDLEY, R. (2000) The evolutionary physiology of animal flight: paleobiological and present perspectives. *Annual Review of Physiology* **62**, 135–155.

DUMOND, D. E. (1987) *The Eskimos and Aleuts. Revised Edition*. London: Thames and Hudson.

DUNBAR, R. I. M. (1998) The social brain hypothesis. *Evolutionary Anthropology* **6**, 178–190.

ELDREDGE, N., GOULD, S. J. (1972) Punctuated equilibria: an alternative to phyletic gradualism. In: SCHOPF, T. J. M. (Ed.), *Models in Paleobiology*. San Francisco: Freeman.

EPICA COMMUNITY MEMBERS (2004) Eight glacial cycles from an Antarctic ice core. *Nature* **429**, 623–628.

GALEF, B. G. JR. (1996) Social enhancement of food preferences in Norway rats: a brief review. In: HEYES, C. M., GALEF, B. G. JR. (Eds.), *Social Learning in Animals: The Roots of Culture*. San Diego: Academic Press.

GOULD, S. J. (2002) *The Structure of Evolutionary Theory*. Cambridge, MA: Harvard University Press.

GRAHAM, J. B., DUDLEY, R., AGUILAR, N. M., GANS, C. (1995) Implications of the Late Paleozoic oxygen pulse for physiology and evolution. *Nature* **375**, 117–120.

HAUG, G. H., GUNTHER, D., PETERSON, L. C., SIGMAN, D. M., HUGHEN, K. A., AESCHLIMANN, B. (2003) Climate and the collapse of Maya civilization. *Science* **299**, 1731–1735.

HENRICH, J. (2004) Demography and cultural evolution: why adaptive cultural processes prod uced maladaptive losses in Tasmania. *American Antiquity* **69**, 197–214.

HENRICH, J., BOYD, R. (1998) The evolution of conformist transmission and the emergence of between-group differences. *Evolution and Human Behavior* **19**, 215–241.

HOLLAND, H. D. (1984) *The Chemical Evolution of the Atmosphere and Oceans*. Princeton, NJ: Princeton University Press.

JERISON, H. J. (1973) *Evolution of the Brain and Intelligence*. New York: Academic Press.

JOHNSEN, S. J., CLAUSEN, H. B., DANSGAARD, W., FUHRER, K., GUNDESTRUP, N., HAMMER, C. U., IVERSEN, P., JOUZEL, J., STAUFFER, B., STEFFENSEN, J. P. (1992) Irregular glacial interstadials recorded in a new Greenland ice core. *Nature* **359**, 311–313.

JONES, T. L., BROWN, G. M., RAAB, L. M., MCVICKAR, J., SPAULDING, W. G., KENNETT, D. J., YORK, A., WALKER, P. L. (**1999**) Environmental imperatives reconsidered. *Current Anthropology* **40**, 137–170.

JORDAN, R. H. (**1984**) Neo-Eskimo Prehistory of Greenland. In: DAMAS, D. (Ed.), *Handbook of North American Indians: Arctic*. Washington, DC: Smithsonian Institution.

KENNETT, D. J., KENNETT, J. P. (**2000**) Competitive and cooperative responses to climatic instability in coastal Southern California. *American Antiquity* **65**, 379–395.

KLEIN, R. G. (**1999**) *The Human Career: Human Biological and Cultural Origins*, 2nd ed. Chicago: University of Chicago Press.

LENSKI, G. E., LENSKI, J. (**1982**) *Human Societies: An Introduction to Macrosociology*, 4th ed. New York: McGraw-Hill.

LINDSAY, L. M. (**1986**) Fremont fragmentation. In: CONDIE, C. J., FOWLER, D. D. (Eds.), *Anthropology of the Desert West: Essays in Honor of Jesse D. Jennings*. Salt Lake City: Anthropological Papers University of Utah.

MARYANSKI, A., TURNER, J. H. (**1992**) *The Social Cage: Human Nature and the Evolution of Society*. Stanford: Stanford University Press.

MAYNARD SMITH, J., SZATHMÁRY, E. (**1995**) *The Major Transitions in Evolution*. Oxford: Freeman/Spectrum.

MCBREARTY, S., BROOKS, A. S. (**2000**) The revolution that wasn't: a new interpretation of the origin of modern human behavior. *Journal of Human Evolution* **39**, 453–563.

MCMANUS, J. F., OPPO, D. W., CULLEN, J. L. (**1999**) A 0.5-million-year record of millennial-scale climate variability in the North Atlantic. *Science* **283**, 971–974.

MOORE, B. R. (**1996**) The evolution of imitative learning. In: HEYES, C. M., GALEF, B. G. JR. (Eds.), *Social Learning in Animals: The Roots of Culture*. San Diego: Academic Press.

NATIONAL RESEARCH COUNCIL (**2002**) *Abrupt Climate Change: Inevitable Surprises*. Washington, DC: National Academy Press.

NITECKI, M. H. (**1988**) *Evolutionary Progress*. Chicago: University of Chicago Press.

OPDYKE, N. D. (**1995**) Mammalian migration and climate over the last seven million years. In: VRBA, E. S., DENTON, G. H., PARTRIDGE, T. C., BURCKLE, L. H. (Eds.), *Paleoclimate and Evolution, with Emphasis on Human Origins*. New Haven, CT: Yale University Press.

POTTS, R. (**1996a**) Evolution and climate variability. *Science* **273**, 922–923.

POTTS, R. (**1996b**) *Humanity's Descent: The Consequences of Ecological Instability*. New York: Avon Books.

POTTS, R. (**1998a**) Environmental hypotheses of hominin evolution. *Yearbook of Physical Anthropology* **41**, 93–136.

POTTS, R. (**1998b**) Variability selection in hominid evolution. *Evolutionary Anthropology* **7**, 81–96.

PRICE, P. W. (**2003**) *Macroevolutionary Theory and Macroecological Patterns*. Cambridge: Cambridge University Press.

RAMIREZ ROZZI, F. V., BERMUDEZ DE CASTRO, J. M. (2004) Surprisingly rapid growth in Neanderthals. *Nature* 428, 936–939.

READER, S. M., LALAND, K. N. (2002) Social intelligence, innovation, and enhanced brain size in primates. *Proceedings of the National Academy of Sciences USA* 99, 4436–4441.

RENDELL, L., WHITEHEAD, H. (2001) Culture in whales and dolphins. *Behavioral & Brain Sciences* 24, 309–382.

RICHERSON, P. J., BOYD, R. (2001) Institutional evolution in the Holocene: the rise of complex societies. In: RUNCIMAN, W. G. (Ed.), *The Origin of Human Social Institutions*. Oxford: Oxford University Press.

RICHERSON, P. J., BOYD, R. (2004) *Not By Genes Alone: How Cultural Transformed Human Evolution*. Chicago: University of Chicago Press.

RICHERSON, P. J., BOYD, R., BETTINGER, R. L. (2001) Was agriculture impossible during the Pleistocene but mandatory during the Holocene? A climate change hypothesis. *American Antiquity* 66, 387–411.

RUDDIMAN, W. F. (2003) The anthropogenic greenhouse era began thousands of years ago. *Climate Change* 61, 261–293.

SCOTESE, C. R. (2003) *Paleomap Project* [website]. Available at http://www.scotese.com/.

SOLTIS, J., BOYD, R., RICHERSON, P. J. (1995) Can group-functional behaviors evolve by cultural group election? An empirical test. *Current Anthropology* 36, 473–494.

STEWART, J. (1997) Evolutionary progress. *Journal of Social and Evolutionary Systems* 20, 335–362.

STINE, S. (1994) Extreme and persistent drought in California and Patagonia during Medieval time. *Nature* 369, 546–549.

TOMASELLO, M. (1996) Do apes ape? In: HEYES, C. M., GALEF, B. G. JR. (Eds.), *Social Learning in Animals: The Roots of Culture*. San Diego: Academic Press.

TOOBY, J., DEVORE, I. (1987) The reconstruction of hominid behavioral evolution through strategic modeling. In: KINZEY, W. (Ed.), *Primate Models of Hominid Behavior*. New York: SUNY Press.

VALENTINE, J. W. (1973) *Evolutionary Paleoecology of the Marine Biosphere*. Englewood Cliffs, NJ: Prentice-Hall.

VALENTINE, J. W. (1985) Biotic diversity and clade diversity. In: VALENTINE, J. W. (Ed.), *Phanerozoic Diversity Patterns: Profiles in Macroevolution*. Princeton, NJ: Princeton University Press.

VEKUA, A., LORDKIPANIDZE, D., RIGHTMIRE, G. P., AGUSTI, J., FERRING, R., MAISURADZE, G., MOUSKHELISHVILI, A., NIORADZE, M., PONCE DE LEON, M., TAPPEN, M., TVALCHRELIDZE, M., ZOLLIKOFER, CH. (2002) A new skull of early *Homo* from Dmanisi, Georgia. *Science* 297, 85–89.

VERHEYEN, E., SALZBURGER, W., SNOEKS, J., MEYER, A. (2003) Origin of the superflock of cichlid fishes from Lake Victoria, East Africa. *Science* 300, 325–329.

VERMEIJ, G. J. (1987) *Evolution and Escalation: an Ecological History of Life*. Princeton, NJ: Princeton University Press.

VRBA, E., DENTON, G. H., PARTRIDGE, T. C., BURCKLE, L. H. (**1995**) *Paleoclimate and Evolution, with Emphasis on Human Origins*: New Haven, CT: Yale University Press.

WALKER, T. D., VALENTINE, J. W. (**1984**) Equilibrium models of evolutionary species diversity and the number of empty niches. *American Naturalist* **124**, 887–899.

WHITE, T. D., ASFAW, B., DEGUSTA, D., GILBERT, H., RICHARDS, G. D., SUWA, G., HOWELL, F. C. (**2003**) Pleistocene *Homo sapiens* from Middle Awash, Ethiopia. *Nature* **423**, 743–747.

WHITEN, A. (**2000**) Primate culture and social learning. *Cognitive Science* **24**, 477–508.

ZACHOS, J., PAGANI, M., SLOAN, L., THOMAS, E., BILLUPS, K. (**2001**) Trends, rhythms, and aberrations in global climate 65 Ma to present. *Science* **292**, 686–693.

8
The Human Impact *

Bernhard Verbeek

8.1
Introduction

Thanks to their mental abilities, humans are not merely an interesting novelty in their particular evolutionary line, they intervene dramatically in the ecology of the whole evolutionary structure as well. These abilities have created a culture that initially improved the living conditions of humans. Later on, however, this culture has developed a doubtful dynamics of its own. After more than two million years of modest human history we witness today a technologically supported civilization ensuring enormous population growth.

In general, all living organisms make use of resources wherever they find them. The less efficient organisms have left no trace of their existence. The supply of resources (including energy) has always regulated the population of organisms. Today's humans, however, do not depend only on the physiological energy of nutrition produced by the ecosystems themselves. Modern humans have learned to also use fossil energy, thus implementing gigantic projects which, without energy input, would have been impossible even for intelligent planning beings.

Culture is undergoing a breathtakingly rapid evolution on a new level. This evolution depends on gene-controlled living beings. However, even the most decisive innovations do not require changes in the genetic program. Nevertheless, there are also conservative elements programmed in the human genome – mainly the perception of culture contents – which function like instincts. An adequate conservatism ensures complexity and reliability.

Competition mechanisms intrinsic to evolution, which occur for animals and plants as well, when technically supported, could take excessive forms. The exploitation strategies are constantly getting better, which – having in mind the possibilities of today – could lead to irreversible consequences to the environment. Continuation of the modern tendency of development may replace regional cultures with a unified global cultural system. In its turn, the latter would make impossible

---

* Translated from the German by Maria Wuketits.

Handbook of Evolution, Vol. 2: The Evolution of Living Systems (Including Hominids)
Edited by Franz M. Wuketits and Francisco J. Ayala
Copyright © 2005 Wiley-VCH Verlag GmbH & Co. KGaA, Weinheim
ISBN: 3-527-30838-5

the success-principle of evolution, namely, the selection among different 'offerings'. There would be no more cultural alternatives.

The intelligent human being is the first species intervening in the evolutionary environment that can evaluate the consequences of its deeds. Why these consequences today are evaluated negatively is a major point for discussion. However, there are cultures that effectively use their limited ecological resources. An example to be presented are island communities in the Pacific that make a living from fishing. Here, sustainability does not result from rational ecological considerations but is rather based on social traditions.

Both organic and cultural evolution develop in particular conditional contexts, e.g., climate. It seems that culture has brought the evolution of life to a critical phase. The alternatives here are either to develop new conservation mechanisms or to take a qualitative step backwards.

8.2
How an Extraterrestrial Would Perceive Modern Humans

Today, the human being plays the most decisive role in the formation of the global environment. The present is a strange phenomenon: strictly speaking, it is infinitely short. At the same time, it shapes the future, i.e., the past to come. Being so infinitely short, the present can be revealed only through the past; the same can also be said of the future.

The past leading to the future can be traced in endlessly interwoven causalities back to the Big Bang. The causality network is so complex that no brain could ever really encompass it. One should concentrate on those essential instances that are recognized only retrospectively and assessed as logically compelling. Actually, without knowledge of later developments, nobody confronted with the worm-like chordates would be able to foresee the evolution of dinosaurs, mammals, or human beings, together with their impact on the planet. Thus, even the first primates (more than three or four million years ago) that are defined by anthropologists as definitely human could not – at that time – have seemed of significance.

For the sake of objectivity in the self-estimation process, we can imagine hypothetical intelligent aliens witnessing the appearance of life on earth almost four billion years ago. They would have noticed that life had endowed our planet with an unparalleled surface and unique atmosphere, at the same time conquering all possible niches. Our alien observers (assuming they had overcome the limits of time) would have noticed some global catastrophes induced by cosmic forces. Life had to fight against them, winning in the end.

Their observation of the history of humans, compared to the whole history of life, would have taken but a split second. The species of apes that later led to contemporary human beings separated from other ape species some four million years ago. The first humans in the true sense of the word appeared about two and a half million years ago. The so-called modern human being, *Homo sapiens*, now driving cars, splitting atoms, cooking on electrical stoves, and surfing the Internet has been around more than 100 000 years (for details, Chapter 6).

However, for some four million years, humans and their ancestors simply drifted through world history like all other living creatures. Actually, nothing in those early humans would have caught the attention of our extraterrestrial observers. They were, in fact, particularly unspectacular. They did not destroy whole landscapes as locust swarms do, and they did not trample down trees like the elephants. They did not flood forests like beavers building their dams. They did not erect sea barriers as the corals did. Their housing (if they had any) was primitive – one could compare them only to the air-conditioned structures of termites or the nests of small birds like weaver finches or wrens. Finally, they caused no mass extinctions as some microorganisms did. For a comparatively long period of time, human beings were not important. The human population was small – with a density of one person per four square kilometers (almost the same as that of contemporary hunter–gatherers in Congo (Heymer 1996)). In a word, the early humans did what other organisms have always done – they used the available ecological resources as well as they could. Perhaps our extraterrestrial observers would have noticed a particular versatility in the human use of the environment – some 700 000 years ago, at the level of *Homo erectus*, humans began to make use of fire. However, from the perspective of an extraterrestrial, there were more spectacular beings to be studied.

Thus it would be understandable if our alien observers, after some hundred thousand years, lost interest in 'modern' man and decided to take a sabbatical lasting a few more tens of thousands of years. If they chose to continue their observations now, they would immediately notice that the human species has long been the most important species on the earth. It has changed its behavior dramatically: it reforms the planet according to its momentary needs to an extent not known before. It can actually change the planet's whole appearance in the shortest of times. At the same time, it seems that the human population is growing immeasurably (Figure 8.1). An earlier law says that, despite some deviations, populations tend to remain constant for long periods of time. The present situation negates this law: on the one hand, we have the steady growth of the human population, and on the other a dramatic decrease in the numbers of other species. Statistically speaking, every 20 minutes we witness the extinction of one species. Coral reefs die, forests melt away, the number of useful plants covering large territories diminishes. Enormous water reservoirs are constructed while elsewhere whole regions dry out. The extraterrestrial observers would now perceive a mass extinction of plant and animal species (e.g., Wuketits 2003).

The land areas covered with buildings and sealed with asphalt grow alarmingly. People and resources are obsessively transported here and there in all possible types of vehicles and with enormous waste of energy. Energy resources are mercilessly exploited even in the remotest corners of the world. From time to time the alien observer is compelled to believe that humans are manipulated and kept as slaves by large and small machines. It seems that the only goal of human existence is to satisfy the needs of their masters – the machines.

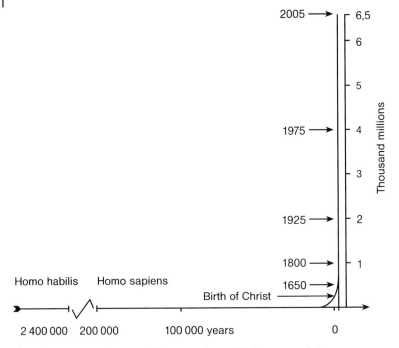

Figure 8.1 For more than two million years, the world's human population remained below ten million individuals (from Verbeek 1998a, actualized).

In a ludicrously short period of time – no more than several decades – humankind is changing the equilibrium of the earth's atmosphere, causing a constant warming of the climate – a situation not experienced for millions of years (compare Gassmann 1994, Bengtsson 1997, Döös 1997). The long-lasting inconspicuousness of humans has come to an end. *Homo sapiens* has all of a sudden become as important for further evolution as a gigantic meteorite smashing into the earth's surface. Our planet is turning more and more into an inhospitable place for a growing number of species; they are leaving forever – by extinction. In the end, humans themselves will be confronted with a global desolation caused by their own activities.

And our imaginary alien observers would be wondering why the most intelligent living being on the earth diligently works toward its own destruction. Among others, this question is discussed in this chapter.

8.3
Pre-established Harmony

One should accept that each living being, including humans, perceives the world from its own situation or perspective. The denial of this fact inevitably leads to a narrower world picture: egocentric, ethnocentric, and anthropocentric. What is more, one tends to perceive everything through the perspective of one's own time.

That is why it is small wonder that the nonbiologist tends to see evolutionary research as a classification of bones demonstrating (as expected) the gradual development of a human being. These naive anthropocentrists consider the appearance of contemporary humans as the climax and fulfillment of evolution.

The readers of this book are of course aware that such a childish approach is untenable. Evolution did not come to an end with the rise of humans; they are a rather recent factor in the history of life, a factor with an enormous and un-precedented impact on the drama of evolutionary happenings. Evolution is an inevitable process that is not to be limited only to life. This process is a part of the cosmic 'essence' and as such is an intrinsic characteristic of our planet and cannot be stopped, not even through the evolution of man.

Einstein used to joke that one shouldn't worry about the future – it will come inevitably. Actually, thoughts and worries about the future are a constant and essential feature of humans. We are sensible enough, considering the fact that although we indeed cannot prevent the future, we are very able to affect it actively. Animals' (and plants') activities influence the future as well – however, they don't worry about it. Nevertheless, they act teleonomically – for example, when they gather and store food for the cold season. Thus, they *form* their future – in the sense of aiding their survival and the survival of their offspring. They have their answers pat – instinctively – to the normal challenges of nature, i.e., to the succession of seasons. In this sense, instincts are well tested answers to questions accompanying the whole phylogeny of the organisms.

These (normally) correct answers are not the result of insight but of genetically conserved programs. Developed on the basis of previous mutation and selection processes, these programs reflect situations that have regularly repeated. Programs that do not adequately answer such situations have no chance of survival. The outcome of this merciless selection looks like something planned and intended. In earlier times, philosophers interpreted nature in exactly such a way; Leibniz used the famous notion of pre-established harmony in this context. What one experiences as ingenious planning is nevertheless to be perceived evolutionarily as post-established, as the result of a simple logic which – although trivial – is obviously not generally recognized, namely: systems that do not adequately respond to the environment, so that they are able to function, simply cannot exist; systems that do not reproduce cannot survive.

Naturally, this is valid also for humans – for their biological existence and their cultural creativity. Humans are able to reflect about the future and to make necessary decisions in a way animals do not. Thanks to this ability we transgress, so to speak, instinct-guided provision for the future. At the same time, because of this very ability we experience a permanent awareness of insecurity and apprehension.

Sure, there have always existed salvation hopes and visions of the end of the world. In 1798 Malthus proclaimed his concern about limited resources opposed to population growth. This concern later provoked Darwin (according to his own words) to formulate his theory of evolution by means of natural selection. The main point was that the potential for over-reproduction is more strongly developed in other species than in humans: one should think about the thousands of eggs laid by the female frog. Nevertheless, only very small parts of the offspring survive – the fittest ones. The fittest are those individuals that (under the particular circumstances) can most successfully use for themselves the spare resources and are able to find sexual partners and live long enough to transfer their genes to the next generation.

The same is also true of humans: 'survival of the fittest' has practically formed their perception and constrained their social behavior and moral conduct. I shall return to this problem later in the chapter.

8.4
The Progress of Civilization

The unexpected progress in industry and agriculture, and especially the development of artificial fertilizers by Justus Liebig, pushed away the gloomy visions of Malthus. There came instead a rose-colored vision of the development of the world: *progress* was defined not as a keeping pace with the times (which would have been the correct way to define it) but rather as general improvement, as unchecked movement toward something better. Such progress was considered to be the feature of both organic evolution and civilization.

This positive evaluation has been opposed by the critical approach of recent decades (Rapp 1992, Wuketits 1998). Indeed, one cannot but agree that our life has become in many ways much more comfortable, at least for the well-off citizens of industrially highly developed countries and affluent societies. At the same time, however, we are also dramatically confronted with the negative consequences of human activities – enough reason to worry. Toxics have entered the food chain; clean air, pure drinking water, and fertile agricultural lands have become rare in many places. The ozone layer protecting the earth from the dangerous ultraviolet rays is disappearing. This leads not only to an increase in instances of skin cancer; it affects the ecosystems and thus the whole evolutionary process to an unknown extent. Human activities are apparently responsible also for the higher concentration of various gases in the atmosphere which influence the radiation impinging on the earth in an unpredictable way (e.g. Bengtsson 1997). It seems that the harmony between living beings and their environment, as well as the ecological framework of the earth, has been endangered by humans. As a result, our planet is developing into a more and more inhospitable place for humans themselves.

Today, mainly under the pressure of the electorate, these problems have acquired a place of importance on the agenda of politicians in developed industrial nations. A number of international conferences have the reduction of greenhouse gases as their goal (Stieger 1995). The dramatic rise of insurance payouts resulting from human-made catastrophes and the unimaginable damage potential of industrial installations and atomic power stations make us, as well the financial world, aware of these problems. The Chernobyl accident in 1986 made parts of Ukraine uninhabitable. "The Global 2000 Report to the President" (Barney 1980) expected the extinction of 15%–20% of the existing species by the end of the twentieth century.

However, most living species are still unknown to biologists. For this reason it is not possible to say whether this prognosis was correct. It seems that, meanwhile, the situation has become even worse (Nentwig 1995, Wilson 1997, Wuketits 1999, 2003). What is nevertheless sure is that the number of existing species is decreasing much quicker than – according to our knowledge of earth's history – new species appear (Kaplan 1989). It seems evident that the presence of man on earth has had a dramatic impact on all bioevolutionary processes.

In this sense, the human being is to be seen not only as the wonderful ascent of consciousness from the dull ocean of unconscious evolution, a being endowed with the light of knowledge, but also as a cosmic catastrophe threatening biological evolution. Humans as a natural catastrophe (Wuketits 1998) influence their environment to an extent that is unprecedented in the long history of the earth. It is to be compared with the extermination of most of the species at the end of the Permian or the extinction of the large dinosaurs at the end of the Cretaceous. Those events, caused most probably by asteroids, led in both instances to a turn in the evolutionary processes.

The new element in the human impact thus has to be looked for not in the size of this impact but in the fact that it is the result of conscious activity of a living being. One cannot but wonder that, despite our consciousness, we are not able to break away from the practice of destroying the ecological conditions of our own existence. The origins of this behavior (as of everything else) are to be traced to the past, i.e., to our evolutionary history. This aspect is further discussed in this chapter, aiming not only at revelation of the respective interrelations but also at the formulation of possible solutions.

8.5
Ecology and the Human Being

For a long time, ecology, i.e., the economy of nature, was not considered to be a particularly important field of interest. What is more, those who dealt with it had always excluded humans from their studies. This tendency has changed. With the publication of *The Silent Spring* (1962), Rachel Carson was one of the first to point out (with enormous impact) the anthropogenic impoverishment of the world. A further important step in this direction proved to be the Club of Rome study *The Limits to Growth* (Meadows 1972), initiated by MIT. The result was a passionate and controversial discussion and an ever-growing number of publications focusing on human ecological problems, e.g., environmental–toxicological and climatologic, as well as problems of humans in a narrower sense. The sheer amount of such publications does not allow me to summarize them in the context of this chapter (for a review, see Verbeek 1998a).

Descriptions of particular – well functioning ecosystems or of the global ecological system normally underline their sustainability. Indeed, the general framework of animated nature represents a system that has not broken down for almost four billion years but has rather persisted. Hence, one should presume that this system is stable and conservative. Such a successful conservation could have been achieved precisely because the system elements are highly dynamic and alterable. Ecological systems are exposed (as is any particular living being) to extremely dramatic processes. One could perhaps even say that biological systems are processes. Even a forest that has survived thousands of years has undergone periods of breakdown and renovation in mosaic-like zones. It is primarily these dramatic changes that make the forest stable.

Some people believe that life has to be explained through a specific force that cannot be reduced to physical principles. Vitalists and neovitalists claim the existence of a *vis vitalis* or *élan vital*. And indeed, it seems that life permanently violates the second law of thermodynamics – the law of entropy. Popularly expressed, this law says: left alone, everything goes downstream and would inevitably lead into self destruction; everything aims at the greatest disorder, at the highest probability, at thermal death. However, life exists although its existence is highly improbable.

Today we are aware that life is not to be explained through some vital force. Nevertheless, it is something special: it possesses structures and information that allow it to achieve order in spite of the otherwise unchecked progress of entropy (Schrödinger 1944). It is thus possible to fight against the destructive forces of entropy, to sail, so to say, against the wind. Life can use the downward current of entropy as an impetus for its upward movement. Figure 8.2 illustrates this possibility with the example of a mechanical analog. To sustain the necessary entropy increase, each organism needs a constant supply of energy that is later released as heat. Naturally, this is valid also for humans. We get our energy in the form of food. Without its constant delivery we inevitably (and relatively quickly) starve to death.

Energy accumulates in food through the sun – directly in plants, and indirectly in animals and humans. Plants acquire energy by photosynthesis. With the help of evolutionarily originated cell structures and through a complicated process (which is not discussed here), CO_2 and water are combined to form carbohydrates. In the course of this process oxygen is produced as a byproduct. The process itself requires energy (to be compared, e.g., to the way water is pumped in a pumping plant).

Figure 8.2 Upward movement is possible with the help of a downward current.

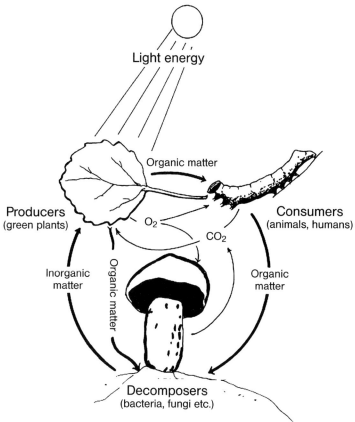

Figure 8.3 The ecological cycle.

In plants this energy is obtained from light photons. Thus, photosynthesis is a process that charges the ecosystem with energy supplied by the sun. To support their energy-consuming life processes, the plants (or other organisms – human beings as well – using plants as nourishment) can, when necessary, use this reservoir by respiring carbohydrates. The latter combine with oxygen to produce the original substrates – carbon dioxide and water.

Decomposers – organisms that obtain their energy from waste products or dead plants or animals – also act in this way. They decompose the complex previously living matter into their former mineral elements, to be used again by plants. This activity is extremely important for sustainability and is carried out by small animals as well, but primarily by microorganisms and fungi.

Thus, there exists a permanent cycle characterizing matter which regenerates itself and which requires a constant supply of energy. The whole earth can be perceived as an ecosystem with a seemingly perfect sustainability. It exists on the basis of atomic fusion processes taking place in the sun, where hydrogen is transformed into helium, releasing energy. The system will remain intact as long as the sun exists (Figure 8.3).

8.6
Evolution of Sustainability

Unfortunately, this illustration is grossly simplified. On the one hand, today humans influence the local and global cycles enormously (which is further discussed in this chapter). On the other hand, the illustration completely ignores the evolutionary aspect of ecology. Most probably the first extremely simple bacteria-like living organisms appeared some 3.8 billion years ago under the infernal conditions existing at that time. Humans would not have had the smallest chance of surviving – the atmosphere did not even contain free oxygen but instead huge quantities of carbon dioxide.

The 'discovery' of photosynthesis by the cyanobacteria proved to be a decisive step in the guarantee of a sustainable energy supply. Nevertheless, one could hardly speak of ecological cycles existing at that time (the situation could be compared to the contemporary non-recycling economic system). The growth-oriented cyanobacteria used their specific advantage 'without consideration'. A generation of cells overgrowing its ancestors cut the latter from the necessary light supply. Huge quantities of carbon had been extracted from the atmosphere through carbon dioxide fixation and had been stored in the lithosphere. The free oxygen produced in the process had initially oxygenated the richly present bivalent iron, which then – as rust – sank to the bottom, thus enriching it. The lack of carbon dioxide and the more aggressive oxygen gradually posed a problem. However, in spite this ecological crisis, the cyanobacteria continued to do what life had always done: they used all available recourses for their own growth. Later – for some 1.5 billion years – after the oxidization of the soluble iron in the oceans, the oxygen ascended in the atmosphere as well (Schidlowski 1981).

Continuing these developments, the photoautotrophic organisms would have been left without the necessary carbon dioxide, which would have inevitably caused the end of life at its very beginning. The evolutionary creativeness, however, found a way to use the free oxygen to gain energy through oxidation. Aerobic metabolism is approximately 18 times more efficient than the process of anaerobic fermentation. The availability of free oxygen eventually made a sustainable producer–consumer–decomposer cycle possible (Figure 8.3).

The freely available oxygen supported the rich development of life on the continents as well. Having in mind the present impact of humans on the biosphere, we can say that the carbon era – some 300 million years ago – is of special importance. At that time, long before the emergence of mammals, a kind of technological revolution was already anticipated. Although a balance between photosynthesis and respiration was firmly established, the principle of ecological cycling was violated at various places on the earth. Repeatedly, plants that became isolated (e.g., because of floods) underwent mass extinction, so that the decomposers had no access to the organic material and could not energetically use it by means of oxidation. Reduced carbon, therefore, became fossilized and remained an energy resource. Mineral oil developed in a rather similar way: probably in highly productive aquatic biotopes with oxygen-free deep zones.

These processes took place long before the appearance of humans. Nevertheless, they laid the foundation for the human ecological impact today. The reduced organic matter remained intact not because of any future-oriented rationality of the organisms. Rather, for most of the time no systems able to use this energy potential existed. The situation has however changed.

8.7
Back to the Past

Our civilization is based on a more or less unlimited energy supply. This has not always been that way. During most of their million-year history humans were able to use only the energy stored through photosynthesis in the ecosystem a (relatively) short time before: their physical power was based on the energy of metabolism: the power of domestic animals, the biomass of plants, wood for cooking and heating, etc. In the Bronze Age humans began to use wood also for smelting metals and other technical activities.

The production of foodstuffs – which mostly consumes energy – in antiquity led to enormous exploitation of the ecosystem. Huge forests disappeared by exploitation or to make room for the development of agriculture. A retrospectively logical causal chain had been established: improved techniques for food production made it possible to feed more people; more people meant more intensive and accelerated cultural development, which facilitated technological progress and made even larger projects possible. However, such gigantic structures like the pyramids, the Great Wall of China, roads, and canals in antiquity were built in the name of a few rulers who possessed other human beings as slaves. (Who of course had to be fed as well.)

In modern times, the invention of the heat engine fundamentally changed the situation. Projects about which one could have only dreamed earlier became reality. Efficient dams, canals, roads, and bridges were built; riverbeds were changed. Mountains were flattened; tunnels were built. Almost every transport problem was soluble. The whole surface of the earth was gradually turning into a construction site.

Heat engines work – analogous to living organisms – by consuming energy in high-quality form and releasing low-quality energy in the form of heat. Using the intelligence of man they ensure their 'nourishment' themselves: with the help of machines and with the help of machine-produced energy, technology can use the accumulated fossil resources – something unknown in the history of any living being. In terms of chemistry, the 'pyrotechnical stage' (Rifkin 1985) in the development of machines recalls the processes of respiration in organisms: the energy required for the maintenance and enlargement of machinery is produced through the oxidation of carbon-containing matter. In living organisms these processes are regulated by enzymes. In machines they are technically organized and accompanied by extreme heat production.

The reduced carbon accumulated in the depths of the earth is being brought to the surface and burned in the process of releasing energy. In this way the carbon

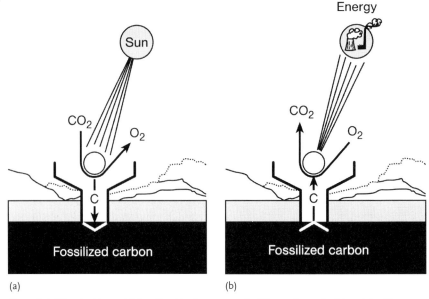

Figure 8.4 What had been (a) fossilized in the course of millions of years is (b) exploited in a year.

dioxide that had been bound in earlier geological periods is released back into the atmosphere. At the same time, an equivalent quantity of oxygen is withdrawn from the atmosphere. The evolutionary–ecological process of carbon fossilization and oxygen release is thus reversed, and the energy stored in earlier times is freed once again (Figure 8.4).

Of particular importance in this context is the following generalization: living beings use resources wherever they find them. The same can be said about technical civilization as well. At the same time, living beings are limited to their own time period – also with respect to their energy sources (time travel is possible only in science fiction). Technology however can use the past to the full. In a year it burns the energy fossilized and accumulated over the course of perhaps one million years. Important is also the fact that the fossilized carbon and the hydrogen usually accompanying it are used not only for producing heat and technical energy but also for preserving the iron ore accumulated on the bottom of the oceans during the process of biotic release of oxygen. To produce metallic iron carbon is needed not only to generate heat for the technical process, it is also an acceptor for the oxygen, which has to be reduced from the ore.

Today we are witnessing a dramatic increase in carbon dioxide levels due, on the one hand, to the combustion of fossil fuels and, on the other to the clearing of forests. In the course of a few decades the concentration of carbon dioxide in the atmosphere will be twice the concentration we would have had without human activity (compare Barney 1980, Rifkin 1985, Gassmann 1994, Nentwig 1995, Brune et al. 1997). Naturally, this enormously influences the climate and thus the ecology and evolution. Vegetation belts will be shifted, as will the community structure of ecological systems. Thus, plants that are productive under conditions of lower carbon dioxide concentration will be replaced by plants that are more efficient at higher carbon dioxide concentrations.

8.8
Culture: Evolution at a Meta Level

Technology, which has made all this possible, is an important aspect of our culture and civilization and is the result of human activity. Thus, analysis of the human impact of technology is intrinsically connected with the study of human cultural capacities, which are – as is everything else in the world – a product of evolution.

It was previously mentioned (Section 8.3) that each living being is endowed with information, tools, and algorithms, which have guaranteed its and its predecessors' survival in an environment hardly to be described as a friendly one. The evolution of life is characterized by ever-growing complexity: from the regulation of metabolism at the molecular level to the instinctively guided capacity for learning, to intentional behavior, and up to culturally imprinted conduct. The lower levels, e.g., metabolism, occur in all organisms, the highest-level capabilities only if a sufficiently developed complex nervous system exists.

Under 'culture' we should subsume not only arts, music, and theatre. Rather, the notion of culture should encompass all forms of expression of human minds that transcend time and generation: built in stone, composed in oral or written word, painted on walls, carved in rocks, cast in brass or in paragraphs. Most of all however, culture is realized through everyday traditions and habits: the mother tongue; methods of food production, supply, and storage; customs of dressing, breeding, education, mutual help (or harm), as well as value systems. Of course, one should not forget the development of items ranging from handicrafts to high technologies. In other words: *our entire civilization.*

Without doubt, much of what we call culture is also found in animals. Titmice traditionalize the opening of milk bottles, red-faced macaques developed a habit of washing foodstuffs or of enjoying a thermal bath. Chimpanzees catch termites and ants with the help of prepared branches; they open nuts with hammers and anvils and are known to use herbs for medical treatments. To the things traditionalized by animals one should add even the recognition of social status and the image of an enemy (Kawai 1965, Boesch 1981, Bonner 1983, Goodall 1991, Paul 1998, Sommer 1989, 1999). Different kinds of birds use different 'dialects'.

It is really just a question of definition whether trans-generational teaching and learning among animals can be considered culture (Bonner 1983) or – to avoid disputes with anthropocentrically inclined opponents – precultural behavior (Kawai 1965). Such disputes are really of minor importance as long as they reflect semantics and not the facts themselves. It is, however, important to point out that only human cultural activity has demonstrated an explosion-like impact, especially in the present time. One has the feeling that in humans a 'critical mass' of a highly explosive material has been surpassed, thus releasing a self-maintaining chain of reactions.

This critical mass refers, on the one hand, to the biologically programmed (DNA) human brain capacity and, on the other, to the intensive interactions between humans and, not least, to the phenomenon of culture itself. If we want to draw a limit between the traditions in the animal world and human culture, then we could use exactly this qualitative leap over the limit of the critical mass, which is not possible in animals until now (Verbeek 1999).

The brain capacity of humans has not significantly changed for the last 120 000 years – since the appearance of *Homo sapiens*. Culture has existed somewhat longer. Unfortunately, the most ancient cultures are poorly documented. This is at least partially because most of the evidence is extremely perishable. Thus, language, which is perhaps the most important cultural carrier, has been documented only in some cases for the past ~5000 years – since the invention of writing (the oldest known writing system dates back to around 5500 BC, see Haarmann 2004, Section 3.2). Then again, the prehistoric culture should have been much less developed. Nevertheless, the cultural thread – like the genealogical thread – has never been totally broken. Otherwise, the consequences would have been fatal also for the genealogical thread. Traditions have been indeed been preserved through generations. At the same time they were spreading only slowly and not explosively. This was partly because the population density was extremely low and the separate human groups were relatively isolated. This situation could be compared to the radioactive bars in an atomic reactor which in the normal working regime are kept well separated from one another to prevent the formation of a critical mass.

Improvements in food production (mainly in agriculture and animal husbandry) supported the growth of population. It became possible to maintain larger populations in towns. The previously unknown high concentrations of people facilitated interactions and feedback between ideas and innovations. The most important cultural achievements (positive as well as negative) – i.e., trade, religion, writing, art, science, war, technology, and the use of fossil resources – are well known. Today it is mainly the information and communication technologies that contribute to the acceleration of cultural evolution. They enable the accumulation and processing of enormous quantities of data. Distances are no longer relevant in hindering telecommunications. Today an idea born in America becomes known almost immediately in Europe or Japan. In pre-Columbian times the Europeans did not even know of the existence of America and vice versa.

8.9
Genome and Culture

Traditionally, one tends to oppose nature and culture, matter and mind, evolution and history as completely different concepts; however, such distinctions are artificial. Nature originated the phenomenon of life, which resulted in the appearance of humans and the human mind. Culture is the consequence of the development of the mind. So all this of course takes place in the framework of nature.

Culture is indeed a part of nature and a part of the evolution (Verbeek 2000). Nevertheless, it has developed a dynamics of its own which dramatically influences evolutionary processes. Its importance is perhaps as powerful as the great events in the past evolution: the origin of life, the inventions of photosynthesis and of respiration, and the development of the nervous system. What is more, with respect to the efficiency and speed of its impact and further realization, culture considerably surpasses these events. Hence it is not only fascinating but of fundamental importance to study the interrelations between culture on the one hand and its evolutionary genetic carriers on the other.

Before the discovery of genes, the theorists developed a view that to a certain extent still exists today: in the course of the earth's history organisms were changing under the pressure of selection like plasticine under the pressure of a sculptor – hominids developed upright posture, their teeth became smaller, their brain increased in size, and a kind of human being emerged (Figure 8.5). Like single frames in an animated cartoon, the different developmental stages merge in our perception into a linear process.

The fact that this idea still dominates the imagination of many is due mainly to our way of thinking, which is rather oriented toward the concrete although biological evolution has always taken place at an abstract level – so obscured like the hidden text of a computer program. One could call this virtual level of abstract genetic programs, written in the language of DNA, the 'level of the primary ideas' – in reference to Plato. The tangible organisms could be described as a test run of these primary ideas.

The primary level is – biologically formulated – the level of the genes (Figure 8.5). This level carries the information necessary to structure a particular organism and to determine its functioning – naturally, with reference to its environment.

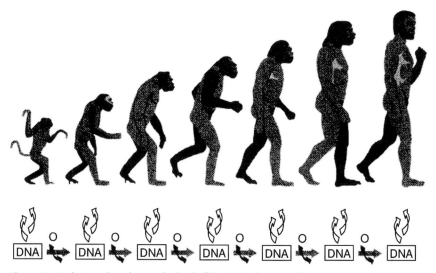

Figure 8.5 Evolution takes place at the level of the DNA, the 'level of the primary ideas' (from Verbeek 2004).

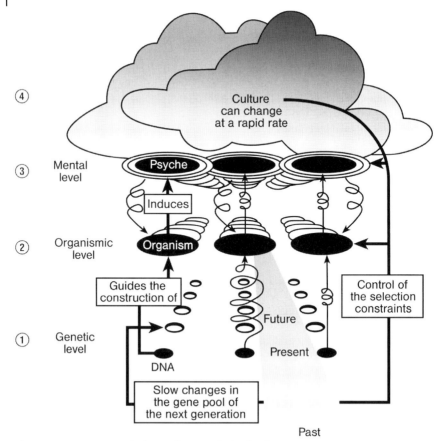

④

③ Mental
level

② Organismic
level

① Genetic
level

Figure 8.6 Connections and relations between the levels of genes,
organisms, psyches and culture (from Verbeek 2004).

Mutations and recombinations take place on the level of the DNA (level 1, Figure 8.6)
in each generation. Characteristics that have proved successful in two organisms
that mate are combined in a third organism i.e. the next generation. These mechanisms for implementing innovation have supported the adaptation abilities of the
genealogical lines of all recent organisms to such an extent that these lines can still
be traced today. In this way they have been able (under all possible circumstances)
to produce enough drafts that could be realized – and thus pass the selection test –
on the level of organisms (level 2). The requirements have been not easy to meet –
life has often demanded the most improbable structures and very specific information. Most of living beings could not stand the pressure of selection and their line
became extinct. The genetic level of higher organisms contains the necessary
information for structuring an efficient brain, inducing a new level: the mental
one (level 3, Figure 8.6). This level opens again new dimensions of self-regulation
and feeds back to the organism. Hence, the carrier of this mind gets access to new
resources, and the genetic program obtains better chances for reproduction.

However, if a nervous system is not sufficiently adapted to the environmental conditions, the mental level can cause somatic diseases or even lead to the death of the organism. Of course, mere mental interaction can also cure the organism. Such a psyche increases the degrees of freedom and thus permits the illusion of total independence of the organism from its material substrate. Here we may find the reason for humanity's hubris, i.e., the belief that we were created in the image of an omnipotent God outside of – and independent of – nature. Although they may have some ecological understanding, many people therefore still tend to forget that they are firmly rooted in nature.

In social beings like humans, there is always intense mental interaction. Humans use the same means of interaction that we know from the animal kingdom, namely, means of nonverbal communication: the visual, the olfactory, and the nonverbal body language and its perception. However, in modern humans these phenomena exist in the shadow of a new and unique communication system: *verbal language*. Its enormous efficiency and possibility to adapt to changing requirements can be regarded as the precondition of culture. Language allows us to transgress, so to speak, the already existing tradition of learned content.

Changes at the organismic level were closely connected with changes at the genetic level, although adaptive modifications at early stages of evolution allowed some freedom of reaction to changing environmental circumstances. Adaptive modification relies on genetically prepared alternative programs that, if required, start to act harmonizing with the environmental conditions. This can concern metabolism, for example, in bacteria that produce the expensive enzymatic tools for the reduction of lactose only when lactose is available in the nutrient medium. In the absence of lactose, the genes for these enzymes remain inactive so that the bacteria do not waste resources producing the enzymes. Adaptive modification can also be of some significance in plants that grow on roadsides and thus run the risk of being trodden down. If they do get run over, they can survive by remaining comparatively small and growing horizontally rather than vertically. Another example is the human skin which, in response to solar radiation, usually gets darker to avoid being damaged by this radiation. Significantly, the skin of many white people, when less exposed to the sunlight, remains fair. The adaptive advantage here initially was that in the winter, enough ultraviolet light has to pass through the skin to support the synthesis of vitamin D. The same modifying ability is observed also in people with darker skin; however, their skin-darkening begins at a higher intensity of solar irradiation.

These modifications have already been established at the phylogenetic level because of the reappearance of the respective situations in the course of a species' history. The establishment of a totally new feature requires the long and multi-generational process of trial-and-error adaptations on the genetic level. At the mental level (level 3, Figure 8.6) the processes are clearly more complicated. Nevertheless, genetically prepared alternative programs that have been preserved by selection exist, to be used if necessary – *instincts* (see Section 8.3).

The cultural level (level 4) works in a completely different way. In humans, adequate 'brain ware' already exists, thanks to their efficient genome. Hence the breathtaking rate of evolution on the level of culture – the 'level of secondary ideas' – requires no changes on the genetic level – the 'level of primary ideas'. As already shown, this new level of evolution has developed its own self-accelerating dynamics. It changes dramatically the selection conditions on our planet at ever smaller time intervals. These changes lead inevitably to the extinction of species, the number of which steadily increases.

8.10
How Do We Learn Culture?

A question that is not discussed frequently is: How do we come to know culture?, i.e.: How does culture, this phenomenon so important to us, enter our brains? According to behaviorists, this takes place in small learning steps that can be learned as in a Skinner box. Rewarded knowledge is stabilized and repeated, punished knowledge is avoided in the future. Behaviorism has indeed been successful in particular fields. Nevertheless, with regard to important cultural phenomena like speech and morality, behaviorism has proved insufficient and wrong. Cultural beings should be endowed with other learning abilities. This can be efficiently illustrated by the example of language acquisition. Language has proved quite suitable for this purpose due first to the fact that it is an important component of culture and second to the fact that, unlike other cultural developments, it is especially easy to study and consequently can serve as an interesting model.

If the only way to learn a language with its complicated grammar, specific phonetics, and rich nuances of intonation were to reward a linguistically correct word and to punish an incorrect one, then no language culture would have developed and nobody would have learned to speak adequately. However, especially children learn a language in a relatively short time, seemingly effortlessly and without an accent. On the other hand, adults confronted with a new language experience serious and often even insoluble difficulties. At the same time they generally are able to understand the use of a technical device or the issues of a physical theory significantly more easily and quickly than children do.

A plausible explanation, which is interesting for evolutionary biology, is Chomsky's (1959) generative-grammar theory, which posits a genetically programmed basic language structure in the brain. However, this potential has to be syntactically and lexically supported for it to be realized as a language.

Meanwhile, several studies have shown that learning the mother tongue differs significantly from learning how to use a slot machine. Thus, it is known that infants adjust their audition to their mother tongue (Eimas 1985). Here, rewards play hardly any role – since young infants do not yet speak. A syntax test is automatically performed by adults demonstrated on the basis of electroencephalographic potentials while hearing correctly or falsely constructed sentences. However, this is true only if the language is learned before the age of four (Neville 1994, Friederici 1996).

Learning without reinforcement, but limited to a sensible phase is typical of learning processes defined in ethology as imprinting. In some distinct cases these learning processes are irreversible. Such specific learning processes are evolutionarily prepared under selection pressure, and they function almost autonomously.

Genetic programs that have been able, according to a 'plan', to supply the functional structures to a brain have proved to possess an advantage compared to less 'able' ones, the latter being superseded with time. Many birds learn the songs of their species (the 'dialect' of their population), or what their sex partners look like in such a specifically prepared way long before they can sing or copulate. Here, no reinforcement of the learning is necessary (Lorenz 1935, Hess 1973, Eibl-Eibesfeldt 1978).

Obviously, humans demonstrate similar abilities in learning culture in a broad sense. Positive or negative reinforcing actions can hardly act as the only mechanism, keeping in mind the complexity of content and the speed of learning. What is more important is that in most cultural contents – morality included – a reward or punishment follows much later or never. It seems that the individual absorbs the offered cultural elements like a sponge absorbs water without waiting for success. Analogous to the sponge, once the young brain is 'satiated' with particular content, there is hardly place for new conflicting or incompatible content. This fact is valid for the acquisition the grammatical structure and the articulation of the mother tongue as well as for more self-evident and elementary phenomena: first of all methods of food providing and food processing, including such macabre practices as cannibalism, which has not always been as objectionable as today (Harris 1970). Clothes, gender roles, manners, ways of living and of child care, education and breeding, religion, witchcraft, and ideology, loyalty towards particular groups of people and moral principles, tolerance or intolerance to other cultures or adversaries, evaluation of honor, power, and their modern surrogate (money) are all culturally dependent in a specific way (Verbeek 1998b, 2004).

Darwin did not know the phenomenon of imprinting. Nevertheless, in *The Descent of Man* (1871) he explicitly stated that any belief constantly impressed upon a brain finally seems to acquire the nature of an instinct, and that the very nature of an instinct is that is followed without being reflected upon.

Learning through imprinting represents a genetically originated ability similar in effect to instincts but different in its ability to adapt to existing ecological and historical situations. In this sense, imprinting and learning through imprinting provide an adequate basis for the establishment and further development of culture. The values, once adopted, show particular resistance to revision. Thus, culture proves to be extremely conservative, to allow the growth of complex, reliable structures. The fact that the cultural contents are not 'written' in the genes makes them – compared to the velocity of evolutionary changes on a genetic level (level 1, Figure 8.6) – able to change very rapidly.

The transmission of new cultural versions and their combinations is no longer bound to the genealogical (vertical) line. It also occurs horizontally – among contemporaries. If these cultural contents correspond to the tendencies of the day and if they appear attractive to other brains then they spread like a prairie fire. In this time of globalization they can overflow the globe in minutes and in this way provoke new developments (Blackmore 1999 has dealt profoundly with this topic, initially discussed by Dawkins). Compared to the rate of evolution of organisms, cultural developments take place like an explosion and prove deadly to many organisms.

It is, however, interesting that new cultural contents are normally limited to individuals in the younger generation. The same can be observed in animals, e.g., in the red-face macaques that traditionalized the washing of sweet potatoes as a new method of food preparation (Kawai 1965). Only a few of the older individuals adopted the new method. Although such conservatism is of hardly any use to the monkeys, in humans it guarantees the reliability of communication as well as of value and reference systems intrinsic to complex civilized society (Voland et al. 1997, Verbeek 1998b, 2000). It is a paradox: the conservative elements seem to be a necessary precondition for the very dynamic development of human culture which, after all, 'infects' evolution at all levels.

8.11
Evolutionarily Formed Learning Dispositions

The attraction of ideas depends on the preceding cultural evolution. For example, a hundred years ago only a few would have accepted the idea of describing the cosmos as an evolutionary process, although today this is more or less common knowledge. In the Middle Ages or even the Renaissance an evolutionary theory would have appeared so wrong that it would not even have been dangerous to propagate it. Actually hardly anyone would have understood it. Not so the heliocentric concept, which did find many supporters and brought, e.g., Giordano Bruno to the stake and Galileo to trial. Cultures and world concepts change. Obviously, the variability of cultures seems to be unlimited.

To think, however, that the brain ware, i.e., genes, can mediate any content, would be a fallacy. Rather, we should consider the existence of particular dispositions with respect to culture-specific contents that determine the material to be learned and which, in phylogenesis, have proved to promote the fitness of the species.

Imprinting experiments with animals have shown that – contrary to the broadly accepted idea – the imprinting process takes place much more quickly and easily if it is induced with natural objects (e.g., baby ducks imprinting on the mother duck) and not with a cardboard cylinder or a human researcher (Hess 1973). With the natural object the imprinting process takes 20 min; with some other object, 20 h. It is small wonder that little ducks are confronted with something like this only in the laboratory. In nature, chicks that spontaneously choose to follow another object and not their mother are usually endangered. Normally, only those survive that, as it were, already almost 'know' what is to be learned. There are even histological data from Guinea fowls (Scheich 1987) which demonstrated that brain circuits harmonize with the natural imprinting object. Under natural conditions very few has to be rectified in the neuronal structure. With artificial imprinting contents however there followed something like a clear-cut on the level of neuronal connections. Many of the primarily offered spines did not survive the process of imprinting. Only a few proved to fit.

In humans everything is, quite naturally, more complex. Here, one should count not only attachment to parents and the mother tongue as fitness-promoting characteristics, but also parental love, altruism, public spirit, and friendship. It is not surprising that in the course of evolution genetic constellations that correspond more adequately to these characteristics replaced the less adequate ones. Some of the useful moral attitudes and skills that are supported by genetically secured rapid-learning programs do possess, however, a bad image, e.g., egoism. This is explained by the fact that the egoism of others is disadvantageous to the individual, which causes a kind of a defense reaction. On the other hand, egoism proves to be extremely important for individual survival – it doesn't allow the total exploitation of that individual. Of particular importance is 'hidden egoism'.

Machiavelli (1513) won his dubious reputation by advising the sovereign that, while demonstrating virtue, he should not be limited by it in his judgments and activities. More successful than any theorist are those politicians who, while openly criticizing Machiavelli, still follow his advise in practice. For illustration an example from history: The Prussian king Friedrich II wrote an "Antimachiavelli", but in practice, however, he was actually led by the maxims of Machiavelli and not by the principles he himself advocated. This de facto machiavellism won him the surname *der Große*, 'the Great".

However, in our everyday life we are usually not confronted with kings and presidents but rather with normal citizens who are not even authors like Machiavelli. However, like the successful politicians they emotionally reject socially unwanted evolutionarily generated properties of human behavioral repertoire and their positive potential. Although we normally have these mental barriers trying to keep particular features of human nature in the subconscious, we do have to recognize them because of their enormous importance for current and future evolution – and not only for cultural evolution.

Almost nobody denies the fact that our animal and human ancestors have always been endangered by natural catastrophes, hunger, biological and social rivals from their own species, predators, and diseases. The following considerations demonstrate clearly the perilous situation of humankind: if we assume an average of four children per woman, the descendants of one couple will be sufficient to regenerate the current world population within less than thousand years. However, for more than million years the world population did not surpass 10 million humans. This shows that selective pressure had been extremely severe and that most individuals born were not able to withstand it. As already mentioned in reference to Malthus and his influence to Darwin, this situation is not specific to humans (Section 8.3).

To pass the evolutionary fitness test that is repeated in each generation, living beings had to use all possible advantages offered by the environment. Those equipped with a more adequate genetic program had better chances in the struggle for survival. Together with such 'positive' features as cooperation and reliability, other properties like egoism, self assertion, dominance, and maximal use of nature proved to be essential. In critical situations these properties facilitated the appropriation of the barely sufficient resources. Competition and selfishness are important driving forces of evolution (e.g., Dawkins 1976, Wickler and Seibt 1977, Voland 1996). Only those who successfully competed could survive. This is a general feature of living beings and not a specifically human trait.

All humans living today – no matter how long their personal documented family tree – descend from individuals who successfully overcame the hazards of the world at least until they could reproduce. Realistically seen, it is not the altruistically minded, the agreeable, the powerless ones who successfully reproduced. According to our personal inclinations we have to acknowledge proudly or with shame that we descend from the successful ones. Realizing that these successful ones were not morally better, many of us are prepared to close our eyes and suppress such knowledge. This reminds me of the anecdote about the wife of an Anglican bishop who, when confronted with Darwin's theory that humans descended from apes, said, Let's hope that this is not true, and if it is true, let's pray that it does not become publicly known.

We know that humans who, thanks to their intelligence and technical progress, are able to look for resources on the moon, are not able to adopt a morality that will prevent the collective plundering of the planet (and of humans as well). The hindrances on the way to the 'new human being' come from the depths of evolution (Verbeek 1992). That is why we have to find a way to get by with the 'old' one.

8.12
The Ecological–Social Dilemma

It is interesting that, despite the imperatives of fitness, e.g. to make use of momentary advantages, animals are prepared in particular situations to dispense with these advantages. Thus, the European field hare does not graze the vegetation near its living place. It is as if it knows that this would reveal its hiding place to predators.

In comparison, rabbits and prairie dogs behave in an egoistic and shortsighted way: they graze recklessly also near their living places. The reason for the different behaviors is that the field hare is a solitary animal and avoids other hares (except in mating season); thus, there are normally no rivals that would graze the grass that it has left untouched. An altruistically inclined rabbit might refrain from eating the plants in front of its 'door' and look for food further away, but that would be much more dangerous. And if so, the plants near its living place would be grazed by the rest of the rabbits in its colony. The altruist would most probably be eaten by a fox while looking for food, and the egoists would get the better grass without endangering themselves. Because to this particular selective pressure, the 'grazing altruists' among the rabbits have almost all disappeared, having been replaced by the 'egoists'.

The exploitation of, for example, commons for grazing livestock confronts humans with similar problems. Unregulated utilization of the meadow would work as long as its usage does not surpass its capacity. One individual alone could hardly hinder any over-utilization even when aware of the danger. An economically rational shepherd will bring to the common meadow as many animals as possible – each animal contributes to his personal enrichment. If, however, everybody acts in this way the system is bound to collapse.

The human being is nevertheless a moral being. Is it not possible to remedy the situation by appealing to their consciences? Unfortunately, this does not happen very often. And if one of the shepherds dispenses with using the commons as a result of this appeal, the consequences for him will be extremely hard. In the extreme case of a pure grazing economy, the shepherd will have to dispense with the herd. At the same time, such a sacrifice by one will hardly improve the overall situation. Hardin (1968) described this phenomenon as the 'tragedy of the commons' and reformulated the conventional moral appeals into two contradictory messages (the first intended, the second unintended):

> If you don't do as we ask, we will openly condemn you for not acting as a responsible citizen.

> If you *do* behave as we ask, we will secretly condemn you as a simpleton who can be shamed into standing aside while the rest of us exploit the commons.

Those who participate in the run for the last resources experience the whole 'fun' of a ruinous exploitation. Obviously, the rabbits have 'understood' (under selective pressure) the second message. It seems that humans understand it equally well. Meanwhile, in the age of 'pyrotechnics', telecommunications, and mass- and long-distance traffic, the whole planet has turned into a 'tragedy of the commons'. Evolution, however, has had no possibility to work out a strategy for sustainable resource use. Recent experiments from environmental psychology have shown that a shortage of common resources (e.g., fish) increases the plundering pressure in humans if no controlling game rules are introduced as opposing factors (Ernst et al. 1998). This experiment was conducted as part of an environmental conflict game dealing with fish stocks in the North Sea. How do humans behave in their real lives?

8.13
Sustainability Through Faka-Tonga

Concerning this problem there exists a study about the use of fish resources in the Kingdom of Tonga (Ernst et al. 1998). The population of these Pacific islands traditionally believe that the fish resources are given by God and that they are not to be exhausted. If the catch is smaller the fishers are convinced that the fish have either migrated or escaped the net by submerging into deeper waterness.

However, progress has come to the island of Uiha. Two large entrepreneurs have introduced modern catching and marketing methods. The island economy is more successful and the people living there have become more diligent. According to our way of thinking these changes are for the better. Nevertheless, this new economy has led to exhaustion of the fishing reserves, and the fishers are forced to expand their fishing range. At the same time, the new methods of work have increased the prosperity of the population in the modern sense of the word. The traditional forms of coexistence founded on informal security have become unnecessary and are slowly disappearing. In earlier times those with poorer catches were helped by the more successful. Today the island inhabitants have to ensure their survival monetarily.

On the nearby island of Lofanga the situation is different – here life runs according to the traditions on a subsistence basis. The resources are still looked upon as given by God and as unlimited. Nevertheless, the fishing resources are used in a sustainable way. Sustainability is however not guaranteed by means of fishing rules but rather by the 'Faka-Tonga' – the way of life of Tonga. According to it, those who do not have luck in fishing or are simply too old or are in some other kind of predicament receive help from the more successful ones. This is the security one receives in this culture. Those who have plenty are supposed to share with the less lucky ones. This is an obligation no one could ignore – the constraint of the local tradition. One is permitted to sell only what is left after this obligation has been met. But this also ensures nobody will try to catch much more than what he and his family actually need.

As already mentioned, it is not some specific economic rule that stabilize the marine ecology in the region. Rather it is the particular culture and the social system, e.g., the generally accepted obligation to share the surplus. This obligation affects the moderate fishers only mildly, the excessive ones, however, seriously. On Lofanga the people are not really ecologically minded. It is simply not profitable to catch more fish than absolutely necessary. The more one catches, the greater the share of social obligation. Already on the way home, fishers with a surplus are coerced into giving away what is not considered necessary for their families. The men on Lofanga spend only one fifth of the time fishing as do the 'modern' entrepreneur-guided fishers on Uiha. This has proved enough for their needs and at the same time advantageous for the local ecology.

Faka-Tonga – the traditional way of life – could have, it seems, lasted forever. But it is affected by the more successful, more diligent neighbors who are invading the territory of the Lofanga people with their modern motor boats and salaries.

8.14
Cultural Comparison

As already shown, culture can change the conditions of selection for all possible organisms. However, it cannot abolish selection as such. Rather, culture itself is affected by the process of selection: if for a long period of time a given cultural design does not adequately meet the requirements of the real world (including the social world), it has to be replaced by a new design. Nevertheless, the differentiation between various organisms' lines is much clearer than the differentiation between separate cultural units, with their mutual exchange of ideas and goods. The different cultures, however, compete with each other. Thus, a culture whose carrier causes the death of people (e.g., on the basis of a dangerous diet theory) would have hardly any chance of survival. Nevertheless, like any deadly virus, it does find new victims for a certain period of time.

It gets more complicated when we consider that different cultural designs do interact and compete among themselves. It seems logical that the more attractive designs would have a better chance to ensure their place in the brain. This would not be really disturbing if these were designs that offered better ecological sustainability. Actually it is – and this is to be expected – usually the opposite.

Empirical studies and logical reasoning about sustainability show that humility, future-oriented reasoning, or other valued properties do not qualify as winners in the fitness competition, but rather the opposite qualities. Such properties are also not the favorite goal of imprinting, but instead such characteristics as power, dominance, glamour, recognition, and immediate success are promoted. One of the behavioral algorithms actually is 'in case of doubt, follow the example of the more successful'. This does not come as a surprise if we keep in mind that all organisms were formed in the past and live in the present. The future (as long as it is not a repetition of the past like the alteration of seasons) plays no role in their repertoire.

The communication networks, the economic systems, and globalization inevitably lead to a worldwide unification of culture. Nevertheless, in the face of the evolutionary competition we can hardly expect everlasting peace or protection of environmental resources.

Like many biologists who have tried to point out evolutionary progress, generations of historians have believed in a particular purpose in the history of humankind. The consequences of this belief – or, rather, false doctrine – have been terrifying. Whole nations have been misled by this belief and, in the worst cases, been thrown into horrid wars (Wuketits 1998). Meanwhile, we are forced to abandon the idea of purpose intrinsic to history. History has always moved forward without a recognizable goal. As long as humans exist, history will also exist – and will go nowhere. In relation to purpose, history does not really differ from organismic evolution. Blind competition and maximum use of resources have been the driving forces in evolutionary development. Nevertheless, and despite all setbacks, the complexity of organisms has continually developed, up to the appearance of human beings. Why, then, should we be so pessimistic with reference to the contemporary determination of cultural history?

The explanation for the principle of success is diversity, which has always ensured that something fit and then something fitter appears without any recognizable planning. Today, however, the diversity of organisms as well as the variety of cultures are rapidly decreasing. This has resulted – notwithstanding the growing efficiency – in a decrease in evolutionary resources at all levels.

Obviously, what is hitherto lacking – also in cultures – is a guiding intelligence that would steer us toward sustainability. Unfortunately, we are confronted with the opposite situation. If, for example, the ecological basis is destroyed and one discovers that the only surviving cultural plan of humankind is a failure, then evolution would have no more possibility to replace it with another one.

We know that new designs developed blindly – through mutation and recombination – and were rarely better than the older ones. Most of them are actually lethal. One could expect a greater win only if the number of attempts at luck is large enough. Should we stake everything on one card and play roulette with the earth – the only one we have?

8.15
The Power of Constraints

Generally, living organisms are not only playthings in the ocean of ecological factors, rather, they actively mount its waves, guided through selection-verified algorithms. The wave to come is also (mostly) reasonably and successfully anticipated by their behavior programs, but only because the new wave has had its predecessors.

There are no evolutionary algorithms for completely new, never-before-experienced challenges. Thanks to their intelligence, human beings are able – at least to a limited extent – to evaluate completely new consequences of their conduct in the future. The prognosis for an earth determined by humans is, as already intimated, not necessarily favorable. Some even see themselves or their descendants wandering to other planets after the destruction of the earth. This will however prove more difficult than inhabiting new parts of the earth in earlier times.

As I have tried to show, culture is indeed bound to its foundation at the DNA level – however, not to the changes taking place in this foundation but rather to its conservatism. The genetic programs, including those of the cultural carrier, have developed to fit a particular environment. Today, the rapid dynamics of culture endangers this very same environment. In Section 8.8 an analogy was made between culture and a controlled nuclear chain reaction; this analogy could be continued further. If the control mechanisms are overtaxed and the chain reaction gets out of control, one would urgently need adequate brakes. However, what kind of brakes?

Perhaps evolutionary theory could be of some help in shaping future perspectives. We speak about selection pressure when the evolution of a particular organism proceeds in a particular direction under the influence of a given factor. The selection pressure is the result of the sum of the existing constraints. Some of the important components of nonhuman nature are, e.g., the climate, food resources, or the presence of enemies. Thus, under the influence of such components arctic mammals acquired thick fur, birds ended up with a particular beak, and hares developed into swift runners. Natural constraints can be affected considerably by living beings and especially by humans. This is illustrated by the development of domestic animals, which differ considerably from their wild relatives. The extinction of numerous species and the man made alterations of the environment have been already mentioned. Additionally, today genetic engineering opens new possibilities for purposeful but also extremely precarious intervention in evolution.

Certainly, it is not easy to reach agreement about all the details of a desirable direction of development. One particular aspect, however, seems relatively clear and undisputable: culture should be allowed to develop freely. Nevertheless, this development should not seriously endanger the existence of organisms and humans, making culture itself impossible. Rather one should pursue a culture of sustainability that will help to define and preserve the environment in a way adequately corresponding also in the future to the genetic programs and requirements of organisms. This will not happen as long as all actors follow the cold logic of the 'tragedy of the commons' and constantly increase the exploitation pressure on the basis of selection-validated algorithms.

The alien observer that we supposed at the beginning would without doubt wonder why *Homo sapiens* does not apply its reflective reasoning abilities and theoretical knowledge in future-oriented planning of social development. Actually, there is a possibility to preserve both civilization and its ecological foundation: the selection pressure for the products of civilization has to be guided from the standpoint of sustainability. We need a 'cultural climate' that supports the development only of those products and ways of conduct that do not destroy the ecological foundations of most living beings and especially of humans.

If we want to influence future evolution along the above-mentioned lines, then we should be defining the selection constraints of civilization in a way that will guarantee everybody a personal advantage from environmental preservation. The self-denial of excess should be rewarded in the long run and not the wasteful exploitation of natural goods or the use of nonrenewable resources greater than personal needs. The story about the 'lazy' fishers from Lofanga compared to those from Uiha illustrates the dependence of our world – its destruction or preservation – on culture.

In the modern world, dominated by mankind the use of fossil energy has a fatal ecological and evolutionary importance. Nations with a cheap access to energy use more of it. At the same time, nations that tax nonrenewable resources show a drastic decrease in their use. Such constraints influence the behavior and stimulate the construction of, e.g., energy-efficient cars and houses. The basic evolutionary law – what is not fit disappears – is valid also in the world of humans. What has proved to be successful remains. What remains is successful. This statement is probably trivial and tautological. Nevertheless it is not without importance.

8.16
Critical Phase of Evolution

It seems that evolution has reached a critical phase that is characterized by the fact that all political measures worldwide introduced (or not introduced) on the basis of future-oriented intelligence have dramatic effects on the evolutionary processes, comparable to the development of photosynthesis or to the disastrous impact of an asteroid. In all cases, on the basis of our abilities to reflect, we have the possibility to determine future developments. This however makes us also responsible for the outcome (Jonas 1984), which in itself is a novelty in the evolution of life on earth.

Unconscious evolution has succeeded in creating life from an ocean of nothing. The cosmic flow of entropy is guided in a way that supports the upward development of life (symbolized in Figure 8.2). Evolution endowed with consciousness and knowledge is now challenged by a stream of destructive selfishness (a kind of social entropy). On the basis of adequate culturally defined constraints, this stream could be forced to work constructively and for preservation. Only our future will show whether – and to what extent – rational beings are able to sail against the wind of their own dispositions.

8.17
References

BARNEY, G. O. (Ed.) (**1980**) *The Global 2000 Report to the President.* Washington, DC: US Government Printing Office.

BENGTSSON, L. (**1997**) Climate Modeling and Prediction. In: BRUNE, D., CHAPMAN, D. V., GWYNNE, M. D., PACYNA, J. M. (Eds.), *The Global Environment: Science, Technology and Management*, Vol. 1. Weinheim: Wiley-VCH, 31–55.

BLACKMORE, S. (**1999**) *The Meme Machine.* Oxford: Oxford University Press.

BOESCH, CH. & BOESCH, H. (1981) Sex differences in the use of natural hammers by wild chimpanzees. A preliminary report. *J. Hum. Evol.* **10**, 585–593.

BONNER, J. T. (**1983**) *The Evolution of Culture in Animals.* Princeton, NJ: Princeton University Press.

BRUNE, D., CHAPMAN, D. V., GWYNNE, M. D., PACYNA, J. M. (Eds.) (**1997**) *The Global Environment: Science, Technology and Management.* 2 Vols. Weinheim: Wiley-VCH.

CARSON, R. (**1962**) *The Silent Spring.* New York: Crest Books.

CHOMSKY, N. (**1959**) Review of Skinner's Verbal Behavior. *Language* **35**, 26–58.

DARWIN, CH. (**1871**) *The Descent of Man.* London: Murray.

DAWKINS, R. (**1976**) *The Selfish Gene.* Oxford: Oxford University Press.

DÖÖS, Bo R. (**1997**) Greenhouse Gases and Climate Change In: BRUNE, D., CHAPMAN, D. V., GWYNNE, M. D., PACYNA, J. M. (Eds.), *The Global Environment: Science, Technology and Management*, Vol. 1. Weinheim: Wiley-VCH, 319–351.

EIBL-EIBESFELDT, I. (**1978**) *Grundriss der vergleichenden Verhaltensforschung.* Munich: Piper.

ERNST, A. (**1997**) *Ökologisch-soziale Dilemmata.* Weinheim: Psychologie Verlagsunion.

ERNST, A., EISENTRAUT, R., BENDER, A., KÄGI, W., MOHR, E., SEITZ, S. (**1998**) Stabilisierung durch Kooperation im Allmende-Dilemma durch institutionelle und kulturelle Rahmenbedingungen. *Gaia* **7**, 271–278.

FRIEDERICI, A. D. (**1996**) Auf der Suche nach den neuronalen Grundlagen der Sprache. *Universitas* **51**, 583–596.

GASSMANN, F. (**1994**) *Was ist los mit dem Treibhaus Erde?* Stuttgart: Teubner.

GOODALL, J. (**1990**) *Through a Window: Thirty Years With the Chimpanzees of Gombe.* London: George Weidenfeld & Nicolson.

HAARMANN, H. (**2004**) Evolution, Language, and the Construction of Culture. In: WUKETITS, F. M. & ANTWEILER, CH. (Eds.), *Handbook of Evolution*, Vol. 1. Weinheim: Wiley-VCH, 77–119.

HARDIN, G. (**1968**) The tragedy of the commons. *Science* **162**, 1243–1248.

HARRIS, M. (**1970**) *Cannibals and Kings.* New York: Knopf.

HESS, E. (**1975**) *Imprinting: Early Experience and the Developmental Psychology of Attachment.* New York: Nostrand.

HEYMER, A. (**1996**) Der Aktionsraum der Pygmäen als Sammler und Jäger im afrikanischen Regenwald. *Ethnogr.-Archäol. Z.* **38**, 479–493.

JONAS, H. (**1984**) *Das Prinzip Verantwortung. Versuch einer Ethik für die technologische Zivilisation.* Frankfurt a. M.: Suhrkamp.

KAPLAN, R. W. (**1989**) Organismenvielfalt und unser Weltbild. *Naturw. Rdsch.* **42**, 354–359.

KAWAI, M. (**1965**) Newly acquired precultural behavior of the natural troop of Japanese monkeys on Koshima Islet. *Primates* **6**, 1–30.

LORENZ, K. (**1935**) Der Kumpan in der Umwelt des Vogels. *J. Ornithol.* **83**, 137–215, 289–413.

LORENZ, K. (**1978**) *Vergleichende Verhaltensforschung.* Vienna: Springer.

MEADOWS, D. H., MEADOWS, D. L., RANDERS, J., BEHRENS, W. W. (**1972**) *The Limits to Growth.* New York: Universe Books.

NENTWIG, W. (**1995**) *Humanökologie. Mensch und Umwelt.* Berlin: Springer.

NEVILLE, H. (**1994**) Development specificity in neurocognitive development in humans. In: GAZZANIGA, M. S. (Ed.), *The Cognitive Neuroscience.* Cambridge, MA: Harvard University Press, 38–52.

PAUL, A. (**1998**) *Von Affen und Menschen. Verhaltensbiologie der Primaten.* Darmstadt: Wissenschaftliche Buchgesellschaft.

RAPP, F. (**1992**) *Fortschritt. Entwicklung und Sinngehalt einer philosophischen Idee.* Darmstadt: Wissenschaftliche Buchgesellschaft.

RIFKIN, J. (**1985**) *Declaration of a Heretic.* Boston: Routledge.

SCHEICH, H. (**1987**) Neural correlates of auditory filial imprinting. *Journal of Comparative Physiology* **A 161**, 605–619.

SCHIDLOWSKI, M. (**1981**) Die Geschichte der Erdatmosphäre. *Spektrum der Wissenschaft*, April, 16–27.

SCHRÖDINGER, E. (**1944**) *What is Life? The Physical Aspect of the Living Cell.* Cambridge: Cambridge University Press.

SOMMER, V. (**1989**) *Die Affen. Unsere wilde Verwandtschaft.* Hamburg: Gruner & Jahr.

SOMMER, V., AMMAN, K. (**1998**) *Die großen Menschenaffen.* Munich: BLV.

STIEGER, R. (**1995**) *Internationaler Umweltschutz. Eine politisch-ökonomische Analyse der Verträge zum Schutz der Ozonschicht.* Bern: P. Lang.

VERBEEK, B. (**1992**) Evolutionsfalle oder: Die ewige Hoffnung auf den 'Neuen Menschen'. *Universitas* **47**, 424–434.

Verbeek, B. (**1998a**) *Die Anthropologie der Umweltzerstörung. Die Evolution und der Schatten der Zukunft*, 3rd ed. Darmstadt: Wissenschaftliche Buchgesellschaft.

Verbeek, B. (**1998b**) Organismische Evolution und kulturelle Geschichte. Gemeinsamkeiten, Unterschiede, Verflechtungen. *Ethik und Sozialwiss.* **9**, 269–280.

Verbeek, B. (**1999**) Kultur als kritische Phase der Evolution. Ethik als Richtschnur. In: Engels, E.-M. (Ed.), *Biologie und Ethik*. Stuttgart: Reclam, 71–99.

Verbeek, B. (**2000**) Kultur: Die Fortsetzung der Evolution mit anderen Mitteln. *Natur und Kultur* **1**, 3–16.

Verbeek, B. (**2004**) *Die Wurzeln der Kriege: zur Evolution ethnischer und religiöser Konflikte*. Stuttgart: Hirzel.

Voland, E., Dunbar, R. I. M., Engel, C., Stephan, P. (**1997**) Population increase and sex-biased parental investment: evidence from 18th and 19th century Germany. *Current Anthropology* **38**, 129–135.

Wickler, W., Seibt, U. (**1977**) *Das Prinzip Eigennutz. Ursachen und Konsequenzen sozialen Verhaltens*. Hamburg: Hoffmann und Campe.

Wilson, E. O. (**1985**) The biological diversity crisis: a challenge to science. *Issues Sci. Technol.* **2**, 265–267.

Wilson, E. O. (1992) *The Diversity of Life*. Cambridge, MA: Harvard University Press.

Wilson, E. O. (**2002**) *The Future of Life*. New York: Knopf.

Wuketits, F. M. (1997) The status of biology and the meaning of biodiversity. *Naturwiss.* **84**, 473–479.

Wuketits, F. M. (**1998**) *Naturkatastrophe Mensch. Evolution ohne Fortschritt*. Düsseldorf: Patmos.

Wuketits, F. M. (**1999**) *Die Selbstzerstörung der Natur. Evolution und die Abgründe des Lebens*. Düsseldorf: Patmos.

Wuketits, F. M. (**2003**) *Ausgerottet: Ausgestorben: über den Untergang von Arten, Völkern und Sprachen*. Stuttgart: Hirzel.

Index

a

Acheulean 228
actual causes 62
adaptation(s) 2, 13, 76, 95, 105, 124, 227
– cultural 233
– functional 117
– gradual 18
– structural 117
adaptationism 76
adaptive advantage 39
adaptive landscapes 223
adaptive radiation 20, 52
– of cultures 237
aDNA 130, 188, 199
Aegyptopithecus 135
African origin 202, 205
Afropithecus 137
Afrotarsius 135
aggregation 50
aggression 38
agriculture 234, 248, 256
altruism 54
altruistic behavior 17
amino acids 3
anagenesis 21
analogistic tradition 89
analogy 22, 63
analytical method 64
Ante-Neanderthals 192
"Anthropocene" 236
anthropology 102, 119
"Anthropomorpha" (Linné) 71
ants 53

Apidium 135
Archaeopteryx 62
archetype 63
Ardipithecus 145
Aristotle 58
artifacts 227
artificial selection 76
Australopithecines 157, 226
Australopithecus 138, 143, 145, 228
Australopithecus afarensis 16, 151
Australopithecus africanus 153
autapomorphic features 123
autonomy of biology 72

b

Baer, Karl E. von 62
barnacles 8
Bateson, William 29
behaviorism 260
binominal nomenclature 72
biodiversity 47, 55
biogeochemistry 224
biogeography 6, 8, 59
biological classification 72
biological development 8
biological systematics 71
biomass 47, 51
biometrics 101
biosphere 125, 252
bipedality 150
body plans 96
Bonner, John 93
Bonnet, Charles 60

Handbook of Evolution, Vol. 2: The Evolution of Living Systems (Including Hominids)
Edited by Franz M. Wuketits and Francisco J. Ayala
Copyright © 2005 Wiley-VCH Verlag GmbH & Co. KGaA, Weinheim
ISBN: 3-527-30838-5